Lecture Notes in Computer Scie

Edited by G. Goos, J. Hartmanis and J. van Le

T0238009

Springer
Berlin
Heidelberg
New York
Barcelona
Hong Kong
London
Milan
Paris
Singapore
Tokyo

Jeffrey S. Vitter   Christos D. Zaroliagis (Eds.)

Algorithm Engineering

3rd International Workshop, WAE'99
London, UK, July 19-21, 1999
Proceedings

Springer

Series Editors

Gerhard Goos, Karlsruhe University, Germany
Juris Hartmanis, Cornell University, NY, USA
Jan van Leeuwen, Utrecht University, The Netherlands

Volume Editors

Jeffrey S. Vitter
Duke University, Department of Computer Science
Box 90129, Durham, NC 27708-0129, USA
E-mail: jsv@cs.duke.edu

Christos D. Zaroliagis
University of London, King's College, Department of Computer Science
Strand, London WC2R 2LS, UK
E-mail: zaro@dcs.kcl.ac.uk

Cataloging-in-Publication data applied for

Die Deutsche Bibliothek - CIP-Einheitsaufnahme

Algorithm engineering : 3rd international workshop ; proceedings /
WAE '99, London, UK, July 19 - 21, 1999. Jeffrey S. Vitter ;
Christos D. Zaroliagis (ed.). - Berlin ; Heidelberg ; New York ;
Barcelona ; Hong Kong ; London ; Milan ; Paris ; Singapore ; Tokyo
: Springer, 1999
   (Lecture notes in computer science ; Vol. 1668)
   ISBN 3-540-66427-0

CR Subject Classification (1998): F.2, C.2, G.2

ISSN 0302-9743
ISBN 3-540-66427-0 Springer-Verlag Berlin Heidelberg New York

Typesetting: Camera-ready by author
SPIN: 10705385   06/3142 – 5 4 3 2 1 0   Printed on acid-free paper

Preface

This volume contains the papers accepted for presentation at the *3rd Workshop on Algorithm Engineering* (WAE'99) held in London, UK, on July 19–21, 1999, together with the extended or short abstracts of the invited lectures by Andrew Goldberg, Bill McColl, and Kurt Mehlhorn. WAE is an annual meeting devoted to researchers and developers interested in the practical aspects of algorithms and their implementation issues. Previous meetings were held in Venice (1997) and Saarbrücken (1998).

Papers were solicited describing original research in all aspects of algorithm engineering including:

- Implementation, experimental testing, and fine-tuning of discrete algorithms
- Development of software repositories and platforms which allow use of and experimentation with efficient discrete algorithms
- Methodological issues such as standards in the context of empirical research on algorithms and data structures
- Methodological issues involved in the process of converting user requirements into efficient algorithmic solutions and implementations

The program committee selected 24 papers from a total of 46 submissions. The program committee meeting was conducted electronically from May 6 to May 16, 1999. The criteria for selection were perceived originality, quality, and relevance to the subject area of the workshop. Considerable effort was devoted to the evaluation of the submissions and to providing the authors with helpful feedback. Each submission was reviewed by at least three program committee members (occasionally assisted by subreferees). However, submissions were not refereed in the thorough way that is customary for journal papers, and some of them represent reports of continuing research. It is expected that most of the papers in this volume will appear in finished form in scientific journals. A special issue of the *ACM Journal on Experimental Algorithmics* will be devoted to selected papers from WAE'99.

We would like to thank all those who submitted papers for consideration, as well as the program committee members and their referees for their invaluable contribution. We gratefully acknowledge the dedicated work of the organizing committee (special thanks to Tomasz Radzik and Rajeev Raman, who did most of the work), the support of the Department of Computer Science at King's College, and the generous help of various volunteers: Gerth Brodal, Sandra Elborough, Ulrich Endriss, Viren Lall, José Pinzón, and Naila Rahman. We thank them all for their time and effort.

July 1999

Jeff Vitter
Christos Zaroliagis

Invited Lecturers

Andrew Goldberg   InterTrust STAR Lab, Sunnyvale, USA
Bill McColl   Oxford University and Sychron Ltd, UK
Kurt Mehlhorn   Max-Planck-Institut für Informatik,
                              Saarbrücken, Germany

Program Committee

Lars Arge   Duke University
Thomas Cormen   Dartmouth Colledge
Giuseppe Italiano   Università di Roma "Tor Vergata"
Bill McColl   Oxford University and Sychron Ltd
Bernard Moret   University of New Mexico
Stefan Näher   Universität Halle
Esko Ukkonen   University of Helsinki
Jeff Vitter, Co-Chair   Duke University & INRIA Sophia-Antipolis
Christos Zaroliagis, Co-Chair King's College, University of London

Organizing Committee

Costas Iliopoulos   King's College, University of London
Tomasz Radzik, Co-Chair   King's College, University of London
Rajeev Raman, Co-Chair   King's College, University of London
Christos Zaroliagis   King's College, University of London

Referees

Rakesh Barve   Dimitrios Gunopulos   Cliff Stein
Julien Basch   Octavian Procopiuc
Gerth Brodal   Tomasz Radzik

Table of Contents

Invited Lectures

Contributed Papers

Selecting Problems for Algorithm Evaluation

Andrew V. Goldberg*

InterTrust STAR Lab., 460 Oakmead Parkway, Sunnyvale, CA 94086, USA
goldberg@intertrust.com

Abstract. In this paper we address the issue of developing test sets for computational evaluation of algorithms. We discuss both test families for comparing several algorithms and selecting one to use in an application, and test families for predicting algorithm performance in practice.

1   Introduction

Experimental methodology is important for any experimental science. Several recent papers [1, 16, 23] address methodology issues in the area of computational evaluation of algorithms. A recent CATS project [13] is aimed at standardizing computational test sets and making them readily available to the research and user communities. In this paper we discuss how to design and select problems and problem families for testing general-purpose implementations, *i.e.*, implementations designed to be robust over a wide range of applications.

Major goals of experimental evaluation of algorithms are to

- determine relative algorithm performance.
- explain observed performance and find bottlenecks.
- facilitate performance prediction in applications.

We address the issue of designing computational experiments to achieve these goals.

A comparative study determines the best algorithm or algorithms for the problem. When no single algorithm dominates the others, one would like to gain understanding of which algorithm is good for what problem class. The best algorithms warrant more detailed studies to gain better understanding of their performance and to enable prediction of their performance in applications.

Theoretical time and space bounds are usually worst-case and hold in abstract models. For many algorithms, their typical behavior is much better than worst-case. Practical considerations, such as implementation constant factors, locality of reference, and communication complexity, are difficult to analyze exactly but can be measured experimentally. Experimentation helps to understand algorithm behavior and establishes real-life bottlenecks. This can lead to improved implementations and even improved theoretical bounds.

---

* Part of this work has been done while the author was at NEC Research Institute, 4 Independence Way, Princeton, NJ 08540.

J.S. Vitter and C.D. Zaroliagis (Eds.): WAE'99, LNCS 1668, pp. 1–11, 1999.
© Springer-Verlag Berlin Heidelberg 1999

Significant investment of resources may be required to develop a software system. Thus one would like to be able to determine in advance if the system performance will meet the requirements. Usually at the system design stage one has an idea of the size, and sometimes the structure, of the subproblems which need to be solved by the system. To obtain performance estimate, one needs to estimate the performance of subroutines for these subproblems. This task is simple if one can generate the subproblems and has good implementations of the subroutines available. Getting the system to the state when it can produce the subproblems, however, may require implementing a large part of the system and may be time-consuming. Obtaining subroutine implementations may also be hard: for example, one may be unwilling to purchase an expensive program until it is clear that the system under development is feasible. Computational studies may allow one to obtain good performance estimates before a system is build.

As algorithms and hardware improve and new applications arise, computational tests need to evolve. Although one can predict some of the future trends (e.g. increased memory and problem size) and prepare for these, one cannot predict other developments (e.g. radically new algorithms).

Designing a good computational study is an art that requires detailed knowledge of the problem in question as well as experimental feedback. Although we cannot give a formula for such a design, we give general principles and helpful ideas, and clarify the experiment design goals.

This paper is based on our experience with computational study of graph algorithms. However, most of what we say should apply to other algorithms as well.

In this paper we restrict the discussion to the test set design. Many important issues, such as test calibration and machine-independent performance measures, fall out of scope of this paper. We also do not discuss architecture dependence, although algorithm performance and the relative performance may depend on the architectural features such as cache size, floating point performance, number of processors, etc. However, we feel these issues are of somewhat different nature than the ones we address here, except for the problem size selection issues. For a discussion of caching issues, see e.g. [9, 19, 26].

This paper is organized as follows. We start with a brief discussion of correctness testing in Section 2. Then we discuss real-life vs. synthetic problems in Section 3. In Section 4 we discuss why and how to generate problems with a desired solution structure. Next, in Section 5, we focus on problem structure and its relationship to randomization. In Section 6, we discuss the importance of using both hard and easy problems in computational tests. In Section 7 we introduce the concept of an algorithm separator that is important for comparative evaluation of algorithms. We discuss parameter value selection in Section 8. In Section 9 we discuss standard test families. In Section 10 we point out how to use an experimental study to select the algorithm for most appropriate for an application. Section 11 gives a summary of the proposed experiment design process.

2 Testing Correctness

The first step in testing a subroutine is to establish its correctness. A good collection of test problems is helpful here. Such a collection should include problems of different types. Problems that demonstrated buggy behavior during algorithm development process may be especially useful is subsequent projects.

A related set of tests is for implementation limits. The limits can be on the problem size, parameter values, or a combination of problems size and parameter values. For example, in the context of the shortest path problem, if one uses 16-bit integers for the input arc lengths and 32-bit integers for the distances, overflows are possible if the input graph has a path with more than 2^{16} arcs.

3 Real-Life and Synthetic Problems

A good library of real-life problem instances can be very useful. Such a library should cover as many different applications as possible. For each application, it is desirable to have a wide range of problem sizes for asymptotic performance estimation. It is also desirable to have enough problems to be able to make meaningful statistical conclusions.

Real-life problem structure can differ significantly from the structure of natural synthetic problems. This can lead to different algorithm performance. Often, real-life problems are very simple. For example, in a study the minimum test set problem, Moret and Shapiro [27] observe that real-life instances from several applications are easy while random instances are hard.

Some real-life instances are hard for some algorithms. For example, for the multicommodity flow implementation of [20] that uses Relax [3] as a minimum-cost flow subroutine, synthetic problems were easy, but a small real-life problem turned out to be hard. This was due to the very poor performance of Relax on some subproblems of the real-life problem.

In some application areas, real-life problems are difficult to find. Many companies consider their problem instances proprietary and do not make them publically available. Only limited number of large problem instances can be maintained due to storage limitations. When only a small number of real-life problems is available, perturbations of these problems can be used as "almost" real-life problems.

Even when real-life problems are available, there are good reasons for using synthetic problems as well. Problems can be generated with desired structure and parameter values. Synthetic problems also anticipate future applications and explore algorithm strengths and weaknesses. In the rest of the paper we discuss how to generate interesting problem instances.

4 Solution Structure

Problem solution structure is very important. This observation is implicit in many computational experiments, but first stated explicitly by Gusfield [15]. For some algorithms, solution structure is closely related to algorithm performance. For example, the Bellman-Ford-Moore algorithm [2, 11, 25] for the shortest path problem usually takes much more time on networks with deep shortest path trees than on the same size networks with shallow trees.

One can generate a problem with the desired solution structure by starting with a solution and then perturbing it and adding data to "hide" the solution. This approach is usually easy to implement and can lead to interesting problem classes. Starting from a solution that ends up as an optimal solution to the generated problem also guarantees that the problem is feasible and facilitates code correctness testing.

For example, one can generate a single-source shortest path problem as follows. First, generate a tree rooted at the source s, assign tree arc lengths, and compute, for every vertex v, the distance $d(v)$ from s to v in the tree. Then generate arcs (x, y) to obtain a desired graph structure. Assign the lengths of (x, y) to be $d(y) - d(x) + \epsilon(x, w)$, where $\epsilon(x, y) > 0$. For any positive function ϵ, the tree is the shortest path tree. If the values of ϵ are very small compared to the tree arc lengths and the number of additional arcs is large, then many trees will give distances close to the shortest path distances, "hiding" the optimal solution. Note that the tree arc lengths and distances may be negative, but the network we construct does not have negative cycles. This is a very simple way to construct a network with cycles, negative-length arcs, but no negative cycles.

Sometimes a natural problem structure implies a solution structure. For example, a shortest path tree in a random graph with independent identically distributed arc lengths is small. One needs to check the solution structure to determine solution properties. For example, if solutions to a problem family are simple, one should be aware of it.

5   Problem Structure and Randomness

Problems with a simple and natural structure, such as random graphs and grids, are easy to generate. See [18] for more natural graph classes. Such problems also can resemble real-life applications. For example, minimum s-t cut problems on grids have been used in statistical physics (e.g. [24, 28]) and stereo vision [31]. Shortest path problems on grids arise in an inventory application [10], one about which we learned only after using such problems in a computational study [7].

If the problem structure is simple, algorithm behavior is easier to understand and analyze. Synthetic problems with simple structure often are computationally simple as well. Not all problems with a simple structure are computationally simple. For example, the bicycle wheel graphs [4, 21] are difficult for all current minimum cut algorithms. When natural problems are easy, this does not necessarily make these problems uninteresting: many real-life problem instances are computationally simple, too.

Sometimes one needs to use more complicated problem structures. In some cases, a complicated structure may better model a real-world application. Simple structures may also allow special-purpose heuristics that take advantage of this particular structure but do not work well on other problem types. An example of a generator that produces fairly complicated problems is the GOTO generator for the minimum-cost flow problem [12]. Problems produced by this generator seem to be difficult for all currently known combinatorial algorithms for the problem.

Some synthetic problem families, especially worst-case families, are deterministic. At the other extreme, some problems are completely random. Most problem families, however, combine problem structure with randomness. For example, the Gridgen generator [7] outputs two-dimensional grid graphs with random arc weights.

Randomness can come into a problem in several ways: it may be part of the problem structure (e.g. random graphs), parameters (such as arc weights in a graph) may be selected at random from a distribution, and a problem can be perturbed at random.

The latter can be done for several reasons. One reason is to hide an optimal solution. Another is to use one problem instance (e.g. a real-life instance) to get many similar instances. Finally, one can perturb certain parameters to test algorithm's sensitivity to these parameters.

6   Hard and Easy Problems

Hard problem families, in particular worst-case families, are interesting for several reasons. They show limitations of the corresponding codes and give upper bounds on how long the codes may take on a problem of a certain size. When a hard problem for one code is not hard for another, one gets an "algorithm separator." See Section 7.

Studying a code on hard problems may give insight into what makes the code slow and may lead to performance improvements. Profiling localizes bottlenecks to be optimized at a low level. The ease of designing hard problem families for a certain code is sometimes closely related to the code's robustness.

Next we discuss easy problems. We say that a problem family is general if problems in this family cannot be solved by a special-purpose algorithm. By easy problem families (for a given code) we mean problem families on which the code's performance compares well with it's performance on other problem instances of the same size. Easy problem families which are also general are more interesting than the non-general ones.

Easy problems give a lower bound on the code's performance. Bottlenecks for easy problems often are different from the bottlenecks for hard problems and need to be optimized separately. In some cases practical problems are easy, so optimizing for easy problems is important.

Consider, for example, a heap-based implementation of Dijkstra's algorithm [8]. Dense graphs are easy for this implementation. Both in theory and in practice, scans of the input graph's arcs dominate algorithm's running time on dense graphs (see e.g. [7]). On these graphs, internal graph representation, that allows to scan arcs going out of a vertex efficiently, is crucial. Large sparse graphs are hard for this implementation. In this case, heap operations are the bottleneck, and efficiency of the heap implementation is crucial for performance.

Designing hard problem families is not easy. Often, worst-case problem families for an algorithm are developed to demonstrate analysis tightness. The goal is to design a problem family that causes as many bottleneck operations as the upper bound of the algorithm analysis allows. See, for example, [5] for a worst-case example of an algorithm for the maximum flow problem. Another example is a graph family of [26], where the authors use a graph family on which Prim's minimum spanning tree algorithm performs a large number of decrease-key operations.

However, some worst-case examples require a sequence of bad choices on algorithm's part which is unlikely in practice. Furthermore, implementations may contain heuristics which improve performance on known worst-case examples. In particular, the above-mentioned maximum flow problems of [5] are not that hard for the better maximum flow codes studied in [6]. This motivated the authors of the latter paper to developed a family of problems which are hard for all the codes in their study.

The design of easy problem families is similar to that of hard families. However, it is usually easier to design simple problem families. In particular, special cases of a problem often lead to simple problem families.

Hard and easy problems give insight into the problem structure and parameters which are unfavorable or favorable for a code, and help in predicting the code's performance in practice.

7   Algorithm Separators

We say that a problem instance *separates* two implementations if there is a big difference in the implementation performance on this instance. Note that performance of two codes is indistinguishable unless one finds problems that separates them. Such algorithm separator problems (or problem families) are important for establishing relative algorithm performance. Ideally, one would like to find problem families on which algorithm performance difference grows with the problem size.

Solution structure, problem structure, and certain choices of parameter values may lead to separator instances. Easy and hard problems are often separators as well.

Natural algorithm separators are of special interest. Knowing that a natural problem structure is hard for one algorithm, but easy for another one, greatly contributes to understanding of these algorithms. For example, an experimental study of [7] shows that certain algorithms, in particular incremental graph algorithms [22, 29, 30], perform poorly on acyclic graphs with negative arc lengths.

Note that if the performance of two algorithms is robust, finding a problem family that separates these algorithms is hard. If one fails to find a separator for two algorithms, this is also interesting, especially if the algorithms are based on different approaches.

8   Parameter Values

One would like to understand how algorithm performance depends on the input parameter values. Since parameters may be interrelated, one is tempted to test all combination of parameters. However, this may be infeasible unless the number of parameters is very small. One should try to find out which parameters are independent and study performance dependency on parameter subsets.

Note that one does not need to report in detail on all experiments performed. For example, if the results for certain parameter combinations were similar to another combination, one can give one set of data and state that the other data was similar. The fact that an algorithm performance is robust with respect to certain parameter values is interesting in itself.

Size is one of the parameters of a problem. One would like to know how algorithm performance scales as the size goes up, with the problem structure and other parameter values staying the same. In such tests, one has to go to as big a size as feasible, because for bigger sizes, low-order terms are less significant and asymptotic bottlenecks of an algorithm are more pronounced.

Memory hierarchy is an issue for problem size selection. Small problems may fit in cache, making reference locality unimportant. Large problems may not fit in the main memory and cause paging. The context of a study determines if one should consider such small or large problems.

9   Commonly Used Problems

In some areas, computational research produced widely accepted data set for algorithm evaluation. Such data sets are important because they make it possible to compare published results and to evaluate new algorithms.

A bad data set, however, may lead to wrong conclusions and wrong directions for algorithm development. If the problem classes represented in the data set are not broad enough, codes tested exclusively on this data set may become tuned to this data set. This may lead to inaccurate prediction of practical performance and the wrong choice of the algorithm for an application.

The 40 Netgen problems [17] provide an example for such an event. These problems were generated in the early 70's. For over a decade, most computational studies of minimum cost flow algorithms used either these problems exclusively, or generated similar problems using Netgen. In the early 90's, Relax [3], a minimum-cost flow subroutine, was used in a multicommodity flow implementation [20], due to its availability and its excellent performance on Netgen-generated problems. (By this time the results of [14] suggested that Relax is not robust.) The authors of [20] discovered that on the only real-life problem in their tests (with only 49 vertices), Relax performed hundreds of times slower than several other codes. This is a good example of the danger of using a non-robust code in a new application.

Designing good common data sets is a very important part of computational study of algorithms. Such data sets, however, need to evolve. With bigger and faster computers and faster algorithms, bigger test problems may be required. New algorithms many require new easy, hard, and separator problem families. As some algorithms become obsolete, some problem families may become obsolete as well.

With the data set evolution, one has to maintain history and justification for the changes, so that old results can be compared to the new ones.

10   Choosing an Algorithm for an Application

How does one use a computational study to find the best algorithm for an application, or to estimate algorithm's performance? This is easy to do given both codes and typical problems. As we discussed earlier, however, one may want to get an estimate without the problems. We describe how to make this decision using a computational study.

First one has to read the computational study and to determine which problems or problem family most resembles the application. If the codes and generators for the computational study are available and no problem family in the study closely resembles the application, one may consider selecting parameter values and generating problems to model the applications as closely as possible.

Then one chooses an algorithm that works well for this problem family. This may involve some compromises. In particular, a more robust algorithm may be preferable over a slightly faster one, especially if the problem family models the application only approximately. Programming ease and code availability may also affect the choice.

To predict the running time on an application problem, one takes the running time of the closest available test instance. Note that the time may need adjustments to account for the difference in hardware and in problem size.

As a case study, we consider two papers. The first paper, by Cherkassky et. al [7], compares different shortest path algorithms. The second one, by Zhan and Noon [32], studies performance of the same codes on real road networks. The latter study contains two problem families, one low detail and another one – high detail. These families differ in the road levels included in the graph.

Road networks are planar, low degree graphs, and among the problems studied in [7], grid graphs model them best. Although this data is not available in [32], it appears that the low detail problems have relatively shallow shortest path trees and the high detail problems have deeper trees. We compare the former with the wide grid family of [7] and the latter with the square grid family. Clearly grids do not model road networks in full detail, and because of this the algorithm rankings differ.

Tables 1 and 2 give rankings of algorithms on the corresponding problem families. In the first table, all rankings are within one of each other except for GOR1 and THRESH. Even though the rankings of the two codes differ substantially, their performance is withing a factor of two. The rankings are closer in the second table: they match except the two fastest codes are switched

	BFP	GOR	GOR1	DIKH	DIKBD	PAPE	TWO_Q	THRESH
Wide grids	4	5	8	7	6	1	2	3
Road networks	3	4	5	8	6	1	2	7

Table 1. Low detail road networks vs. wide grid problem family: algorithm ranks.

	BFP	GOR	GOR1	DIKH	DIKBD	PAPE	TWO_Q	THRESH
Square grids	7	4	8	6	5	1	2	3
Road networks	8	4	7	6	5	2	1	3

Table 2. High detail road networks vs. square grid problem family: algorithm ranks.

and the two slowest codes are switched. The grid graphs is a fair model for the road networks, and lead to a moderately good relative performance predictions.

Based on the results of [7], one may chose chose PAPE or TWO_Q for a road network application. This agrees with the results of [32].

11   Summary

In conclusion, we summarize the process of designing a computational study. One starts with several natural, hard, and easy problem families and experiments with various parameter values to find interesting problem families and to gain a better understanding of algorithm performance. Additional problem families may be developed to clarify certain aspects of algorithm performance and to separate algorithms not yet separated. As a final step, one selects the problem families to report on in a paper documenting the research.16680001.ps The writeup should state unresolved experimental issues. For example, if the authors were unable to separate two algorithms, they should state this fact.

As we have mentioned, algorithm test sets needs to evolve as computers, algorithms, and applications develop. Feedback from applications, both in the form of comments and real-life instances, is important for evolution and for improved cooperation between algorithm theory, computational testing, and applications.

Acknowledgments

We would like to thank Boris Cherkassky, Dan Gusfield, Bernard Moret, and Cliff Stein for helpful discussions and comments.

References

1. W. Barr, B. Golden, J. Kelly, M. Resende, and W. Stewart. Designing and Reporting on Computational Experiments with Heuristic Methods. *J. Heuristics*, 1:9–32, 1995.

2. R. E. Bellman. On a Routing Problem. *Quart. Appl. Math.*, 16:87–90, 1958.

3. D. P. Bertsekas and P. Tseng. Relaxation Methods for Minimum Cost Ordinary and Generalized Network Flow Problems. *Oper. Res.*, 36:93–114, 1988.

4. C. S. Chekuri, A. V. Goldberg D. R. Karger, M. S. Levine, and C. Stein. Experimental Study of Minimum Cut Algorithms. In *Proc. 8th ACM-SIAM Symposium on Discrete Algorithms*, pages 324–333, 1997.

5. J. Cheriyan and S. N. Maheshwari. Analysis of Preflow Push Algorithms for Maximum Network Flow. *SIAM J. Comput.*, 18:1057–1086, 1989.

6. B. V. Cherkassky and A. V. Goldberg. On Implementing Push-Relabel Method for the Maximum Flow Problem. *Algorithmica*, 19:390–410, 1997.

7. B. V. Cherkassky, A. V. Goldberg, and T. Radzik. Shortest Paths Algorithms: Theory and Experimental Evaluation. *Math. Prog.*, 73:129–174, 1996.

8. E. W. Dijkstra. A Note on Two Problems in Connexion with Graphs. *Numer. Math.*, 1:269–271, 1959.

9. N. Eiron, M. Rodeh, and I. Steinwarts. Matrix Multiplication: A Case Study of Algorithm Engineering. In *Proc. 2-nd Workshop on Algorithm Engineering, Saarbruken, Germany*, pages 98–109, 1998.

10. R.E. Erickson, C.L. Monma, and A.F. Veinott Jr. Send-and-split method for minimum-concave-cost network flows. *Math. of Oper. Res.*, 12:634–664, 1979.

11. L. R. Ford, Jr. and D. R. Fulkerson. *Flows in Networks*. Princeton Univ. Press, Princeton, NJ, 1962.

12. A. V. Goldberg and M. Kharitonov. On Implementing Scaling Push-Relabel Algorithms for the Minimum-Cost Flow Problem. In D. S. Johnson and C. C. McGeoch, editors, *Network Flows and Matching: First DIMACS Implementation Challenge (Refereed Proceedings)*, pages 157–198. AMS, 1993.

13. A. V. Goldberg and B. M. E. Moret. Combinatorial Algorithms Test Sets (CATS): The ACM/EATCS Platform for Experimental Research. In *Proc. 10th ACM-SIAM Symposium on Discrete Algorithms*, pages S913–S914, 1999.

14. M. D. Grigoriadis and Y. Hsu. The Rutgers Minimum Cost Network Flow Subroutines. *SIGMAP Bulletin of the ACM*, 26:17–18, 1979.

15. D. Gusfield. University of california at davis. personal communication. 1996.

16. D. B. Johnson. How to do Experiments. Unpublished manuscript, AT&T Laboratories, 1996.

17. D. Klingman, A. Napier, and J. Stutz. Netgen: A Program for Generating Large Scale Capacitated Assignment, Transportation, and Minimum Cost Flow Network Problems. *Management Science*, 20:814–821, 1974.

18. D. Knuth. *The Stanford GraphBase*. ACM Press, New York, NY. and Addison-Wesley, Reading, MA, 1993.

19. A. LaMarca and R. E. Ladner. The Influence of Caches on the Performance of Heaps. *ACM Jour. of Experimental Algorithmics*, 1, 1996.

20. T. Leong, P. Shor, and C. Stein. Implementation of a Combinatorial Multicommodity Flow Algorithm. In D. S. Johnson and C. C. McGeoch, editors, *Network Flows and Matching: First DIMACS Implementation Challenge (Refereed Proceedings)*, pages 387–406. AMS, 1993.

21. M. S. Levine. Experimental Study of Minimum Cut Algorithms. Technical Report MIT-LCS-TR-719, MIT Lab for Computer Science, 1997.

22. B. Ju. Levit and B. N. Livshits. *Nelineinye Setevye Transportnye Zadachi*. Transport, Moscow, 1972. In Russian.

23. C. McGeoch. Towards an Experimental Method for Algorithm Simulation. *INFORMS J. Comp.*, 8:1–15, 1996.

24. A. A. Middleton. Numerical Result for the Ground-State Interface in Random Medium. *Phys. Rev. E*, 52:R3337–R3340, 1995.

25. E. F. Moore. The Shortest Path Through a Maze. In *Proc. of the Int. Symp. on the Theory of Switching*, pages 285–292. Harvard University Press, 1959.

26. B. M. E. Moret and H. D. Shapiro. An Empirical Assessment of Algorithms for Constructing a Minimum Spanning Tree. In *DIMACS Monographs in Discrete Mathematics and Theoretical Computer Science 15*, pages 99–117, 1994.

27. B. M.E. Moret and H. D. Shapiro. On Minimizing a Set of Tests. *SIAM J. Sceintific & Statistical Comp.*, 6:983–1003, 1985.

28. A. T. Ogielski. Integer Optimization and Zero-Temperature Fixed Point in Ising Random-Field Systems. *Physical Review Lett.*, 57:1251–1254, 1986.

29. S. Pallottino. Shortest-Path Methods: Complexity, Interrelations and New Propositions. *Networks*, 14:257–267, 1984.

30. U. Pape. Implementation and Efficiency of Moore Algorithms for the Shortest Root Problem. *Math. Prog.*, 7:212–222, 1974.

31. S. Roy and I. Cox. A Maximum-Flow Formulation to the Stereo Correspondence Problem. In *Proc. Int. Conf. on Computer Vision (ICCV '98), Bombay, India*, pages 492–499, 1998.

32. F. B. Zhan and C. E. Noon. Shortest Path Algorithms: An Evaluation using Real Road Networks. *Transp. Sci.*, 32:65–73, 1998.

BSP Algorithms – "Write Once, Run Anywhere"

Bill McColl

Oxford University Computing Laboratory
mccoll@comlab.ox.ac.uk
and
Sychron Ltd
mccoll@sychron.com

Scalable computing is rapidly becoming the normal form of computing. In a few years time it may be difficult to buy a computer system which has only one processor. Scalable systems will come in all shapes and sizes, from cheap PC servers and Linux clusters with a small number of processors, up to large, expensive supercomputers with hundreds or thousands of symmetric multiprocessor (SMP) nodes. An important challenge for the research community is to develop a unified framework for the design, analysis and implementation of scalable parallel algorithms.

In this talk I will describe some of the work which has been carried out on BSP computing over the last ten years. BSP offers a number of important advantages from the perspective of scalable algorithm engineering: applicability to all architectures, high performance implementations, source code portability, predictability of performance across all architectures, simple analytical cost modelling, globally compositional design style based on supersteps which facilitates development, debugging and reasoning about correctness. The BSP cost model can be used to characterise the capabilities of a scalable architecture in a concise "machine signature" which accurately describes its computation, communication and synchronisation properties. It can also be used to analyse and explore the space of possible scalable algorithms for a given problem, taking account of their differing computation, communication, synchronisation, memory and input/output requirements. Moreover, it can be used to investigate the various possible tradeoffs amongst these resource requirements.

References

1. A V Gerbessiotis and L G Valiant. Direct bulk-synchronous parallel algorithms. *Journal of Parallel and Distributed Computing*, 22:251–267, 1994.
2. J M D Hill, B McColl, D C Stefanescu, M W Goudreau, K Lang, S B Rao, T Suel, T Tsantilas, and R H Bisseling. BSPlib: The BSP programming library. *Parallel Computing*, 24(14):1947–1980, 1998.
3. W F McColl. Scalable computing. In J van Leeuwen, editor, *Computer Science Today: Recent Trends and Developments. LNCS Volume 1000*, pages 46–61. Springer-Verlag, 1995.
4. W F McColl. Foundations of time-critical scalable computing. In K Mehlhorn, editor, *Fundamentals - Foundations of Computer Science. Proc. 15th IFIP World*

J.S. Vitter and C.D. Zaroliagis (Eds.): WAE'99, LNCS 1668, pp. 12–13, 1999.

Computer Congress, 31 August - 4 September 1998, Vienna and Budapest, (Invited Paper), pages 93–107. Osterreichische Computer Gesellschaft, 1998.

5. W F McColl and A Tiskin. Memory-efficient matrix multiplication in the BSP model. *Algorithmica*, 24(3):287–297, 1999.
6. D B Skillicorn, J M D Hill, and W. F. McColl. Questions and answers about BSP. *Scientific Programming*, 6(3):249–274, Fall 1997.
7. L G Valiant. A bridging model for parallel computation. *Communications of the ACM*, 33(8):103–111, 1990.

Ten Years of LEDA
Some Thoughts

Kurt Mehlhorn

Max-Planck-Institut für Informatik
Im Stadtwald
66123 Saarbrücken
Germany
http://www.mpi-sb.mpg.de/~mehlhorn

Stefan Näher and I started the work on LEDA [LED] in the spring on 1989. Many collegues and students have contributed to the project since then. A first publication appeared in the fall of the same year [MN89]. The LEDAbook [MN99] will appear in the fall of 1999 and should be available at WAE99. *In my talk I will discuss how the work on LEDA has changed my research perspective.*

References

[LED] LEDA (Library of Efficient Data Types and Algorithms). www.mpi-sb.mpg.de/LEDA/leda.html.

[MN89] K. Mehlhorn and S. Näher. LEDA: A library of efficient data types and algorithms. In Antoni Kreczmar and Grazyna Mirkowska, editors, *Proceedings of the 14th International Symposium on Mathematical Foundations of Computer Science (MFCS'89)*, volume 379 of *Lecture Notes in Computer Science*, pages 88–106. Springer, 1989.

[MN99] K. Mehlhorn and S. Näher. *The LEDA Platform for Combinatorial and Geometric Computing*. Cambridge University Press, 1999.

J.S. Vitter and C.D. Zaroliagis (Eds.): WAE'99, LNCS 1668, pp. 14–14, 1999.
© Springer-Verlag Berlin Heidelberg 1999

Computing the K Shortest Paths: A New Algorithm and an Experimental Comparison*

Víctor M. Jiménez and Andrés Marzal

Departamento de Informática, Universitat Jaume I, Castellón, Spain
{vjimenez,amarzal}@inf.uji.es

Abstract. A new algorithm to compute the K shortest paths (in order of increasing length) between a given pair of nodes in a digraph with n nodes and m arcs is presented. The algorithm recursively and efficiently solves a set of equations which generalize the Bellman equations for the (single) shortest path problem and allows a straightforward implementation. After the shortest path from the initial node to every other node has been computed, the algorithm finds the K shortest paths in $O(m + Kn \log(m/n))$ time. Experimental results presented in this paper show that the algorithm outperforms in practice the algorithms by Eppstein [7, 8] and by Martins and Santos [15] for different kinds of random generated graphs.

1 Introduction

The problem of enumerating, in order of increasing length, the K shortest paths between two given nodes, s and t, in a digraph $G = (V, E)$ with n nodes and m arcs has many practical applications [8] and has received considerable attention in the literature [1, 2, 4, 6–10, 14, 18–22].

The asymptotically fastest known algorithm to solve this problem is due to Eppstein [7, 8]. After computing the shortest path from every node in the graph to t, the algorithm builds a graph representing all possible deviations from the shortest path. Building this graph takes $O(m + n \log n)$ time in the basic version of the algorithm, and $O(m + n)$ time in a more elaborate but "rather complicated" [7, 8] version. Once this deviations graph has been built, the K shortest paths can be obtained in order of increasing length, taking $O(\log k)$ time to compute the kth shortest path, so that the total time required to find the K shortest paths after computing the shortest path from every node to t is $O(m + n + K \log K)$. (Eppstein also solves in [7, 8] the related problem of computing the unordered set of the K shortest paths in $O(m + n + K)$ time.)

Martins' *path-deletion* algorithm [14] constructs a sequence of growing graphs $G_1, G_2, \ldots, G_K$, such that the (first) shortest path in G_k is the kth shortest path in G. In [1], Azevedo *et al.* avoid the execution of a shortest path algorithm on each graph G_k, for $k > 1$, by properly using information already computed for

---

* This work has been partially supported by Spanish CICYT under contract TIC-97-0745-C02.

G_{k-1}. In [2], a further computational improvement is proposed that reduces the number of nodes for which new computations are performed. More recently, Martins and Santos [15] have proposed a new improvement that avoids the number of arcs growing up when building G_k from G_{k-1}, thus reducing the space complexity. The total time required by the resulting algorithm to find the K shortest paths in order of increasing length after computing the shortest path from s to every node is $O(Km)$. However, in experiments reported by Martins and Santos, their algorithm outperforms in practice Eppstein's algorithm [15].

In this paper, a new algorithm that finds the K shortest s-t paths is proposed. The so-called *Recursive Enumeration Algorithm* (REA), which is described in detail in Section 3, efficiently solves a set of equations that generalize the Bellman equations for the (single) shortest path problem and allows a straightforward implementation. The algorithm recursively computes every new s-t path by visiting at most the nodes in the previous s-t path, and using a heap of candidate paths associated to each node from which the next path from s to the node is selected. In the worst case all these heaps are initialized in $O(m)$ time. Once these heaps are initialized, for $k > 1$ the kth shortest path is obtained in $O(\lambda_{k-1} \log d)$ time, where λ_{k-1} is the minimum between n and the number of nodes in the $(k-1)$th shortest path and d is the maximum input degree of the graph. The total time required to find the K shortest paths in order of increasing length after computing the shortest path from s to every node is $O(m + Kn \log(m/n))$. However, this worst case bound corresponds to a rather exceptional situation in which the one-to-all K shortest paths need to be computed. Experimental results, reported in Section 4, show that the REA outperforms in practice the algorithms by Eppstein [7,8] and by Martins and Santos [15] for different kinds of random generated graphs.

2   Problem Formulation

Let $G = (V, E)$ be a directed graph, where V is the set of nodes and $E \subseteq V \times V$ is the set of arcs. Let n be the number of nodes and m the number of arcs. Let $\ell : E \to \mathbb{R}$ be a weighting function on the arcs. The value of $\ell(u, v)$ will be called *length* of (u, v). The set of nodes $u \in V$ for which there is an arc (u, v) in E will be denoted by $\Gamma^{-1}(v)$. The nodes in $\Gamma^{-1}(v)$ are called predecessors of v. The number of nodes in $\Gamma^{-1}(v)$ is known as the input degree of v and will be denoted by $|\Gamma^{-1}(v)|$.

Given $u, v \in V$, a path from u to v is a sequence $\pi = \pi_1 \cdot \pi_2 \cdot \ldots \cdot \pi_{|\pi|} \in V^+$, where $\pi_1 = u$, $\pi_{|\pi|} = v$, and $(\pi_i, \pi_{i+1}) \in E$, for $1 \leq i < |\pi|$ (the notation $|\cdot|$ will be used both for the number of nodes in a path and the cardinality of a set). The length of a path π is $L(\pi) = \sum_{1 \leq i < |\pi|} \ell(\pi_i, \pi_{i+1})$, and $L(\pi) = 0$ if $|\pi| = 1$.

Let us consider the problem of computing the K shortest paths from a starting node $s \in V$ to a terminal node $t \in V$ in order of increasing length. For every $v \in V$, the kth shortest path from s to v will be denoted by $\pi^k(v)$, and $L^k(v) = L(\pi^k(v))$ will denote its length. Self-loops, cycles and positive and negative arc-lengths are allowed, but it will be assumed that G does not contain negative length cycles which can be reached from s (in which case the problem would have no solution). For the sake of notational simplicity we will present the new algorithm in next section assuming that there are not parallel arcs between the same pair of nodes, although it should be noted that this is not a requisite of the algorithm.

3   Recursive Enumeration Algorithm

3.1   Derivation and Correctness

The following theorem formulates the computation of the K shortest paths from s to v as the resolution of a set of recursive equations, which for $k = 1$ are the well-known Bellman equations [3] for the (single) shortest path problem.

Theorem 1 *For all $v \in V$,*

$$L^k(v) = \begin{cases} 0, & \text{if } k = 1 \text{ and } v = s; \\ \min_{\pi \in C^k(v)} L(\pi), & \text{otherwise;} \end{cases} \tag{1a}$$

$$\pi^k(v) = \begin{cases} s, & \text{if } k = 1 \text{ and } v = s; \\ \arg\min_{\pi \in C^k(v)} L(\pi), & \text{otherwise;} \end{cases} \tag{1b}$$

where if $k = 1$ and $v \neq s$, or $k = 2$ and $v = s$, then

$$C^k(v) = \{\pi^1(u) \cdot v : u \in \Gamma^{-1}(v)\}; \tag{1c}$$

otherwise, if u and k' are the node and index, respectively, such that $\pi^{k-1}(v) = \pi^{k'}(u) \cdot v$ then

$$C^k(v) = \left(C^{k-1}(v) - \{\pi^{k'}(u) \cdot v\}\right) \cup \{\pi^{k'+1}(u) \cdot v\}, \tag{1d}$$

assuming that $\{\pi^{k'+1}(u) \cdot v\}$ denotes the empty set if $\pi^{k'+1}(u)$ does not exist, which happens when $C^{k'+1}(u)$ is empty.

Proof: Let $\mathcal{P}^k(v)$ denote the set of the k shortest paths from s to v. Each path in $\mathcal{P}^k(v)$ reaches v from some node $u \in \Gamma^{-1}(v)$. In order to compute $\pi^k(v)$ we could consider, for every $u \in \Gamma^{-1}(v)$, *all paths* from s to u that do not lead to a path in $\mathcal{P}^{k-1}(v)$. However, considering that $k_1 < k_2 \implies L(\pi^{k_1}(u)) + \ell(u, v) \leq L(\pi^{k_2}(u)) + \ell(u, v)$, only *the shortest* of these paths needs to be taken into account when computing $\pi^k(v)$. Thus, we can associate to v a set of candidate paths $C^k(v)$ among which $\pi^k(v)$ can be chosen, that contains at most one path from each predecessor node $u \in \Gamma^{-1}(v)$ and is recursively defined by equation (1c-1d). $\square$

An alternative but equivalent formulation of these equations was initially given by Dreyfus [6], who also proposed an iterative algorithm for solving them in which, for every $k \geq 2$, the kth shortest path from s is computed node by node by first computing it at nodes that are one arc distant from s in the shortest path, then two arcs, etc. Fox [9, 10] pointed out the convenience of using heaps in Dreyfus' algorithm. In this way, the K shortest paths from s to *every* node in the graph are computed in $O(m + Kn \log(m/n))$ time once the shortest path from s to every node has been computed. However, this algorithm computes the K shortest paths from s to *every* node in the graph, even when we are interested only in paths between s and t.

The following algorithm, which we will refer to as the *Recursive Enumeration Algorithm* (REA), computes the K shortest paths from s to t, and constitutes a direct recursive resolution of equations (1a–1d) for $2 \le k \le K$ and $v = t$, once the shortest path from s to every other node in the graph has been computed in step A.1:

A.1 Compute $\pi^1(v)$ for all $v \in V$ by means of any adequate one-to-all shortest path algorithm and set $k \leftarrow 1$.

A.2 Repeat until $\pi^k(t)$ does not exist or no more paths are needed:

A.2.1 Set $k \leftarrow k + 1$ and compute $\pi^k(t)$ by calling $NextPath(t, k)$.

For $k > 1$, and once $\pi^1(v)$, $\pi^2(v),\dots,$ $\pi^{k-1}(v)$ are available, $NextPath(v, k)$ computes $\pi^k(v)$ as follows:

B.1 If $k = 2$, then initialize a set of candidates to the next shortest path from s to v by setting $C[v] \leftarrow \{\pi^1(u) \cdot v : u \in \Gamma^{-1}(v)$ and $\pi^1(v) \ne \pi^1(u) \cdot v\}$.

B.2 If $v = s$ and $k = 2$, then go to B.6.

B.3 Let u and k' be the node and index, respectively, such that $\pi^{k-1}(v) = \pi^{k'}(u) \cdot v$.

B.4 If $\pi^{k'+1}(u)$ has not already been computed, then compute it by calling $NextPath(u, k' + 1)$.

B.5 If $\pi^{k'+1}(u)$ exists, then insert $\pi^{k'+1}(u) \cdot v$ in $C[v]$.

B.6 If $C[v] \ne \emptyset$, then select and delete the path with minimum length from $C[v]$ and assign it to $\pi^k(v)$, else $\pi^k(v)$ does not exist.

In the case that different paths from s to v with the same length exist, any of them can be chosen first, but we assume (without loss of generality) that if Step A.1 obtains $\pi^1(v_j) = v_1 \cdot v_2 \cdot \dots \cdot v_j$, then it obtains $\pi^1(v_i) = v_1 \cdot v_2 \cdot \dots \cdot v_i$ for $1 \le i \le j$. For graphs with zero length cycles, this implies that the particular $\pi^1(v)$ computed in Step A.1 does not contain cycles.

The enumeration of paths by loop A.2 can be finished once the K shortest paths from s to t have been obtained. However, the number of paths to be computed does not really need to be fixed *a priori*, and the algorithm can be used in general to enumerate the shortest paths until the first one satisfying some desired restriction is obtained.

Note that $C[v]$ in the algorithm corresponds to $C^k(v)$ in Theorem 1. Steps B.1–B.5 compute $C^k(v)$ from $C^{k-1}(v)$, according to equations (1c–1d) (Step B.1 initializes $C[v]$ when the second shortest path from s to v is required). In Step B.6, $\pi^k(v)$ is selected from $C^k(v)$ according to (1b).

The following theorems prove that the recursive procedure terminates and indicate interesting properties of the algorithm. Theorem 2 proves that the recursive calls to *NextPath* to compute $\pi^k(t)$ visit, in the worst case, all the nodes in $\pi^{k-1}(t)$. Theorem 3 proves that, in the case that $\pi^{k-1}(t)$ contains a loop, the recursive calls finish before visiting the same node twice.

Theorem 2 *For $k > 1$ and for all $v \in V$, the computation of $\pi^k(v)$ by means of* NextPath(v, k) *may recursively generate calls to* NextPath(u, j) *only for nodes u in $\pi^{k-1}(v)$.*

Proof: Let us suppose that $\pi^{k-1}(v) = u_1 u_2 \ldots u_p$ (where $u_1 = s$ and $u_p = v$). For every $i = 1, \ldots, p$, let k_i be the index such that $\pi^{k_i}(u_i) = u_1 u_2 \ldots u_i$. Since $\pi^{k-1}(v) = \pi^{k_{p-1}}(u_{p-1}) \cdot v$, *NextPath*$(v, k)$ may require a recursive call to *NextPath*$(u_{p-1}, k_{p-1} + 1)$ in case $\pi^{k_{p-1}+1}(u_{p-1})$ has not been already computed; since $\pi^{k_{p-1}}(u_{p-1}) = \pi^{k_{p-2}}(u_{p-2}) \cdot u_{p-1}$, *NextPath*$(u_{p-1}, k_{p-1} + 1)$ may require a recursive call to *NextPath*$(u_{p-2}, k_{p-2} + 1)$; and so on. In the worst case, the recursive calls extend through the nodes $u_p, u_{p-1}, \ldots, u_1$. Since $u_1 = s = \pi^1(s)$, if the recursion reaches u_1, that is, if *NextPath*$(s, 2)$ is called to compute $\pi^2(s)$, then the condition in Step B.2 holds and no more recursive calls are performed. □

Theorem 3 *For $k > 1$ and for all $v \in V$, computing $\pi^k(v)$ by* NextPath(v, k) *does not generate a recursive call to* NextPath(v, j) *for any j.*

Proof: Let us suppose that $\pi^{k-1}(v) = u_1 u_2 \ldots u_p$ (where $u_1 = s$ and $u_p = v$) contains $u_i = u_p$, $i < p$. Due to the condition in Step B.4, recursive calls to *NextPath* through the nodes of $\pi^{k-1}(v)$ can reach u_{i+1} only if $u_1 u_2 \ldots u_{i+1}$ is the last computed path ending at u_{i+1}. In that case, the index k' in Step B.3 is the position of the path $u_1 u_2 \ldots u_i$ in the list of shortest paths from s to u_i, since it is obvious that $u_1 u_2 \ldots u_{i+1} = (u_1 u_2 \ldots u_i) \cdot u_{i+1}$. Since at least the path $u_1 u_2 \ldots u_i \ldots u_j$ ending at u_j (where $u_j = u_i$) has already been generated, Step B.4 detects that $\pi^{k'+1}(u_i)$ has already been computed and *NextPath* is not invoked at u_i. □

3.2 Data Structures and Computational Complexity

Representation of Paths. The paths $\pi^1(v), \pi^2(v), \ldots$ ending at v can be dynamically stored (by increasing length) in Step B.6 in a linked list connected to v; on the other hand, each path $\pi^k(v) = \pi^j(u) \cdot v$ can be compactly represented by its length $L^k(v)$ and a back-pointer to the path $\pi^j(u)$ in the predecessor node u (see Fig. 1). In this way, each one of the K shortest paths from s to t can be recovered at the end of the algorithm in time proportional to the number of nodes in the path, following the back-pointers from the list associated with node t, as is usual in (single) shortest path algorithms (see for example [5]).

The following example, depicted in Fig. 1, illustrates how *NextPath* would operate with this representation. Let us assume that $\Gamma^{-1}(t) = \{u, v, w\}$ and

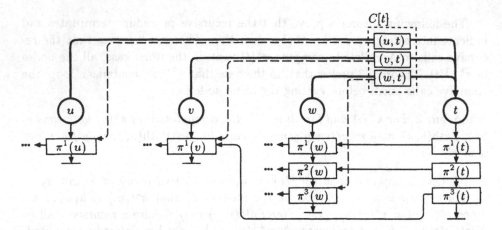

Fig. 1. Representation of paths and sets of candidates in the REA. Thick lines represent nodes and arcs in G. The 3 shortest paths reaching node t have been computed: $\pi^1(t) = \pi^1(w) \cdot t$, $\pi^2(t) = \pi^2(w) \cdot t$, and $\pi^3(t) = \pi^1(v) \cdot t$. These paths are arranged in a linked list hanging from node t. Each path is represented by a back-pointer to another path in a predecessor node. The set of candidates $C[t]$ contains one element for each incoming arc, pointing to the last considered path in the corresponding predecessor node.

$\pi^3(t) = \pi^1(v) \cdot t$ was chosen as the best path in $C[t] = C^3(t) = \{\pi^1(u) \cdot t, \pi^1(v) \cdot t, \pi^3(w) \cdot t\}$. When *NextPath* is called in order to compute $\pi^4(t)$, Step B.3 reaches $\pi^1(v)$ in time $O(1)$ through the back-pointer in $\pi^3(t)$ and Step B.4 checks in time $O(1)$ whether $\pi^2(v)$ has been previously computed by looking for the path following $\pi^1(v)$ in the list of paths associated with v. If it hasn't, *NextPath* is called to compute it. When this recursive call ends, and if we assume that $\pi^2(v)$ exists, Step B.5 inserts $\pi^2(v) \cdot t$ (scored by $L^2(v) + \ell(v,t)$) into the set of candidates associated with t. Then, Step B.6 selects the best candidate from $C[t] = C^4(t) = \{\pi^1(u) \cdot t, \pi^2(v) \cdot t, \pi^3(w) \cdot t\}$ and links it following $\pi^3(t)$ in the list of paths associated with t.

Representation of Sets of Candidates. Several alternative data structures can be used in practice to efficiently handle the sets of candidates. Given that each node keeps one candidate for each incoming arc, the total space required by all the sets candidates is $O(m)$. In each call to *NextPath*, at most one candidate is inserted in Step B.5, and one candidate is extracted in Step B.6, so that the total size of the set of candidates remains constant or decreases by one.

1. If the sets of candidates are implemented with heaps, then Step B.5 requires time $O(\log |\Gamma^{-1}(v)|)$, and step B.6 requires time $O(1)$ by postponing the deletion of the minimum element in the heap until a new insertion is performed by step B.5 on the same heap. Step B.1 can be performed in time $O(|\Gamma^{-1}(v)|)$ with *heap-build* [5].

2. If the sets of candidates are kept unsorted, then Step B.5 requires $O(1)$ time and Step B.6 requires $O(|\Gamma^{-1}(v)|)$ time, for every call to *NextPath*. This

is asymptotically worse than using heaps. However, it is well known that in practice an unsorted array can be more efficient than a heap in certain cases.

3. To benefit from the asymptotic behavior of heaps but avoid a costly initialization that may not be amortized unless a large number of paths is computed, a hybrid data structure can be used in which the set of candidates is implemented as an array that is in part unsorted and in part structured as a heap. Initially, the full array is unsorted; whenever a candidate in the array is selected, the next path obtained from it is inserted in the heap part. The best candidate in Step B.6 must be chosen between the best candidate in the heap and the best candidate in the rest of array (which is computed only when it is unknown). Experimental results that are reported in Section 4 show the convenience of this approach in some cases.

Computational Complexity. Let us analyze the computation time required by the algorithm (once the shortest paths tree is available) in case that the sets of candidates are implemented with heaps. Since Step B.1 is done at most once for each node in the graph, in the worst case all the sets of candidates are initialized in total time $O(m)$. The recursive calls to *NextPath* to compute $\pi^k(t)$ go through the nodes of $\pi^{k-1}(t)$ (Theorem 2), and never visit the same node twice (Theorem 3). Hence the number of recursive calls needed to compute $\pi^k(t)$ is upper bounded by $\lambda_{k-1} = \min(n, |\pi^{k-1}(t)|)$. At any given time, the set $C[v]$ contains at most one candidate for each predecessor node of v and the time required by Steps B.2–B.6, when executed on node v, is $O(\log |\Gamma^{-1}(v)|)$. In the worst case, $\lambda_{k-1} = n$ and then the total running time of Steps B.2–B.6 in the recursive calls to compute $\pi^k(t)$ is $O(n \log(m/n))$, since $\sum_{v \in V} |\Gamma^{-1}(v)| = m$ and $\sum_{v \in V} \log |\Gamma^{-1}(v)|$ is maximized when all the terms are of equal size. Hence, the total time required by the REA to find the K shortest paths in order of increasing length after computing the shortest path from s to every node is $O(m + Kn \log(m/n))$.

We find worth to emphasize that, as a consequence of Theorems 2 and 3, the number of recursive calls generated by *NextPath*(t, k) is at most $\min(n, |\pi^{k-1}(t)|)$. Therefore, once the heaps are initialized, for $k > 1$ the kth shortest path is obtained in $O(\lambda_{k-1} \log d)$ time, where d is the maximum input degree of the graph. The number of recursive calls could even be less than λ_{k-1} if at some intermediate node the next path has already been previously computed, in which case the recursion ends. The worst case bound $O(m + Kn \log(m/n))$ corresponds in the REA to a rather exceptional situation in which *NextPath*(t, k) generates n recursive calls for all $k = 2, \ldots, K$. This can only happen if all the $K - 1$ shortest s-t paths have at least n nodes. However, in many practical situations the shortest paths are composed by a small fraction of the nodes in the graph, in which case the time required by the REA to compute the kth shortest path may be negligible. An experimental study, described in the next section, has been done in order to assess the efficiency of the REA in practice and to compare it with alternative algorithms.

The algorithm can also be extended to find the K shortest paths from s to every node in a set $T \subseteq V$ by just calling *NextPath*(t, k) for each $t \in T$ and for each $k = 2, \ldots, K$, if $\pi^k(t)$ has not been previously computed. In this way, instead of starting the computation from scratch for each t, the previous computations are reused (*NextPath* will also generate recursive calls only if the required paths have not been previously computed). The worst case time complexity of this algorithm is still $O(m + Kn \log(m/n))$. In the particular case $T = V$ it constitutes an alternative to the algorithm by Dreyfus and Fox for the one-to-all K shortest paths computation, with the same time complexity but differing from it in that the nodes do not need to be processed in any particular order. Also, the number of paths to be computed can be different for each node and can be decided dynamically, using the algorithm to obtain alternative paths only when they are needed.

If K is known *a priori*, a further computational improvement is possible: since the algorithm will select at most the K best paths from $C[v]$, in nodes v in which $K < |\Gamma^{-1}(v)|$ the size of the heap that represents $C[v]$ can be limited to K, so that Step B.5 requires time $O(\log K)$ instead of $O(\log |\Gamma^{-1}(v)|)$ and the worst case time complexity of the algorithm becomes $O(m + Kn \log(\min(K, m/n)))$.

4 Experimental Comparison

In this section, experimental results are reported comparing, on three different kinds of random generated graphs, the time required to compute the K shortest s-t paths by the Martins and Santos' algorithm [15] (MSA), the basic (more practical) version of Eppstein's algorithm [7,8] (EA), and the Recursive Enumeration Algorithm with two different representations of the sets of candidates: heaps (REA) and the hybrid data structures (in part heaps and in part unsorted) described in the previous section (HREA).

Implementation of the Algorithms. We have used the implementation of MSA made publicly available by Martins (http://www.mat.uc.pt/~eqvm). EA, REA and HREA have been implemented by the authors of this work and are also publicly available (ftp://terra.act.uji.es/pub/REA). All the programs have been implemented in C.

Experimental Methodology and Computing Environment. Since MSA uses a different implementation of Dijkstra's algorithm and we are interested in comparing the performance of the algorithms once the shortest path tree has been computed, all time measurements start when Dijkstra's algorithm ends. All the implementations share the routines used to measure the CPU running time (with a resolution of 0.01 seconds). Each point in the curves shows the average execution time for 15 random graphs generated with the same parameters, but different random seeds. In all cases the greatest arc length was chosen to be 10000.

The programs have been compiled with the GNU C compiler (version 2.7.2.3) using the maximum optimization level. The experiments have been performed on a 300 MHz Pentium-II computer with 256 megabytes of RAM, running under Linux 2.0.

4.1   Results for Graphs Generated with Martins' General Instances Generator

First, we compared the algorithms using Martins' general instances generator (geraK), the same used in the experiments reported in [15] (and also made publicly available by Martins at http://www.mat.uc.pt/~eqvm). The input to this graph generator are four values: seed for the random number generator, number of nodes, number of arcs, and maximum arc length. The program creates the specified number of nodes and joins them with a Hamiltonian cycle to assure that the start and terminal nodes are connected; then, it completes the set of arcs by randomly choosing pairs of nodes. The arc lengths are generated uniformly distributed in the specified range.

Fig. 2 represents the CPU time required to compute up to 100, up to 10000, and up to 100000 paths by each of the algorithms with this kind of graphs, for different values of the number of nodes and number of arcs.

Comparing the results of REA and HREA, it can be observed that heaps (REA) are preferable for small values of the average input degree, while the hybrid sets of candidates (HREA) are preferable for large values of the average input degree (when initializing the heaps is more costly). The results using totally unsorted sets of candidates were worse than the results with heaps in all the experiments and are not included in the figures for the sake of clarity.

Regarding the behavior of Eppstein's algorithm, it can be observed that, once the graph of deviation paths has been built, the K shortest paths are found very efficiently. However, building this graph requires (in comparison with the other algorithms) a considerable amount of time that can be clearly identified in the figures at the starting point $K = 2$.

4.2   Results for Multistage Graphs

Multistage graphs are of interest in many applications and underlie many discrete Dynamic Programming problems. A multistage graph is a graph whose set of nodes can be partitioned into S disjoint sets (stages), $V_1, V_2, \ldots, V_S$, such that every arc in E joins a node in V_i with a node in V_{i+1}, for some i such that $1 \le i < S$. We have implemented a program that generates random multistage graphs with the specified number of stages, number of nodes per stage, and input degree (which is fixed to the same value for all the nodes). Arc lengths are generated uniformly distributed between 0 and the specified maximum value. The results with these kind of graphs are presented in Figs. 3 and 4.

Fig. 3 shows that the dependency with the number of paths is similar to what has been observed for the graphs generated by geraK. However, in this case EA is the fastest algorithm to compute $K = 10000$ paths when the input degree is $d = 10$ and the number of stages is $S = 1000$ (Fig. 3b), while REA is the fastest algorithm for smaller values of K (Fig. 3a-3b), smaller values of S (Fig. 3c) or larger values of d (Fig. 3d).

Figs. 4a and 4b represent the dependency of the running time with the number of stages, to compute up to 1000 and up to 10000 paths, respectively. These

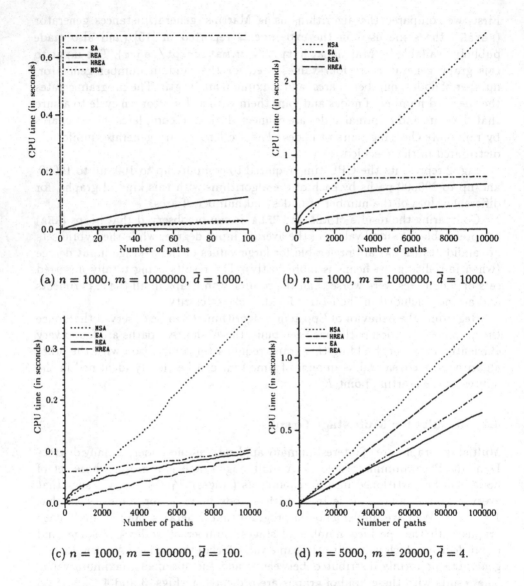

(a) $n = 1000$, $m = 1000000$, $\bar{d} = 1000$.

(b) $n = 1000$, $m = 1000000$, $\bar{d} = 1000$.

(c) $n = 1000$, $m = 100000$, $\bar{d} = 100$.

(d) $n = 5000$, $m = 20000$, $\bar{d} = 4$.

Fig. 2. Experimental results for graphs generated with Martins' general instances generator. CPU time as a function of the number of computed paths. (a) is an enlargement of the initial region of (b) to appreciate more clearly the differences between the algorithms for small values of K. (Parameters: n, number of nodes; m, number of arcs; $\bar{d} = m/n$, average input degree.)

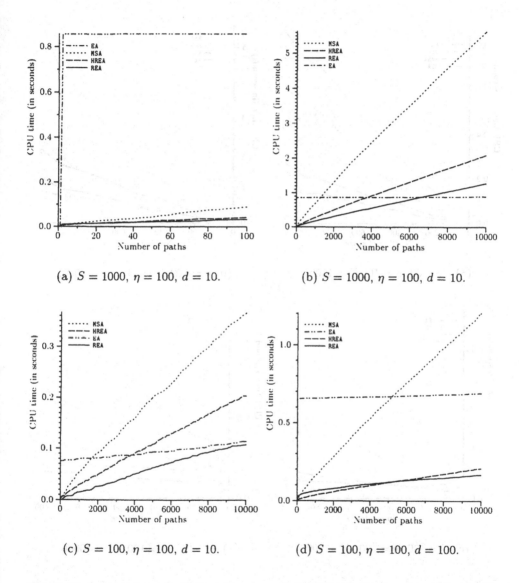

Fig. 3. Experimental results for multistage graphs. CPU time as a function of the number of computed paths. (a) is an enlargement of the initial region of (b). (Parameters: S, number of stages; η, number of nodes per stage; d, input degree.)

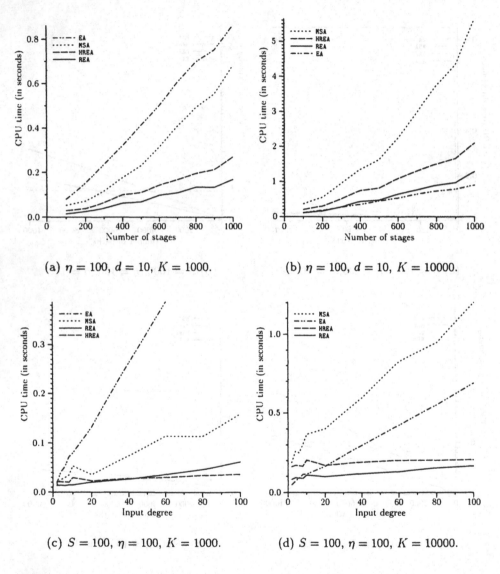

(a) $\eta = 100$, $d = 10$, $K = 1000$.

(b) $\eta = 100$, $d = 10$, $K = 10000$.

(c) $S = 100$, $\eta = 100$, $K = 1000$.

(d) $S = 100$, $\eta = 100$, $K = 10000$.

Fig. 4. Experimental results for multistage graphs. CPU time as a function of the number of stages (a and b) and the input degree (c and d). (Parameters: S, number of stages; η, number of nodes per stage; d, input degree; K, number of computed paths.)

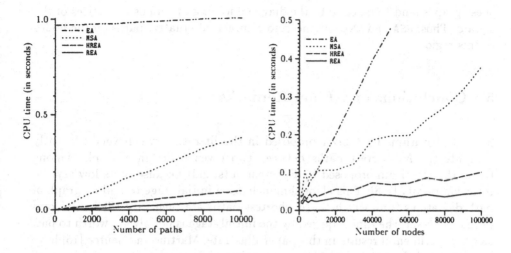

Fig. 5. Results for Delaunay triangulation graphs. CPU time as a function of (a) the number of computed paths (with 100000 nodes) and (b) the number of nodes (to compute 10000 paths).

figures illustrate the evolution from Fig. 3c to Fig. 3b. In the case of MSA and REA, the dependency with the number of stages is due to the fact that $\pi^k(t)$ is computed by visiting, in the worst case, all the nodes in $\pi^{k-1}(t)$, and in multistage graphs the number of nodes in any path is the number of stages.

Figs. 4c and 4d represent the dependency of the running time with the input degree to compute up to 1000 and up to 10000 paths, respectively. In these figures the input degree ranges from 2 to 100, and this includes the evolution from Fig. 3c to Fig. 3d, as well as the behavior for smaller values of the input degree. Observe that the time required by REA and HREA only increases slightly when the input degree increases, while MSA and EA are clearly more affected by this parameter, so that the difference between both algorithms and REA increases as the input degree does.

4.3   Results for Delaunay Triangulation Graphs

Delaunay triangulations are a particular kind of graphs that can model communication networks. We have implemented a graph generator that uniformly distributes a given number of points in a square and then computes their Delaunay triangulation and assigns to each arc the Euclidean distance between the two joined points (nodes). The initial and final nodes are located in opposite vertices of the square.

The results can be observed in Fig. 5. In this case EA is far from being competitive even to compute as many as 10000 paths and even though the average degree in these graphs is typically lower than 10. It can be clearly observed how the difference between EA and the other algorithms increases as the number of nodes does. This can be explained by the fact that the K shortest paths in these graphs tend to be close to the diagonal joining the opposite vertices of the square. Thus, MSA and REA only need to compute alternative paths on the nodes of this region.

5   Conclusions and Final Remarks

Several algorithms have been proposed in the literature which very efficiently compute the K shortest paths between two given nodes in a graph. Among these, the algorithm proposed by Eppstein outstands because of its low asymptotic complexity [7,8]. This algorithm includes a initial stage to build a graph of path deviations from which the K shortest paths are then very efficiently computed. However, the time required by the initial stage is not always worth to pay, as the experimental results in this paper illustrate. Martins and Santos [15] have presented a different algorithm that, under certain circumstances, runs faster in practice. A new algorithm proposed in this paper, the Recursive Enumeration Algorithm (REA), is another useful, practical alternative according to the experimental results. Experiments with different kinds of random generated graphs and for many different settings of the parameters determining the size of the problem have shown the superiority of the new method in many practical situations. The REA is specially well suited for graphs in which shortest paths are composed by a small fraction of the nodes in the graph. On the other hand, the REA can be derived from a set of equations that have been formally proved to solve the problem, it relies on quite simple data structures, and it can be easily implemented.

The REA has been extended to enumerate the N-best sentence hypotheses in speech recognition [11–13,17]. This is a related problem that can be modeled as the search for the K shortest paths in a multistage graph, but with the additional requirement that only paths with different associated word sequences are of interest. The REA, modified to discard at intermediate nodes those partial paths whose associated word sequence has already been generated, has been applied in a task involving the recognition of speech queries to a geographical database in which the underlying multistage graph has (in average per utterance) about $2 \cdot 10^6$ nodes and 450 stages [13], and in a city-name spelling recognition task in which the underlying multistage graph has about $2 \cdot 10^5$ nodes and 272 stages [12]. The REA has also been extended to compute the K best solutions to general discrete Dynamic Programming problems [16].

Acknowledgement

The authors wish to thank E. Q. V. Martins and J. L. E. Santos at the University of Coimbra for making the implementation of their algorithm and their random graph generator publicly available.

References

1. Azevedo, J.A., Costa, M.E.O.S., Madeira, J.J.E.R.S., Martins, E.Q.V.: An Algorithm for the Ranking of Shortest Paths. European J. Op. Res. **69** (1993) 97–106
2. Azevedo, J.A., Madeira, J.J.E.R.S., Martins, E.Q.V., Pires, F.P.A.: A Computational Improvement for a Shortest Paths Ranking Algorithm. European J. Op. Res. **73** (1994) 188–191
3. Bellman, R.: On a Routing Problem. Quarterly Applied Math. **16** (1958) 87–90
4. Bellman, R., Kalaba, R.: On kth Best Policies. J. SIAM **8** (1960) 582–588
5. Cormen, T.H., Leiserson, C.E., Rivest, R.L.: Introduction to Algorithms. The MIT Press, Cambridge, MA (1990)
6. Dreyfus, S.E.: An Appraisal of Some Shortest Path Algorithms. Op. Res. **17** (1969) 395–412
7. Eppstein, D.: Finding the k Shortest Paths. In: Proc. 35th IEEE Symp. FOCS (1994) 154–165
8. Eppstein, D.: Finding the k Shortest Paths. SIAM J. Computing **28(2)** (1999) 652–673
9. Fox, B.L.: Calculating kth Shortest Paths. INFOR - Canad. J. Op. Res. and Inform. Proces. **11(1)** (1973) 66–70
10. Fox, B.L.: Data Structures and Computer Science Techniques in Operations Research. Op. Res. **26(5)** (1978) 686–717
11. Jiménez, V.M., Marzal, A.: A New Algorithm for Finding the N-Best Sentence Hypotheses in Continuous Speech Recognition. In: Casacuberta, F., Sanfeliu, A. (eds.): Advances in Pattern Recognition and Applications. World Scientific (1994) 218–228. Translated from Proc. V Symp. of Spanish AERFAI (1992) 180–187
12. Jiménez, V.M., Marzal, A., Monné, J.: A Comparison of Two Exact Algorithms for Finding the N-Best Sentence Hypotheses in Continuous Speech Recognition. In: Proc. 4th ESCA Conf. EUROSPEECH (1995) 1071–1074
13. Jiménez, V.M., Marzal, A., Vidal, E.: Efficient Enumeration of Sentence Hypotheses in Connected Word Recognition. In: Proc. 3rd ESCA Conf. EUROSPEECH (1993) 2183–2186
14. Martins, E. Q. V.: An Algorithm for Ranking Paths that may Contain Cycles. European J. Op. Res. **18** (1984) 123–130
15. Martins, E.Q.V., Santos, J.L.E.: A New Shortest Paths Ranking Algorithm. Technical report, Univ. de Coimbra, http://www.mat.uc.pt/~eqvm (1996)
16. Marzal, A.: Cálculo de las K Mejores Soluciones a Problemas de Programación Dinámica. PhD thesis (in Spanish) , Univ. Politécnica de Valencia, Spain (1994)
17. Marzal, A., Vidal, E.: A N-best sentence hypotheses enumeration algorithm with duration constraints based on the two level algorithm. In Proc. of the Int. Conf. on Pattern Recognition (1992)
18. Miaou, S.P., Chin, S.M.: Computing K-Shortest Paths for Nuclear Spent Fuel Highway Transportation. European J. Op. Res. **53** (1991) 64–80
19. Shier, D.R.: Iterative Methods for Determining the k Shortest Paths in a Network. Networks **6** (1976) 205–229
20. Shier, D.R.: On Algorithms for Finding the k Shortest Paths in a Network. Networks **9** (1979) 195–214
21. Skicism, C.C., Golden, B.L.: Computing k-Shortest Path Lengths in Euclidean Networks. Networks **17** (1987) 341–352
22. Skicism, C.C., Golden, B.L.: Solving k-Shortest and Constrained Shortest Path Problems Efficiently. Annals of Op. Res. **20** (1989) 249–282

Efficient Implementation of Lazy Suffix Trees

Robert Giegerich[1], Stefan Kurtz[1]*, and Jens Stoye[2]

[1] Technische Fakultät, Universität Bielefeld, Postfach 100 131
D-33501 Bielefeld, Germany
{robert,kurtz}@techfak.uni-bielefeld.de
[2] German Cancer Research Center (DKFZ), Theoretical Bioinformatics (H0300)
Im Neuenheimer Feld 280, D-69120 Heidelberg, Germany
j.stoye@dkfz-heidelberg.de

Abstract. We present an efficient implementation of a write-only top-down construction for suffix trees. Our implementation is based on a new, space-efficient representation of suffix trees which requires only 12 bytes per input character in the worst case, and 8.5 bytes per input character on average for a collection of files of different type. We show how to efficiently implement the lazy evaluation of suffix trees such that a subtree is evaluated not before it is traversed for the first time. Our experiments show that for the problem of searching many exact patterns in a fixed input string, the lazy top-down construction is often faster and more space efficient than other methods.

1 Introduction

Suffix trees are efficiency boosters in string processing. The suffix tree of a text t is an index structure that can be computed and stored in $O(|t|)$ time and space. Once constructed, it allows to locate any substring w of t in $O(|w|)$ steps, independent of the size of t. This instant access to substrings is most convenient in a "myriad" [2] of situations, and in Gusfield's recent book [9], about 70 pages are devoted to applications of suffix trees.

While suffix trees play a prominent role in algorithmics, their practical use has not been as widespread as one should expect (for example, Skiena [16] has observed that suffix trees are the data structure with the highest need for better implementations). The following pragmatic considerations make them appear less attractive:

- The linear-time constructions by Weiner [18], McCreight [15] and Ukkonen [17] are quite intricate to implement. (See also [7] which reviews these methods and reveals relationships much closer than one would think.)
- Although asymptotically optimal, their poor locality of memory reference [6] causes a significant loss of efficiency on cached processor architectures.
- Although asymptotically linear, suffix trees have a reputation of being greedy for space. For example, the efficient representation of McCreight [15] requires 28 bytes per input character in the worst case.

---

* Partially supported by DFG-grant Ku 1257/1-1.

J.S. Vitter and C.D. Zaroliagis (Eds.): WAE'99, LNCS 1668, pp. 30–42, 1999.
© Springer-Verlag Berlin Heidelberg 1999

– Due to these facts, for many applications, the construction of a suffix tree does not amortize. For example, if a text is to be searched only for a very small number of patterns, then it is usually better to use a fast and simple online method, such as the Boyer-Moore-Horspool algorithm [11], to search the complete text anew for each pattern.

However, these concerns are alleviated by the following recent developments:

– In [6], Giegerich and Kurtz advocate the use of a write-only, top-down construction, referred to here as the *wotd*-algorithm. Although its time efficiency is $O(n \log n)$ in the average and even $O(n^2)$ in the worst case (for a text of length n), it is competitive in practice, due to its simplicity and good locality of memory reference.
– In [12], Kurtz developed a space-efficient representation that allows to compute suffix trees in linear time in 46% less space than previous methods. As a consequence, suffix trees for large texts, e.g. complete genomes, have been proved to be manageable.
– The question about amortizing the cost of suffix tree construction is almost eliminated by incrementally constructing the tree as demanded by its queries. This possibility was already hinted at in [6], where the *wotd*-algorithm was called "lazytree" for this reason.

When implementing the *wotd*-algorithm in a lazy functional programming language, the suffix tree automatically becomes a lazy data structure, but of course, the general overhead of using a lazy language is incurred. In the present paper, we explicate how a lazy and an eager version of the *wotd*-algorithm can efficiently be implemented in an imperative language. Our implementation technique avoids a constant alphabet factor in the running time.[1] It is based on a new space efficient suffix tree representation, which requires only $12n$ bytes of space in the worst case. This is an improvement of $8n$ bytes over the most space efficient previous representation, as developed in [12]. Experimental results show that our implementation technique leads to programs that are superior to previous ones in many situations. For example, when searching $0.1n$ patterns of length between 10 and 20 in a text of length n, the lazy *wotd*-algorithm (*wotdlazy*, for short) is on average almost 35% faster and 30% more space efficient than a linked list implementation of McCreight's [15] linear time suffix tree algorithm. *wotdlazy* is almost 13% faster and 50% more space efficient than a hash table implementation of McCreight's linear time suffix tree algorithm, eight times faster and 10% more space efficient than a program based on suffix arrays [13], and *wotdlazy* is 99 times faster than the iterated application of the Boyer-Moore-Horspool algorithm [11]. The lazy *wotd*-algorithm makes suffix trees also applicable in contexts where the expected number of queries to the text is small relative to the

---

[1] The suffix array construction of [13] and the linear time suffix tree construction of [5] also do not have the alphabet factor in their running time. For the linear time suffix tree constructions of [15, 17, 18] the alphabet factor can be avoided by employing hashing techniques, see [15], however, for the cost of using considerably more space, see [12].

of the text, with an almost immeasurable overhead compared to its eager variant *wotdeager* in the opposite case. Beside its usefulness for searching string patterns, *wotdlazy* is interesting for other problems (see the list in [9]), such as exact set matching, the substring problem for a database of patterns, the DNA contamination problem, common substrings of more than two strings, circular string linearization, or computation of the q-word distance of two strings. Documented source code, test data, and complete results of our experiments are available at http://www.techfak.uni-bielefeld.de/~kurtz/Software/wae99.tar.gz.

2   The *wotd*-Suffix Tree Construction

2.1   Terminology

Let Σ be a finite ordered set of size k, the *alphabet*. Σ^* is the set of all strings over Σ, and ε is the *empty string*. We use Σ^+ to denote the set $\Sigma^* \setminus \{\varepsilon\}$ of non-empty strings. We assume that t is a string over Σ of length $n \geq 1$ and that $\$ \in \Sigma$ is a character not occurring in t. For any $i \in [1, n+1]$, let $s_i = t_i \ldots t_n\$$ denote the ith non-empty suffix of $t\$$. A Σ^+-*tree* T is a finite rooted tree with edge labels from Σ^+. For each $a \in \Sigma$, every node u in T has at most one a-edge $u \xrightarrow{av} w$ for some string v and some node w. An edge leading to a leaf is a *leaf edge*. Let u be a node in T. We denote u by $\overline{w}$ if and only if w is the concatenation of the edge labels on the path from the *root* to u. $\overline{\varepsilon}$ is the *root*. A string s *occurs* in T if and only if T contains a node $\overline{sv}$, for some string v. The *suffix tree* for t, denoted by $ST(t)$, is the Σ^+-tree T with the following properties: (i) each node is either a leaf or a branching node, and (ii) a string w occurs in T if and only if w is a substring of $t\$$. There is a one-to-one correspondence between the non-empty suffixes of $t\$$ and the leaves of $ST(t)$. For each leaf $\overline{s_j}$ we define $\ell(\overline{s_j}) = \{j\}$. For each branching node $\overline{u}$ we define $\ell(\overline{u}) = \{j \mid \overline{u} \xrightarrow{v} \overline{uv} \text{ is an edge in } ST(t), j \in \ell(\overline{uv})\}$. $\ell(\overline{u})$ is the *leaf set* of $\overline{u}$.

2.2   A Review of the *wotd*-Algorithm

The *wotd*-algorithm adheres to the recursive structure of a suffix tree. The idea is that for each branching node $\overline{u}$ the subtree below $\overline{u}$ is determined by the set of all suffixes of $t\$$ that have u as a prefix. In other words, if we have the set $R(\overline{u}) := \{s \mid us \text{ is a suffix of } t\$\}$ of *remaining suffixes* available, we can evaluate the node $\overline{u}$. This works as follows: at first $R(\overline{u})$ is divided into groups according to the first character of each suffix. For any character $c \in \Sigma$, let $group(\overline{u}, c) := \{w \in \Sigma^* \mid cw \in R(\overline{u})\}$ be the c-*group of* $R(\overline{u})$. If for some $c \in \Sigma$, $group(\overline{u}, c)$ contains only one string w, then there is a leaf edge labeled cw outgoing from $\overline{u}$. If $group(\overline{u}, c)$ contains at least two strings, then there is an edge labeled cv leading to a branching node $\overline{ucv}$, where v is the longest common prefix (*lcp*, for short) of all strings in $group(\overline{u}, c)$. The child $\overline{ucv}$ can then be evaluated from the set $R(\overline{ucv}) = \{w \mid vw \in group(\overline{u}, c)\}$ of remaining suffixes.

The *wotd*-algorithm starts by evaluating the *root* from the set $R(root)$ of all suffixes of $t\$$. All nodes of $ST(t)$ can be evaluated recursively from the corresponding set of remaining suffixes in a top-down manner.

Example Consider the input string $t = abab$. The *wotd*-algorithm for t works as follows: At first, the *root* is evaluated from the set $R(root)$ of all non-empty suffixes of the string $t\$$, see the first five columns in Fig. 1. The algorithm recognizes 3 groups of suffixes. The a-group, the b-group, and the $\$$-group. The a-group and the b-group each contain two suffixes, hence we obtain two unevaluated branching nodes, which are reached by an a-edge and by a b-edge. The $\$$-group is singleton, so we obtain a leaf reached by an edge labeled $\$$. To evaluate the unevaluated branching node corresponding to the a-group, one first computes the longest common prefix of the remaining suffixes of that group. This is b in our case. So the a-edge from the *root* is labeled by ab, and the remaining suffixes $ab\$$ and $\$$ are divided into groups according to their first character. Since this is different, we obtain two singleton groups of suffixes, and thus two leaf edges outgoing from $\overline{ab}$. These leaf edges are labeled by $ab\$$ and $\$$. The unevaluated branching node corresponding to the b-group is evaluated in a similar way, see Fig. 1.

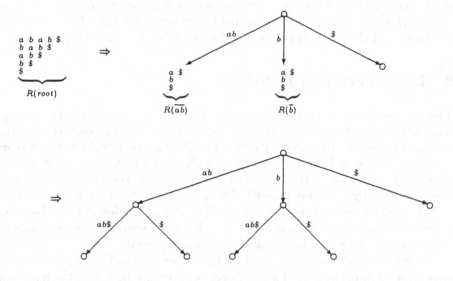

Fig. 1. The write-only top-down construction of $ST(abab)$

2.3   Properties of the *wotd*-Algorithm

The distinctive property of the *wotd*-algorithm is that the construction proceeds top-down. Once a node has been constructed, it needs not be revisited in the construction of other parts of the tree (unlike the linear-time constructions of [5, 15, 17, 18]). As the order of subtree construction is independent otherwise, it may be arranged in a demand-driven fashion, obtaining the lazy implementation detailed in the next section.

The top-down construction has been mentioned several times in the literature [1, 6, 8, 14], but at the first glance, its worst case running time of $O(n^2)$ is
disappointing. However, the expected running time is $O(n \log_k n)$ (see e.g. [6]),
and experiments in [6] suggest that the *wotd*-algorithm is practically linear for
moderate size strings. This can be explained by the good locality behavior: the
wotd-algorithm has optimal locality on the tree data structure. In principle,
more than a "current path" of the tree needs not be in memory. With respect
to text access, the *wotd*-algorithm also behaves very well: For each subtree, only
the corresponding remaining suffixes are accessed. At a certain tree level, the
number of suffixes considered will be smaller than the number of available cache
entries. As these suffixes are read sequentially, practically no further cache misses
will occur. This point is reached earlier when the branching degree of the tree
nodes is higher, since the suffixes split up more quickly. Hence, the locality of
the *wotd*-algorithm improves for larger values of k.

Aside from the linear constructions already mentioned, there are $O(n \log n)$
time suffix tree constructions (e.g. [3,8]) which are based on Hopcroft's partitioning technique [10]. While these constructions are faster in terms of worst-case
analysis, the subtrees are not constructed independently. Hence they do not share
the locality of the *wotd*-algorithm, nor do they allow for a lazy implementation.

3   Implementation Techniques

This section describes how the *wotd*-algorithm can be implemented in an eager
language. The "simulation" of lazy evaluation in an eager language is not a very
common approach. Unevaluated parts of the data structure have to be represented explicitly, and the traversal of the suffix tree becomes more complicated
because it has to be merged with the construction of the tree. We will show,
however, that by a careful consideration of efficiency matters, one can end up
with a program which is not only more efficient and flexible in special applications, but which performs comparable to the best existing implementations of
index-based exact string matching algorithms in general.

We first describe the data structure that stores the suffix tree, and then we
show how to implement the lazy and eager evaluation, including the additional
data structures.

3.1   The Suffix Tree Data Structure

To implement a suffix tree, we basically have to represent three different items:
nodes, edges and edge labels. To describe our representation, we define a total
order $\prec$ on the children of a branching node: Let $\bar{u}$ and $\bar{v}$ be two different nodes
in $ST(t)$ which are children of the same branching node. Then $\bar{u} \prec \bar{v}$ if and only
if $\min \ell(\bar{u}) < \min \ell(\bar{v})$. Note that leaf sets are never empty and $\ell(\bar{u}) \cap \ell(\bar{v}) = \emptyset$.
Hence $\prec$ is well defined.

Let us first consider how to represent the edge labels. Since an edge label v is a
substring of $t\$$, it can be represented by a pair of pointers (i, j) into $t' = t\$$, such

that $v = t_i' \ldots t_j'$. In case the edge is a leaf edge, we have $j = n+1$, i.e., the right pointer j is redundant. In case the edge leads to a branching node, it also suffices to only store a left pointer, if we choose it appropriately: Let $\overline{u} \overset{v}{\longrightarrow} \overline{uv}$ be an edge in $ST(t)$. We define $lp(\overline{uv}) := \min \ell(\overline{uv}) + |u|$, the *left pointer* of $\overline{uv}$. Now suppose that $\overline{uv}$ is a branching node and $i = lp(\overline{uv})$. Assume furthermore that $\overline{uvw}$ is the smallest child of $\overline{uv}$ w.r.t. the relation $\prec$. Hence we have $\min \ell(\overline{uv}) = \min \ell(\overline{uvw})$, and thus $lp(\overline{uvw}) = \min \ell(\overline{uvw}) + |uv| = \min \ell(\overline{uv}) + |u| + |v| = lp(\overline{uv}) + |v|$. Now let $r = lp(\overline{uvw})$. Then $v = t_i \ldots t_{i+|v|-1} = t_i \ldots t_{lp(\overline{uv})+|v|-1} = t_i \ldots t_{lp(\overline{uvw})-1} = t_i \ldots t_{r-1}$. In other words, to retrieve edge labels in constant time, it suffices to store the left pointer for each node (including the leaves). For each branching node $\overline{u}$ we additionally need constant time access to the child of $\overline{uv}$ with the smallest left pointer. This access is provided by storing a reference *firstchild*$(\overline{u})$ to the first child of $\overline{u}$ w.r.t. $\prec$. The *lp-* and *firstchild*-values are stored in a single integer table T. The values for children of the same node are stored in consecutive positions ordered w.r.t. $\prec$. Thus, only the edges to the first child are stored explicitly. The edges to all other children are implicit. They can be retrieved by scanning consecutive positions in table T.

Any node $\overline{u}$ is referenced by the index in T where $lp(\overline{u})$ is stored. To decode the tree representation, we need two extra bits: A *leaf bit* marks an entry in T corresponding to a leaf, and a *rightmost child bit* marks an entry corresponding to a node which does not have a right brother w.r.t. $\prec$. Fig. 2 shows a table T representing $ST(abab)$.

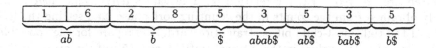

Fig. 2. A table T representing $ST(abab)$ (see Fig. 1). The input string as well as T is indexed from 1. The entries in T corresponding to leaves are shown in grey boxes. The first value for a branching node $\overline{u}$ is $lp(\overline{u})$, the second is *firstchild*$(\overline{u})$. The leaves $\overline{\$}$, $\overline{ab\$}$, and $\overline{b\$}$ are rightmost children

3.2   The Evaluation Process

The *wotd*-algorithm is best viewed as a process evaluating the nodes of the suffix tree, starting at the root and recursively proceeding downwards into the subtrees.

We first describe how an unevaluated node $\overline{u}$ of $ST(t)$ is stored. For the evaluation of $\overline{u}$, we need access to the set $R(\overline{u})$ of remaining suffixes. Therefore we employ a global array *suffixes* which contains pointers to suffixes of $t\$$. For each unevaluated node $\overline{u}$, there is an interval in *suffixes* which stores pointers to all the starting positions in $t\$$ of suffixes in $R(\overline{u})$, ordered by descending suffix-length from left to right. $R(\overline{u})$ is then represented by the two boundaries *left*$(\overline{u})$

and $right(\overline{u})$ of the corresponding interval in *suffixes*. The boundaries are stored in the two integers reserved in table T for the branching node $\overline{u}$. To distinguish evaluated and unevaluated nodes, we use a third bit, the *unevaluated bit*.

Now we can describe how $\overline{u}$ is evaluated: The edges outgoing from $\overline{u}$ are obtained by a simple counting sort [4], using the first character of each suffix stored in the interval $[left(\overline{u}), right(\overline{u})]$ of the array *suffixes* as the key in the counting phase. Each character c with count greater than zero corresponds to a c-edge outgoing from $\overline{u}$. Moreover, the suffixes in the c-group determine the subtree below that edge. The pointers to the suffixes of the c-group are stored in a subinterval, in descending order of their length. To obtain the complete label of the c-edge, the lcp of all suffixes in the c-group is computed. If the c-group contains just one suffix s, then the lcp is s itself. If the c-group contains more than one suffix, then a simple loop tests for equality of the characters $t_{suffixes[i]+j}$ for $j = 1, 2, \ldots$ and for all start positions i of the suffixes in the c-group. As soon as an inequality is detected, the loop stops and j is the length of the lcp of the c-group.

The children of $\overline{u}$ are stored in table T, one for each non-empty group. A group with count one corresponds to a subinterval of width one. It leads to a leaf, say $\overline{s}$, for which we store $lp(\overline{s})$ in the next available position of table T. $lp(\overline{s})$ is given by the left boundary of the group. A group of size larger than one leads to an unevaluated branching node, say $\overline{v}$, for which we store $left(\overline{v})$ and $right(\overline{v})$ in the next two available positions of table T. In this way, all nodes with the same father $\overline{u}$ are stored in consecutive positions. Moreover, since the suffixes of each interval are in descending order of their length, the children are ordered w.r.t. the relation $\prec$. The values $left(\overline{v})$ and $right(\overline{v})$ are easily obtained from the counts in the counting sort phase, and setting the leaf-bit and the rightmost-child bit is straightforward. To prepare for the (possible) evaluation of $\overline{v}$, the values in the interval $[left(\overline{v}), right(\overline{v})]$ of the array *suffixes* are incremented by the length of the corresponding lcp. Finally, after all successor nodes of $\overline{u}$ are created, the values of $left(\overline{u})$ and $right(\overline{u})$ in T are replaced by the integers $lp(\overline{u}) := suffixes[left(\overline{u})]$ and $firstchild(\overline{u})$, and the unevaluated bit for $\overline{u}$ is deleted.

The nodes of the suffix tree can be evaluated in an arbitrary order respecting the father/child relation. Two strategies are relevant in practice: The *eager* strategy evaluates nodes in a depth-first and left-to-right traversal, as long as there are unevaluated nodes remaining. The program implementing this strategy is called *wotdeager* in the sequel. The *lazy* strategy evaluates a node not before the corresponding subtree is traversed for the first time, for example by a procedure searching for patterns in the suffix tree. The program implementing this strategy is called *wotdlazy* in the sequel.

3.3 Space Requirement

The suffix tree representation as described in Sect. 3.1 requires $2q + n$ integers, where q is the number of non-root branching nodes. Since $q = n - 1$ in the worst case, this is an improvement of $2n$ integers over the best previous representation, as described in [12]. However, one has to be careful when comparing the $2q + n$ representation of Sect. 3.1 with the results of [12]. The $2q + n$ representation is tailored for the *wotd*-algorithm and requires extra working space of $2.5n$ integers in the worst case:[2] The array *suffixes* contains n integers, and the counting sort requires a buffer of the width of the interval which is to be sorted. In the worst case, the width of this interval is $n - 1$. Moreover, *wotdeager* needs a stack of size up to $n/2$, to hold references to unevaluated nodes.

A careful memory management, however, allows to save space in practice. Note that during eager evaluation, the array *suffixes* is processed from left to right, i.e., it contains a completely processed prefix. Simultaneously, the space requirement for the suffix tree grows. By reclaiming the completely processed prefix of the array *suffixes* for the table T, the extra working space required by *wotdeager* is only little more than one byte per input character, see Table 1. For *wotdlazy*, it is not possible to reclaim unused space of the array *suffixes*, since this is processed in an arbitrary order. As a consequence, *wotdlazy* needs more working space.

4   Experimental Results

For our experiments, we collected a set of 11 files of different sizes and types. We restricted ourselves to 7-bit ASCII files, since the suffix tree application we consider (searching for patterns) does not make sense for binary files. Our collection consists of the following files: We used five files from the Calgary Corpus: *book1*, *book2*, *paper1*, *bib*, *progl*. The former three contain english text, and the latter two formal text (bibliographic items and lisp programs). We added two files (containing english text) from the Canterbury Corpus:[3] *lcet10* and *alice29*. We extracted a section of 500,000 residues from the PIR protein sequence database, denoted by *pir500*. Finally, we added three DNA sequences: *ecoli500* (first 500,000 bases of the ecoli genome), *ychrIII* (chromosome III of the yeast genome), and *vaccg* (complete genome of the vaccinia virus).

All programs we consider are written in C. We used the *ecgs* compiler, release 1.1.2, with optimizing option –O3. The programs were run on a Computer with a 400 MHz AMD K6-II Processor, 128 MB RAM, under Linux. On this computer each integer and each pointer occupies 4 bytes.

In a first experiment we ran three different programs constructing suffix trees: *wotdeager*, *mccl*, and *mcch*. The latter two implement McCreight's suffix tree construction [15]. *mccl* computes the improved linked list representation, and *mcch* computes the improved hash table representation of the suffix tree, as

---

[2] Moreover, the *wotd*-algorithm does not run in linear worst case time, in contrast to e.g. McCreight's algorithm [15] which can be used to construct the $5n$ representations of [12] in constant working space. It is not clear to us whether it is possible to construct the $2q + n$ representation of this paper within constant working space, or in linear time. In particular, it is not possible to construct it with McCreight's [15] or with Ukkonen's algorithm [17], see [12].

[3] Both corpora can be obtained from http://corpus.canterbury.ac.nz

described in [12]. Table 1 shows the running times and the space requirements. We normalized w.r.t. the length of the files. That is, we show the relative time (in seconds) to process 10^6 characters (i.e., $rtime = (10^6 \cdot time)/n$), and the relative space requirement in bytes per input character. For *wotdeager* we show the space requirement for the suffix tree representation (*stspace*), as well as the total space requirement including the working space. *mccl* and *mcch* only require constant extra working space. The last row of Table 1 shows the total length of the files, and the averages of the values of the corresponding columns. In each row a grey box marks the smallest relative time and the smallest relative space requirement, respectively.

file	n	k	wotdeager			mccl		mcch	
			rtime	stspace	space	rtime	space	rtime	space
book1	768771	82	2.82	8.01	9.09	3.55	10.00	2.55	14.90
book2	610856	96	2.60	8.25	9.17	2.90	10.00	2.31	14.53
lcet10	426754	84	2.48	8.25	9.24	2.79	10.00	2.30	14.53
alice29	152089	74	1.97	8.25	9.43	2.43	10.01	2.17	14.54
paper1	53161	95	1.69	8.37	9.50	1.88	10.02	1.88	14.54
bib	111261	81	1.98	8.30	9.17	2.07	9.61	1.89	14.54
progl	71646	87	2.37	9.19	10.42	1.54	10.41	1.95	14.54
ecoli500	500000	4	3.32	9.10	10.46	2.42	12.80	2.84	17.42
ychrIII	315339	4	3.20	9.12	10.70	2.28	12.80	2.70	17.41
vaccg	191737	4	3.70	9.22	11.07	2.14	12.81	2.56	17.18
pir500	500000	20	3.06	7.81	8.61	5.10	10.00	2.58	15.26
	3701614		2.66	8.53	9.71	2.65	10.77	2.34	15.40

Table 1. Time and space requirement for different programs constructing suffix trees

All three programs have similar running times. *wotdeager* and *mcch* show a more stable running time than *mccl*. This may be explained by the fact that the running time of *wotdeager* and *mcch* is independent of the alphabet size. For a thorough explanation of the behavior of *mccl* and *mcch* we refer to [12]. While *wotdeager* does not give us a running time advantage, it is more space efficient than the other programs, using 1.06 and 5.69 bytes per input character less than *mccl* and *mcch*, respectively. Note that the additional working space required for *wotdeager* is on average only 1.18 bytes per input character.

In a second experiment we studied the behavior of different programs searching for many exact patterns in an input string, a scenario which occurs for example in genome-scale sequencing projects, see [9, Sect. 7.15]. For the programs of the previous experiment, and for *wotdlazy*, we implemented search functions. *wotdeager* and *mccl* require $O(km)$ time to search for a pattern string of length m. *mcch* requires $O(m)$ time. Since the pattern search for *wotdlazy* is merged with the evaluation of suffix tree nodes, we cannot give a general statement about the running time of the search. We also considered suffix arrays, using the original program code developed by Manber and Myers [13, page 946]. The suffix array program, referred to by *mamy*, constructs a suffix array in $O(n \log n)$ time. Searching is performed in $O(m + \log n)$ time. The suffix array requires $5n$ bytes of space. For the construction, additionally $4n$ bytes of working space are required. Finally, we also considered the iterated application of an on-line string searching algorithm, our own implementation of the Boyer-Moore-Horspool algorithm [11], referred to by *bmh*. The algorithm takes $O(n + m)$ expected time per search, and uses $O(m)$ working space.

We generated patterns according to the following strategy: For each input string t of length n we randomly sampled ρn substrings $s_1, s_2, \ldots, s_{\rho n}$ of different lengths from t. The proportionality factor ρ was between 0.0001 and 1. The lengths were evenly distributed over the interval $[10, 20]$. For $i \in [1, \rho n]$, the programs were called to search for pattern p_i, where $p_i = s_i$, if i is even, and p_i is the reverse of s_i, otherwise. Reversing a string s_i simulates the case that a pattern search is often unsuccessful. Table 2 shows the relative running times for $\rho = 0.1$. For *wotdlazy* we show the space requirement for the suffix tree after all ρn pattern searches have been performed (*stspace*), and the total space requirement. For *mamy*, *bmh*, and the other three programs the space requirement is independent of ρ. Thus for the space requirement of *wotdeager*, *mccl*, and *mcch* see Table 1. The space requirement of *bmh* is marginal, so it is omitted in Table 2.

			wotdlazy			*wotdeager*	*mccl*	*mcch*	*mamy*		*bmh*
file	*n*	*k*	*rtime*	*stspace*	*space*	*rtime*	*rtime*	*rtime*	*rtime*	*space*	*rtime*
book1	768771	82	3.17	3.14	7.22	3.37	5.70	3.19	20.73	8.04	413.60
book2	610856	96	2.85	3.12	7.22	3.14	5.17	2.96	21.23	8.06	298.25
lcet10	426754	84	2.65	3.07	7.22	3.07	4.52	2.88	20.55	8.07	206.40
alice29	152089	74	2.04	3.13	7.23	2.43	3.62	2.70	17.10	8.14	74.76
paper1	53161	95	1.50	3.23	7.65	2.07	2.82	2.26	9.41	8.68	23.70
bib	111261	81	1.89	3.06	7.23	2.43	3.06	2.43	14.11	8.24	47.28
progl	71646	87	1.81	2.91	7.24	2.79	2.37	2.37	14.52	8.42	32.24
ecoli500	500000	4	3.60	3.71	8.02	3.82	3.36	3.68	24.84	8.52	724.56
ychrIII	315339	4	3.23	3.84	8.02	3.62	3.17	3.49	26.73	8.83	453.39
vaccg	191737	4	2.97	3.81	8.03	3.39	2.87	3.34	23.73	8.34	291.13
pir500	500000	20	2.46	3.55	7.62	3.66	6.34	3.10	29.38	8.06	248.42
	3701614		2.56	3.32	7.52	3.07	3.91	2.94	20.21	8.31	255.79

Table 2. Time and space requirement for searching $0.1n$ exact patterns

Except for the DNA sequences, *wotdlazy* is the fastest and most space efficient program for $\rho = 0.1$. This is due to the fact that the pattern searches only evaluate a part of the suffix tree. Comparing the *stspace* columns of Tables 1 and 2 we can estimate that for $\rho = 0.1$ about 40% of the suffix tree is evaluated. We can also deduce that *wotdeager* performs pattern searches faster than *mcch*, and much faster than *mccl*. This can be explained as follows: searching for patterns means that for each branching node the list of successors is traversed, to find a particular edge. However, in our suffix tree representation, the successors are found in consecutive positions of table T. This means a small number of cache misses, and hence the good performance. It is remarkable that *wotdlazy* is more space efficient and eight times faster than *mamy*. Of course, the space advantage of *wotdlazy* is lost with a larger number of patterns. In particular, for $\rho \geq 0.3$ *mamy* is the most space efficient program. Figs. 3 and 4 give a general overview, how ρ influences the running times. Fig. 3 shows the average relative running time for all programs and different choices of ρ for $\rho \leq 0.005$. Fig. 4 shows the average relative running time for all programs except *bmh* for all values of ρ. We observe that *wotdlazy* is the fastest program for $\rho \leq 0.3$, and *wotdeager* is the fastest program for $\rho \geq 0.4$. *bmh* is faster than *wotdlazy* only for $\rho \leq 0.0003$. Thus the index construction performed by *wotdlazy* already amortizes for a very small number of pattern searches.

We also performed some tests on two larger files (english text) of length 3 MB and 5.6 MB, and we observed the following:

- The relative running time for *wotdeager* slightly increases, i.e. the superlinearity in the complexity becomes visible. As a consequence, *mcch* becomes faster than *wotdeager* (but still uses 50% more space).
- With ρ approaching 1, the slower suffix tree construction of *wotdeager* and *wotdlazy* is compensated for by a faster pattern search procedure, so that there is a running time advantage over *mcch*.

5   Conclusion

We have developed efficient implementations of the write-only top-down suffix tree construction. These construct a representation of the suffix tree, which requires only $12n$ bytes of space in the worst case, plus $10n$ bytes of working space. The space requirement in practice is only $9.71n$ bytes on average for a collection of files of different type. The time and space overhead of the lazy implementation is very small. Our experiments show that for searching many exact patterns in an input string, the lazy algorithm is the most space and time efficient algorithm for a wide range of input values.

Acknowledgments

We thank Gene Myers for providing a copy of his suffix array code.

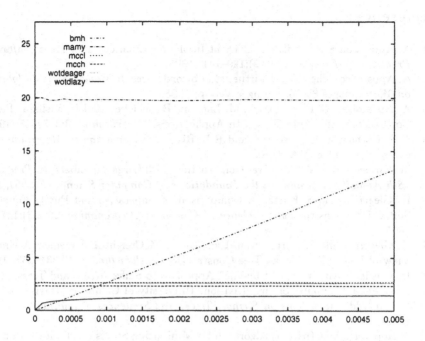

Fig. 3. Average relative running time (in seconds) for different values of $\rho \in [0, 0.005]$

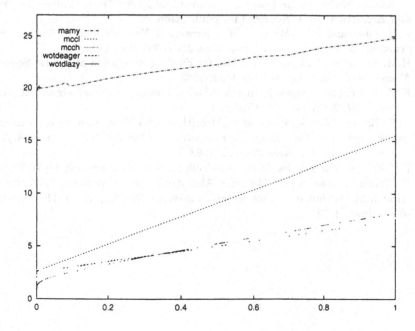

Fig. 4. Average relative running time (in seconds) for different values of $\rho \in [0, 1]$

References

1. A. Andersson and S. Nilsson. Efficient Implementation of Suffix Trees. *Software—Practice and Experience*, 25(2):129–141, 1995.
2. A. Apostolico. The Myriad Virtues of Subword Trees. In *Combinatorial Algorithms on Words*, pages 85–96. Springer Verlag, 1985.
3. A. Apostolico, C. Iliopoulos, G. M. Landau, B. Schieber, and U. Vishkin. Parallel Construction of a Suffix Tree with Applications. *Algorithmica*, 3:347–365, 1988.
4. T.H. Cormen, C.E. Leiserson, and R.L. Rivest. *Introduction to Algorithms*. MIT Press, Cambridge, MA, 1990.
5. M. Farach. Optimal Suffix Tree Construction with Large Alphabets. In *Proc. of the 38th Annual Symposium on the Foundations of Computer Science (FOCS)*, 1997.
6. R. Giegerich and S. Kurtz. A Comparison of Imperative and Purely Functional Suffix Tree Constructions. *Science of Computer Programming*, 25(2-3):187–218, 1995.
7. R. Giegerich and S. Kurtz. From Ukkonen to McCreight and Weiner: A Unifying View of Linear-Time Suffix Tree Constructions. *Algorithmica*, 19:331–353, 1997.
8. D. Gusfield. An "Increment-by-one" Approach to Suffix Arrays and Trees. Report CSE-90-39, Computer Science Division, University of California, Davis, 1990.
9. D. Gusfield. *Algorithms on Strings, Trees, and Sequences*. Cambridge University Press, 1997.
10. J. Hopcroft. An $O(n \log n)$ Algorithm for Minimizing States in a Finite Automaton. In *Proceedings of an International Symposium on the Theory of Machines and Computations*, pages 189–196. Academic Press, New York, 1971.
11. R.N. Horspool. Practical Fast Searching in Strings. *Software—Practice and Experience*, 10(6):501–506, 1980.
12. S. Kurtz. Reducing the Space Requirement of Suffix Trees. *Software—Practice and Experience*, 1999. Accepted for publication.
13. U. Manber and E.W. Myers. Suffix Arrays: A New Method for On-Line String Searches. *SIAM Journal on Computing*, 22(5):935–948, 1993.
14. H.M. Martinez. An Efficient Method for Finding Repeats in Molecular Sequences. *Nucleic Acids Res.*, 11(13):4629–4634, 1983.
15. E.M. McCreight. A Space-Economical Suffix Tree Construction Algorithm. *Journal of the ACM*, 23(2):262–272, 1976.
16. S. S. Skiena. Who is Interested in Algorithms and Why? Lessons from the Stony Brook Algorithms Repository. In *Proceedings of the 2nd Workshop on Algorithm Engineering (WAE)*, pages 204–212, 1998.
17. E. Ukkonen. On-line Construction of Suffix-Trees. *Algorithmica*, 14(3), 1995.
18. P. Weiner. Linear Pattern Matching Algorithms. In *Proceedings of the 14th IEEE Annual Symposium on Switching and Automata Theory*, pages 1–11, The University of Iowa, 1973.

Experiments With List Ranking for Explicit Multi-Threaded (XMT) Instruction Parallelism (Extended Abstract)*

Shlomit Dascal and Uzi Vishkin

Abstract. Algorithms for the problem of list ranking are empirically studied with respect to the Explicit Multi-Threaded (XMT) platform for instruction-level parallelism (ILP). The main goal of this study is to understand the differences between XMT and more traditional parallel computing implementation platforms/models as they pertain to the well studied list ranking problem. The main two findings are: (i) Good speedups for much smaller inputs are possible. (ii) In part, this finding is based on competitive performance by a new variant of a 1984 algorithm, called the No-Cut algorithm. The paper incorporates analytic (*non-asymptotic*) performance analysis into experimental performance analysis for relatively small inputs. This provides an interesting example where experimental research and theoretical analysis complement one another. [1]

Explicit Multi-Threading (XMT) is a fine-grained computation framework introduced in our SPAA'98 paper. Building on some key ideas of parallel computing, XMT covers the spectrum from algorithms through architecture to implementation; the main implementation related innovation in XMT was through the incorporation of low-overhead hardware and software mechanisms (for more effective fine-grained parallelism). The reader is referred to that paper for detail on these mechanisms. The XMT platform aims at faster single-task completion time by way of ILP.

---

* Partially supported by NSF grant CCR-9416890 at U. Maryland.

[1] This paper suggests a possible new utility for the developing field of algorithm engineering; contribute to the interplay between two research questions: how to build new computing systems? and how to design algorithms?
More concretely, reducing the completion time of a single general-purpose computing task by way of parallelism has been one of the main challenges for the design of new computing systems over the last few decades. ILP has been a realm where success has already been demonstrated. A motivation for originally proposing XMT has been to provide a platform for the new utility in the context of the ILP realm. The referees asked us a few general questions: (i) Does it make sense to compare speed-ups for XMT, which for now is nothing but a simulator, with speed-ups for a real machine? Noting that such comparisons are standard in computer architecture research, we agree that caution is required. (ii) Why is the XMT model not more succinctly presented? The XMT model is based on an assembly language and an execution model since this is the way it will be implemented. This is similar to the the underlying assumptions in measuring ILP in Section 4.7 in [HP96]. (iii) How is bandwidth incorporated in the XMT model? Bandwidth does not explicitly appear; within a chip it is generally not a bottleneck, see [DL99]).

1   Introduction

This work considers practical parallel list-ranking algorithms. The model for which programs are written is a single-program multiple-data (SPMD) "bridging model". This model is designated as a programmer's model for a fine-grained computation framework called Explicit Multi-Threading (XMT), which was introduced in [VDBN98]; the XMT framework covers the spectrum from algorithms through architecture to implementation; it is meant to provide a platform for faster single-task completion time by way of instruction-level parallelism (ILP). The performance of XMT programs is evaluated as follow: the performance of a matching optimized XMT assembly code is measured within an XMT execution model. (We use in the current paper the so-called *Spawn-MT* programming model - the easier to implement among the two programming models presented in [VDBN98]). The XMT approach deviates from the standard PRAM approach by incorporating *reduced synchrony* and departing from the lock-step structure in its so-called *asynchronous mode*. Our envisioned platform uses an extension to a standard serial instruction set. This extension efficiently implements PRAM-style algorithms using explicit multi-threaded ILP, which allows considerably more fine-grained parallelism than the previously studied parallel computing implementation platforms/models.

The list ranking problem was the first problem considered as we examined and refined many of the concepts in the XMT framework. The problem arises in parallel algorithms on lists, trees and graphs and is considered a fundamental problem in the theory of parallel algorithms. Experimental results are presented.

Empirical study of parallel list ranking algorithms Implementation of parallel algorithms for list ranking has been considered by many including [HR96], [Re94], [SGS], [Si97a] and [Si97b]; Section 7 gives more detail. The three main observations in this paper are as follows: (i) *Good speedup relative to the corresponding serial algorithm are possible for smaller input sizes than before.* We are not aware of previous results for this problem which are competitive with the counterpart serial algorithm for input size not exceeding 1000; often applicable input sizes have been much larger. By efficiently supporting lighter threads, the XMT envisioned implementation platform made this possible. List ranking is a routine which is employed in many large combinatorial applications; we proceed to illustrate the *importance of achieving good speed-ups for smaller inputs* by comparing two hypothetical parallel computer systems, called A and B. The only difference is that System B allows speed-ups for smaller inputs than System A. *Assumptions Concerning System A (resp. B):* In 95% (resp. 99%) of the time in which the list ranking routine is used speed-ups relative to the serial algorithm is achieved; the speed-up is by a factor of 100. The improvement from System A to B is significant! If a fraction F of computation can be enhanced by a (speed-up) factor of S, Amdahl's Law ([HP96]) implies overall speed-up of $\frac{1}{(1-F)+F/S}$. So, overall speed-up for system A is $\frac{1}{0.05+.95/100} = 1/0.0595 \leq 17$ and for system B ≥ 50; nearly 3 times the speed-up for System A! (ii) *Choosing a fastest list ranking algorithm is more involved than implied by the literature.* A challenge for our experimental work has been to compare some new insights we have with the known experimental work on list ranking. While our results affirm that the RM (Reid-Miller) algorithm is faster for large inputs, for smaller inputs a new variant of [Vi84], called the No-Cut algorithm, offers competitive performance. (iii) *Analytic (non-asymptotic) performance analysis is possible for smaller inputs.*

Due to space limitations, we must refer interested readers to [DV99].

2   List-Ranking Algorithms

The input is an array of N elements, each having a pointer to its successor in a linked list (the successor can be anywhere in the array). The pointer is labeled with the distance it represents. The *list ranking problem* is to find for each element in the list its total (weighted) distance from the end of the list.

To evaluate our list-ranking algorithms, we implemented a serial list-ranking algorithm, with serial assembly code, and several parallel list-ranking algorithms with our new explicitly parallel assembly code. For each of the codes, we aimed at figuring out the best ILP performance, taking into account constant factors and not only asymptotic behavior. The serial algorithm is followed below by three parallel algorithms: No-Cut, Cut-6, and Wyllie's algorithm. See Section 7.1 for RM's "random subset" algorithm.

Serial Algorithm The serial algorithm for the problem consists of first finding the "head" of the list; the head is the only element of the list that is not the successor of any other element in the list. Second ("forward loop"), the list is traversed to find the ranking of the "head", determining for each element by how much closer it is to the "tail" of the list than the head. Finally ("final ranking"), the ranking of all elements in the list are derived.

Parallel Randomized Coin-Tossing Algorithm (No-Cut) The main new parallel algorithm we used is randomized and works in iterations. See Figure 1. Assuming an array with precomputed results of a "randomly tossed coin" (0 or 1) is provided, an iteration first assigns such 0 or 1 to each element of the list. Every element, which is assigned 1 whose successor is assigned 0, is "selected"; other elements are not selected. Note that: (i) on average a quarter of the elements are selected, and (ii) it is impossible to select two successive elements. So far this is the same basic idea introduced in [Vi84]. The new part is: for each selected element, "pointer-jump over" its successor elements, stopping once another selected element is reached. This results in a new list which contains only the selected elements. (*In the next section the length of the "chain" between selected elements is observed to be relatively short, implying that the performance of this "No-Cut" algorithm is attractive.*) The selected element, which pointer-jumped over a non-selected element, is its "ranking parent". All selected elements are included in a smaller array. The compaction into the smaller array is achieved using a new multiple-PS instruction. The smaller array is the input for the next iteration. These "forward iterations" will result in finding the rank of the head of the list. All the elements which have been jumped over (i.e., skipped) during each forward iteration are compacted into another smaller array noting for each its "ranking parent". Playing "backwards" the iterations will extend the ranking from ranking parents to all elements of the list. Our evaluations show that for some input sizes this No-Cut algorithm requires less work (but may need more time) than other versions and algorithms.

Time-Emphasized Parallel Randomized Coin-Tossing Algorithm (Cut-6) A variant of the parallel algorithm of the previous paragraph is presented next. Dubbed "Cut-6", it achieves better time but does more work. That is, it runs faster than the No-Cut algorithm if available (machine) parallelism is above some threshold. The only difference relative to the No-Cut algorithm is that in the forward iterations a selected element stops pointer-jumping after 6 elements even if it has not reached another selected element. Forward iterations also find the rank of the head of the list and playing the iterations "backwards" extends the ranking to all elements of the list.

Wyllie's Parallel Algorithm consists of $log_2 N$ steps, each comprising parallel pointer jumping by all elements, till all elements are ranked.

3   Optimization and Analysis of the Algorithms

The Level of Parallelism denoted P is (for an ILP architecture) *the number of instructions whose execution can overlap in time.* P cannot exceed the product of the number of instructions that can be issued in one clock with the number of stages in the execution pipeline. (The popular text [HP96] suggests in p. 430 that multiple-issue of 64 is realizable in the time frame of the years 2000-2004; so, a standard 5- or 6-stage pipeline leads to P exceeding 300. It is hard to guess why P has hardly increased since 1995; perhaps, this is because vendors have not found a way to harness such an increase for cost effective performance improvement of existing code.)

Our building blocks for parallel list ranking are several parallel algorithms each of which is *non-dominated*, i.e., for each of these algorithms there is some input size and some P value for which it is the fastest. [DV99] describes low- and high-level considerations that were involved in getting the best performance out of each algorithm, and for assembling them in line with the accelerating cascades technique (see [CV86a]). This section discusses particularly noteworthy issues related to work and time analysis of the algorithms.

Serial Algorithm
Loop unrolling is the main concern in analyzing the execution of the serial algorithm. Issues such as the size of the code, the limited number of registers R, and the overhead of leaving too large a "modulo" from the loop, are considered. Most limiting was the value of R, and [DV99] discusses a choice of $R = 128$.

Parallel Randomized Coin-Tossing Algorithm (No-Cut)

Two issues require most attention: (i) the convergence rate of the forward iterations, and (ii) bounding the length of the *longest chain* from one selected element to its successive selected element. The No-Cut algorithm is optimized for less work. It has smaller work constants than the time-emphasized Cut-6 algorithm, but its asymptotic time bound includes a $O(log_2{}^2 N)$ term. The algorithm becomes dominant (in the term $\frac{W}{P} + T$ - a standard way for evaluating PRAM algorithms) for relatively small values of P. The work term is $O(N)$ and its constant is very important. *But, why does the "No-cut" algorithm perform less work?* The answer is: a significant decrease in the number of (forward and backward) iterations. We have longer threads but fewer of them in fewer iterations, so on balance the overall work term is smaller. (An optimization not discussed here is for reducing the number of spawn-join pairs.)

Convergence of Iterations The expected fraction of skipped elements in each iteration is $\frac{3}{4}$. Let N be the initial input size, and let M_i be the expected number of active elements in the i-th iteration (i.e., the input size for the i-th iteration); we make the simplifying assumption that the number of active elements in each iteration will be: $M_0 = N, M_1 \leq \frac{1}{4}N, \cdots, M_i \leq (\frac{1}{4})^i N$. So, $\Sigma M_i \leq 1.34N$. The number of forward iterations $\#iter$ will not exceed $log_4 N = 0.5 log_2 N$.

According to the simulations for various sizes of inputs, $\Sigma M_i \leq 1.33N$ and $\#iter \leq 0.46 log_2 N$; both fully supporting our analysis.

Bounding the Length of the Longest Chain In order to analyze the execution time for the *No-Cut* algorithm we need to find a realistic bound on the longest chain. This bound is important since the critical-time-path length of each forward iteration is determined by the "slowest" thread, i.e., the thread that has to short-cut over the longest chain.

Some definitions follow: (i) A *"chain"* is a series of consecutive (based on pointers) elements in a list starting at "1,0" (coin-tossing selection results) and ending just before the next "1,0" (coin-tossing selection results). (ii) A *chain's "length"* is the number of elements *after* the initial "1" (i.e., the elements which will actually be short-cut). (iii) Given an element, $P(length \geq i)$ is the probability that the length of a chain starting at it is i, or longer.

Claim 1: $P(length \geq log N + log log N) \leq \frac{1}{N}$.

Verbally, this can be described as: the probability for chains whose length is longer than $log N + log log N$ is upper bounded by $\frac{1}{N}$.

To prove Claim 1, we find an i such that $P(length \geq i) \leq \frac{1}{N}$. *The probability of the length of a chain* will be determined as follows: (i) A chain will be of length $\geq l$ if $\forall i : 1 \leq i < l, next^i(x)$ *is not selected* where: (1) x is the selected element that starts the chain, and is considered to be the zero element of the chain. (2) $next^i(x)$ is the i-th successor of element x. And, (3) an element is selected only if its coin is "1" and its successor's coin is "0". (ii) $\frac{1}{4}$ of the $next^2(x)$ elements will be selected (when $next^2(x) = 1$ and $next^3(x) = 0$). $\frac{2}{8}$ of the $next^3(x)$ elements will be selected, since only 2 out of their successors which are still active (8 options) will be $next^4(x) = 0$. (iii) So the series of selected elements will be: $0, \frac{1}{4}, \frac{2}{8}, \frac{3}{16}, \frac{4}{32}, \frac{5}{64}, \ldots, \frac{i-1}{2^i}, \ldots$. The situation is depicted in Figure 2. For the algorithm this means the following: the fraction of threads that will stop after 1 short-cut is $\frac{1}{4}$, after 2 short-cuts is $\frac{2}{8}$, after 3 short-cuts is $\frac{3}{16}$, and so on.

Claim 1 follows from $P(i) = \frac{i+1}{2^{i+2}} \leq \frac{1}{n}$.

[DV99] reports several simulations to support our formal analysis of the longest chain bounds. The simulations resulted in the following: (i) We ran over 200 different N-values in the range: $10^2 \leq N \leq 10^7$. (ii) $Maxlen < 1.23 * log_2 N$; $Maxlen$ denotes the length of the longest chain. (iii) Only 6 simulations resulted in $Maxlen \geq 1.20 * log_2 N$. (iv) Practical conclusion: $E(Maxlen) = 1.2 * log_2 N$.

The Cut-6 Algorithm For analyzing the Cut-6 algorithm we consider the sequence which describes the convergence of the forward iterations. Recall that during the forward iterations, a selected element stops pointer-jumping after 6 elements even if it has not reached another selected element. An analysis which is similar to the proof of Claim 1 applies. The expected fraction of skipped elements in each iteration will be $[\frac{1}{4} + \frac{3}{16} + \frac{4}{32} + \frac{5}{64} + \frac{6}{128} + \frac{7}{256}] = \frac{183}{256}$. Let N be the initial input size, and let M_i be the number of active elements in the i-th iteration. Then, under similar assumptions to the analysis of the no-cut algorithm, the number of active elements in each iteration will be: $M_0 = N, M_1 \leq \frac{73}{256}N, \cdots, M_i \leq (\frac{73}{256})^i N$ and $\Sigma M_i \leq \frac{256}{183}N \leq 1.4N$. The number of iterations necessary for the algorithm to finish is $log_{(256/73)}N = 0.552 log_2 N$.

Why, among the family of cut-k algorithms, did we chose the one that stops after short-cutting at most 6 elements? We experimented with $k = 3, 4, 5, 6, \ldots, 11$. The minimum time constant, for the $O(log_2 N)$ term, was obtained for $k = 6$. The work constant, for the $O(N)$ term, continued decreasing until $k = 10$, but wasn't significantly smaller than for $k = 6$. Thus we chose k to be 6.

Wyllie's Parallel Algorithm is relatively simple. Most of its work and time analysis is clear from the code and the counts presented. The only place where additional assumptions were needed concerns the calculation of $log_2 N$ (for the number of iterations). As discussed in [DV99], the $log_2 N$ calculation adds a large constant to the work term.

General comment on compilation feasibility Translation from the high level code (C and extended-C) to the optimized assembly code, which has actually been done manually, is feasible using known compiler techniques. See [DV99].

4   The Instruction Set

The parallel algorithms presented in this work all use an asynchronous mode of the so-called *Spawn-MT programming model* for the XMT framework; this allows expressing the fine-grained parallelism in the algorithms. Concretely, we extended standard instruction sets (for assembly languages) to enable transition back and forth from serial state to parallel state and allow in the parallel state any number of (virtual) threads. A Spawn command marks a transition from a serial state to parallel state and a Join command marks a transition in the opposite direction. See left side of Figure 3. The semantics of the assembly language, called the *"independence of order semantics (IOS)"*, allows each thread to progress at its own speed from its initiating Spawn command to its terminating Join command, without having to ever wait for other threads. Synchronization occurs only at the Join command; all threads must terminate at the Join command before the execution can proceed.

The assembly language includes *multi-operand* instructions, mainly the **prefix-sum** (PS) instruction, whose specification is similar to the known Fetch-and-Add. The paper [Vi97] explains why if the number of operands does not exceed k, for some parameter k which depends on the hardware, and each of the elements is, say, one bit, multiple-PS can be implemented (in hardware) very efficiently (in "unit time"), even where the base is any integer. The **high-level language** for Spawn-MT (as well as the assembly language) is single-program multiple-data (SPMD). It is an extension of standard C which includes Spawn and multioperand instructions such as prefix-sum. Using standard C scoping, the Joins are implicit. A Join implementation comprises a parallel sum operation for monitoring the number of terminating threads.

The XMT instruction set extends the standard serial MIPS instruction set (see [HP94]). The table describes a few new instructions. A fuller specification for the explicit parallel instruction set (also referred to as spawn-join instruction set) can be found in the extended summary version of [VDBN98]; it describes the new instructions and how they can be efficiently implemented in hardware.

Few XMT Instructions

Notation: $R1\$$ is local register number 1 of thread \$.

Instruction	Instruction Meaning
PS Ri, Rj	Atomically: (i) $Ri := Ri + Rj$, and (ii) $Rj := (original)Ri$. Ri called *base for PS*.
PS Ri, Rj_f for $1 \leq f \leq g$	Multiple prefix-sum: If a sequence of g Prefix-sum commands comes from one thread: follows serial semantics. If the g commands come from different threads: follow some (arbitrary) serial semantics. Execute by special hardware. Assumptions made: if each Rj_f between 0 and 3, then unit time execution if $g \leq k$ (base is any integer).
PSM $15(Ri)Rj$	*Prefix sum to memory*, with base $Memory[15 + Ri]$. Multiple PSM: only inter-thread.
Mark Ri, Rj	If $Ri = 0$, then atomically: $Ri = 1$ and $Rj = 0$; else $Rj = 1$. Also: Multiple-Mark executed by special hardware.
MarkM $15(Ri), Rj$	Mark to memory. Also, multiple MarkM.

5   Execution Model and Execution Times

We are interested in comparing performance results for the serial algorithm in a serial (ILP) execution model with the parallel algorithms in our new execution model. The performance results we obtained are derived by finding the execution critical-time-path T and counting the work W from the assembly code written for each of the algorithms in their respective instruction sets.

Following is an overview of the process by which we obtained the performance results.

(i) Given the code, we tried to place each instruction as early as possible without violating data and control dependences.

(ii) *Speculation.* Generally, we were able to obtain our results without resorting to speculation; however, we allowed speculation for branches which are known to happen with high probability (e.g., branching on whether an element of a linked list is last in the list fails for all but one of the elements of the list).

(iii) *Cache misses.* We examined every memory access. If, due to spatial-locality or temporal-locality, the address could be found in the cache, we assumed unit-time memory access. If a case for spatial or temporal locality could not be made, we applied a penalty of $d = 5$ time units (to simplify our analysis).

This is unquestionably an arbitrary decision which oversimplifies an involved situation. The issue is how to account for memory response time without resorting to detailed memory system simulations; the problem is that to reach the level of detail that merits the use of simulations, we will have to specify the hardware in ways which exceed our more general bridging model and our basic intent to reserve judgement on hardware implementation issues which complement our new ideas. We could only say that picking somewhat bigger or smaller numbers for d did not appear to change our overall comparison of serial and parallel code significantly. For problems, which are not exceedingly large, the memory could fit within the levels of the cache (which reached 16MB for all cache levels in 1995, with a maximum response time of 15 cycle, see [HP96]). So, the number 5 is not unreasonable for average response time. Also: (a) the sensitivity of our speed-up results to replacing 5 by slightly bigger or smaller numbers was marginal; (b) presented and explained in [VDBN98], a Scheduling Lemma justifies the use of a single average number for simulating memory response time.

(iv) *Prefetching.* To alleviate this cache-miss penalty, prefetching could begin prior to a branch or spawn instruction, that precedes the memory access; however, to minimize speculative execution, their commit is deferred to after the resolution of the branch or spawn.

(v) *Spawn and Join instructions.* Our analysis uses the following simplifying assumption: a Spawn instruction contributes one to the operation count, and a Join instruction (in each thread) is counted as one operation. Our rationale follows. The principal goal has been comparing the performance of two kinds of instruction codes: serial and parallel. We were faced with a situation where the extensive hardware used currently in practice in order to extract ILP from serial code and control its execution is not accounted for in the instruction code. This gives an unfair advantage to the serial code since it is likely to require more complicated hardware than for implementing our explicitly parallel code; one reason is that *parallelism mandated by the code frees the hardware from verifying that no dependences occur.* This stresses our case. It explains why in terms of devoted hardware, standard serial code is unlikely to be implemented faster, or using less hardware, than our envisioned parallel code (for the same level of parallelism).

Still, we take the extremely conservative approach of counting the Spawn and Join instructions, while essentially not counting anything on the serial code end. The only case where we applied to serial code the same costs as to the parallel one was where loop parallelism, which can be determined at run-time, is needed. Note that this still reflects a "discount" for the serial code: in practice only some fixed loop-unrolling is possible for serial code, while our Spawn mechanism has no problem handling this for parallel code.

(vi) *Unit time instructions.* All other instructions are assumed to take unit time.

(vii) *Scheduling Lemma, number of registers and code length.* The Scheduling Lemma (in [VDBN98]) assumes that the number of registers is unlimited and that the length of the code can be increased with no penalty (as needed for loop-unrolling). (However, our performance results consider limiting these assumptions, as well.)

The detailed execution assumptions and techniques for figuring out time and work counts are described under http://www.umiacs.umd.edu/~vishkin/XMT/ in *"Explicit Multi-Threading (XMT) - Specifications and Assumptions"*; they are followed there by a simple example, to facilitate understanding before approaching the full algorithms. The proof of the Scheduling Lemma is presented in the full paper of [VDBN98].

A Conservative Comparison Approach We made every effort to avoid unfairly favoring the performance estimates of the new code over standard serial code. Concretely, we followed the following guidelines:

(1) For standard serial code, we took T as the running time, since this is a **lower bound** on the execution time achievable.

(2) For parallel code, we took $W/p+T$ as the running time, since this is an **upper bound** on the execution time achievable. (This bound relies on the Scheduling Lemma in [VDBN98].)

6 Summary of Work and Time Execution Counts and Speed-Up Results

The execution times for the four algorithms were derived by counting the execution time and work of each algorithm, following the execution assumptions and techniques described above. The fully commented assembly code for each of the algorithms can be found through the web site http://www.umiacs.umd.edu/~vishkin/XMT/. Detailed counts are also presented there in a table format for each of the algorithms, and the execution work and time below are derived from that. (For later reference, we also included data on the RM algorithm.)

Algorithm	Work $\leq$	Time
Serial[2]	$15N + 4(\frac{N}{R}) + 5(N\,mod\,R) + 22$	$9N + 4(\frac{N}{R}) + 4(N\,mod\,R) + 22$
No-Cut[3]	$58N + 4log_2 N + 2$	$4.2log_2{}^2 N + 18.5log_2 N + 27$
Cut-6	$80.5N + 2.2log_2 N + 2$	$44.7log_2 N + 27$
Wyllie[4]	$10Nlog_2 N + 12N + 2log_2 N + 268$	$15log_2 N + 30$
RM[5]	$32N + 10log_2 N + 9$	$9log_2{}^2 N + 27log_2 N + 6$

6.1   Speed-Up Results

We compared the execution time for the 4 algorithms (serial, No-Cut, Cut-6 and Wyllie's) for various input-size (N) values and machine parallelism (P).

We considered P values up to 2000, which cover the horizon for the amount of parallelism currently envisioned for explicit ILP.

We considered N values (input size) up to 10^7.

Upper bounds on the execution time for the parallel algorithms ($\frac{W}{P} + T$) are compared with a lower bound on execution time for the serial algorithm (T), to comply with our conservative approach.

We noticed the following: (i) The serial algorithm, under our assumptions and execution counts, is competitive for $P < 7$. (ii) Each of the three parallel algorithms has an area (of P-values and N-values) for which it is the best (see table at the end). When looking at a specific P (along a vertical section) we see, as expected, that:

---

[2] In the serial algorithm: (i) The loop unrolling is bound by R, the number of available registers (which affects the work and time counts). (ii) For the comparison with the parallel algorithms we will only look at the *lower bound* of the serial algorithm, i.e., the time count alone (which amounts to a critical path in a dependence graph).

[3] The $log_2{}^2 N$ term in the time count for the "work-emphasized" No-Cut algorithm is based on the fact that the "longest chain" (in each iteration of the No-Cut algorithm) does not exceed $2log_2 N$ with high probability.

[4] The high constant in Wyllie's algorithm's work is due to the calculation of $log_2 N$ (the number of iterations) at the beginning (assuming the N is 64 bits).

[5] In the RM algorithm: (i) we used a powerful atomic instruction to eliminate multiple selections of the same element (and remain with one copy of each selected element). An alternative setting, where such an atomic instruction is not available, would imply that the the time and work counts increase. (ii) The random selection of elements is counted as one operation per iteration only. This implies that the random numbers should be prepared ahead of time and at a minimum scaled (which requires at least one additional operation) to the relevant list size at each iteration. These assumptions provide wide margins to make sure that we do not favor the new algorithms (No-Cut, Cut-6, etc) to RM's algorithm. As a result the work and time counts for RM's algorithms ended up being a bit optimistic. See later reference in Section 7.1.

- For larger N-values, the work-emphasized No-Cut algorithm is ahead.
- For intermediate N-values, the time-emphasized Cut-6 is ahead.
- For smaller N-values, Wyllie's algorithm is ahead.

Plots of the areas in which each algorithm is dominant are shown in Figure 4. Note that the table at the end of the paper adds data about RM's algorithm and shows that it is ahead for larger N-Values.

7   Other Implementations of List-Ranking Algorithms

In [Re94] list ranking is discussed. After considering several work efficient list ranking algorithms, such as [MR85], [AM90] and [CV89], they implemented a new algorithm which has experimentally optimized smaller constants for the parallel work, but not fast asymptotic time. The algorithm does recursive iterations using a randomized division into $m + 1$ sublists, where $m < \frac{n}{\log n}$. The variance in the size of the sublists (which does not present a problem in our XMT model) is amortized by giving several sublists to each processor. When the number of sublists becomes small enough (determined experimentally) they switch to use either the serial or Wyllie's algorithm. Note that the code requires a careful manual adjusting of parameters to achieve good speedup each time a different number of processors is used. The platform used is the *Cray C90* vector multiprocessor, with 128 vector elements per processor (in Figure 2 there). As explained there, when vectorizing a serial problem that requires gather/scatter operations, the best speedup one can expect on a single processor *Cray C90* is about a factor of 12-18, but when the vectorized algorithm does more work than the serial one then one can expect smaller speedups (the vectorization is used to hide latencies). The results presented are for 1, 2, 4 and 8 processors. The speedups obtained for list ranking (presented in Figure 11 there) using 8 processors (i.e., $8 * 128 = 1024$ vector elements) are approximately 40 for $n = 3 * 10^7$, 35 for $n = 4 * 10^6$ and 4 for $n = 8 * 10^3$. In Section 7.1 below, a detailed comparison between RM's algorithm and our No-Cut algorithm is given.

In [HR96] the implementation of three parallel list ranking algorithms on the massively parallel SIMD (Single Instruction Multiple Data) machine *MasPar MP-1* with virtual processing, using the MPL C-based language is presented. The *MP-1* has 16384 physical processors. The algorithms implemented were: Wyllie's suboptimal deterministic algorithm, a simple optimal randomized algorithm and a combination of the two where initially the randomized algorithm is run and when the number of elements remaining was no more than the number of physical processors they switch to Wyllie's algorithm. A comparison to the sequential algorithm on a *UNIX* system is presented, where the raw computational power of the *MP-1* was at least 63 times larger than the *SPARC* used, and the total amount of main memory available for computation on the massively parallel computer is approximately 32 times larger. A speedup of 2 was achieved when all cells in the linked list fit into the *SPARC*'s main memory ($n \simeq 3 * 10^6$). However, when heavy swapping was needed in the sequential implementation, its performance degraded dramatically.

The platform used in [Si97a], [Si97b] and [SGS] is the *Intel Paragon*, which has a grid like structure. Three algorithms were implemented in [SGS]: Wyllie's pointer-jumping, independent set removal and sparse ruling set. Using 100 processors the best speedups were respectively: 5 for $n \simeq 3 * 10^6$, 14 for $n \simeq 5 * 10^7$ and 27 for $n \simeq 2 * 10^8$. In [Si97b] the main result is an "external algorithm" which achieves a speedup of 25 using 100 processors for $n = 10^8$; for an illustration of the advantage of XMT over these multi-processing results for smaller input, see right side of Figure 3.

Several deterministic work-optimal EREW list ranking algorithms (see [Tr96]) have been implemented using *Fork95* for the PAD library of PRAM algorithms and data structures in connection with the the *SB-PRAM* and *Fork95* projects. *Fork95* is a c-based parallel programming language. All the algorithms switch at some experimentally determined parameter to Wyllie's algorithm. The execution was evaluated on a simulator for the SB-PRAM. The results reported were: with 32 processors for $n = 42968$ there is a break-even between Anderson-Miller's algorithm and the serial one; with 128 processors the best algorithm is Wyllie's with speedups of approximately 3.5 for $n = 42968$ and 5.7 for $n = 10742$. Their practical conclusion was that none of the studied algorithms is fully satisfactory, due to large constants; although some randomized algorithm might achieve better results.

7.1   Detailed Comparison to Reid-Miller's Algorithm

RM's algorithm also has forward and backward iterations. A forward iteration works as follows. A parameter k is specified. In parallel, k random numbers each between 1 and n (the number of current elements in the list) are picked. Each element, whose index is one of these random numbers, is selected (after getting rid of multiple selections of the same element). For each selected element the forward iteration proceeds by pointer jumping over all non-selected elements. This results in linking all selected elements. All selected elements are then included in a smaller array. RM's algorithm performs less work at one interesting point. Unlike the No-Cut and Cut-6 algorithms it need not check for each element (at the beginning of the forward iteration) whether it was selected. We implemented RM's algorithm in the same XMT programming model. Since our main interest was comparison with our No-Cut algorithm, we implemented RM's algorithm in assembly code for the XMT computational framework. (We left the cascading aside as we did for other algorithms we implemented.)

For comparison purposes, we took the number of randomly picked sublists to be $k = \frac{n}{4}$. Note, however, that in the No-Cut algorithm, we could have actually controlled the expected number of selected elements by picking a probability different than $\frac{1}{2}$ for getting a Head at each node. In other words, No-Cut can also be flexible with the number of selected elements similar to RM's algorithm.

We note several things in our implementation of RM's algorithm and in the comparison to the No-Cut algorithm. First, the random selection of k elements, as the heads of the sublists, is presented and counted as only one operation per iteration in the instruction code. This implies that the random numbers should be prepared ahead of time and (at least) scaled (which should require at least one additional operation for each draw of a number in each iteration) to the relevant list-size at each iteration (to achieve uniform distribution for each iteration). Second, we used the markM atomic instruction (which is a simpler variant of the PSM instruction) to eliminate multiplicities among the k selected elements. This instruction is counted as 1 work unit and 5 time units (since it has to read a memory location). Had multiplicities been eliminated differently the time and work necessary per iteration would have been different. These assumptions were used as part of a conservative comparison with RM's algorithm. We expected RM's algorithm to perform less work, since only the threads that are active (i.e., selected) start working after the selection was completed. In the No-Cut implementation all threads are spawned and those who aren't selected terminate immediately (which accounts for some of the additional work). We should remember that the No-Cut algorithm was time-optimized to compete with the Cut-6 algorithm, thus putting most of the effort into decreasing the critical-time-path even when that results in additional work. This included, amongst other things, using an "active-array" and copying into it most of the information in the elements, in order to save a few cache misses (when accessing data through two pointer accesses, which must be done in RM's algorithm). When optimizing for less work a different approach should be used for a better comparison. In such a case, we could save roughly $15 * N$ work units by using a selected-list (similar to the one used in RM's algorithm) at the end of the selection step (coin-tossing), spawning only the selected elements and not storing duplicates of most of the information in an active-array. This, of course, would increase the critical-time-path (by at least $5 * log_2 N$), and both algorithms would be much more similar.

The discussion above explains how our performance results made RM's algorithm look much better than other algorithms. This was done since the comparison was against RM's algorithm. The results of the analysis show that the work term for the RM algorithm ($32N + 10log_2 N + 9$) is smaller than the one for the No-Cut algorithm (see the results in section 6), but the time term for the No-Cut algorithm is smaller than the one for RM's algorithm ($9log_2 N^2 + 27log_2 N + 6$). For smaller list sizes the No-Cut algorithm performs better. A few of the speedups, comparing to the serial algorithm, obtained for RM's list ranking and for the No-Cut algorithm on the XMT are compared to the ones on the Cray-90 in the following table.

| Input Size | RM on Cray-C90 | RM on XMT | No-Cut on XMT |
n	$P = 8 * 128 = 1024$	$P = 1000$	$P = 1000$
$3.5 * 10^7$	speedup $\simeq 40$	speedup $\simeq 280$	speedup $\simeq 155$
$4.2 * 10^6$	speedup $\simeq 35$	speedup $\simeq 270$	speedup $\simeq 154$
$8 * 10^3$	speedup $\simeq 4$	speedup $\simeq 34$	speedup $\simeq 51$

Due to the new mechanism for enhancing thread spawning using PS (as per [VDBN98]), XMT circumvents the main hurdle for implementing RM's algorithm on the Cray, which is the need to group several sublists for amortizing unequal distributions of length. RM's solution used larger input length to overcome this hurdle. However, this, in turn, increased the input length for which good speed-ups are obtained. For XMT one can get good speed-ups already from shorter inputs. We should also remember that we favored the RM implementation on XMT for the comparison with No-Cut; this in turn somewhat biases its comparison to the implementation on the Cray. Detailed speedup results are presented in [DV99]. The analysis and simulations presented there are summarized below. The *most interesting conclusion* in this analysis is that the expected length of the longest chain in RM's algorithm is $\simeq 2.75 * log_2 N$, while for the No-Cut algorithm Section 3 derived the result $E(Maxlen) \simeq 1.2 * log_2 N$. This highlights an interesting advantage of the No-Cut algorithm; this advantage is due to its different randomized mechanism for picking "selected" elements. This explains why the time term for RM in Section 6 is larger than for No-Cut and why No-Cut is ahead for certain input sizes and values of P.

Few more details on the analysis of the RM algorithm A "paper and pencil" analysis of the behavior of RM's algorithm was done picking $k = \frac{n}{4}$ elements randomly at each iteration as heads of the sub-lists. Also, results obtained from simulations of the algorithm's behavior using the same k are reported.

We wish to find the probability that an element is selected; i.e., picked at least once during the $k = \frac{n}{4}$ draws. The probability that a specific element is *not* selected at all during all the draws, denoted as p_{ns}, is:

$$(1 - \frac{1}{n})^{\frac{n}{4}} = ((1 - \frac{1}{n})^n)^{\frac{1}{4}} \xrightarrow[n \to \infty]{} (\frac{1}{e})^{\frac{1}{4}} = 0.779$$

Notice that the limit of $(1 - \frac{1}{n})^n$ goes to $\frac{1}{e}$ very quickly (it is already very close, 0.366 from 0.3679, for $n = 100$). And the probability that an element is selected is $(1 - 0.779) = 0.221$.

This led to an analytical bound of $2.77 * log_2 N$ on the length of the longest chain. This was backed by experimental results whose practical conclusion was that $E(Maxlen) \simeq 2.75 * log_2 N$.

References

[AM90]   Anderson R. and Miller G.L. A Simple randomized parallel algorithm for list-ranking. *Information Processing Letters*, 33(5), 1990, 269-273.

[CV86a]   Cole R. and Vishkin U. Deterministic coin tossing and accelerating cascades: micro and macro techniques for designing parallel algorithms. In *Proc. 18th Annual ACM Symp. on Theory of Computing*, 1986, 206-219.

[CV89]   Cole R. and Vishkin U. Faster optimal parallel prefix sums and list ranking. *Information and Computation* 81(3), 1989, 334-352.

[DL99]   W.J. Dally and S. Lacy. VLSI Architecture: Past, Present, and Future, Proc. Advanced Research in VLSI Conference, Atlanta, March 1999.

[DV99]   S. Dascal and U. Vishkin. Experiments with list ranking for Explicit Multi-Threaded (XMT) instruction parallelism (extended summary), UMIACS-TR-99-33, June 1999.

[HP94]   Hennessy J.L., Patterson D.A., *Computer Organization and Design - The Hardware / Software Interface*. Morgan/Kaufmann. 1994.

[HP96]   Hennessy J.L., Patterson D.A., *Computer Architecture - A Quantitative Approach*, 2nd Edition. Morgan/Kaufmann. 1996.

[HR96]   Hsu T. and Ramachandran V. Efficient massively parallel implementation of some combinatorial problems. *Theor. Comp. Sci.* 6(2), 1996, 297-322.

[MR85]   Miller G.L. and Reif J.H. Parallel tree contraction and its applications. In *Proc. Symposium on Foundations of Computer Science*, 1985, 478-489.

[Re94]   Reid-Miller M. List ranking and list scan on the Cray C90. In *Proc. 6th ACM-SPAA*, 1994, 104-113.

[Si97a]   Sibeyn J.F. Better Trade-Offs for parallel list ranking. In *Proc. 9th ACM-SPAA*, 1997, 221-230.

[Si97b]   Sibeyn J.F. From Parallel to external list ranking. *TR-MPI-I-97-1-021,* Max-Planck-Institut Fur Informatik, Saarbrucken, Germany, 1997.

[SGS]   Sibeyn J.F., Guillaume F. and Seidel T. Practical parallel list ranking. *Preprint*, 1997.

[Tr96]   Traff J.L. List ranking on the PRAM. *The PAD Library Documentation*, Max-Planck-Institut fur Informatik, Saarbrucken, 1996.

[Vi84]   Vishkin U. Randomized speed-ups in parallel computation. In *Proc. 16th ACM-STOC*, 1984, 230-239.

[Vi97]   Vishkin U. From algorithm parallelism to instruction-level parallelism: An encode-decode chain using prefix-sum. In *Proc. 9th ACM-SPAA*, 1997, 260-271.

[VDBN98] Vishkin U. et al. Explicit Multi-Threading (XMT) bridging models for instruction parallelism (extended abstract). In *Proc. SPAA '98*, 1998. For greater detail see the extended summary of this extended abstract under http://www.umiacs.umd.edu/~vishkin/XMT/ .

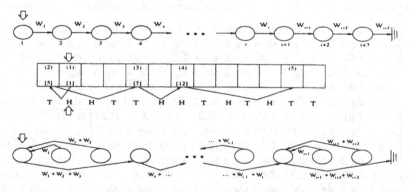

Fig. 1. *Upper figure:* original linked list. ⇓ - head of linked list. W_i - weight of a link. Elements numbering - their order in the linked list. *Middle figure:* input array representation of the linked list. Element numbering is in round brackets; pointer to array index of the next element in square brackets. Selected elements (H followed by T) are shaded. *Lower figure:* the linked list after one forward iteration. Backwards arrow to ranking parent.

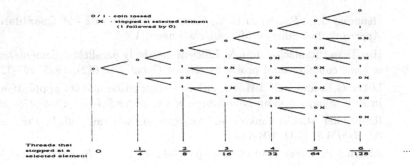

Fig. 2. Probabilities of selected elements and of chain length

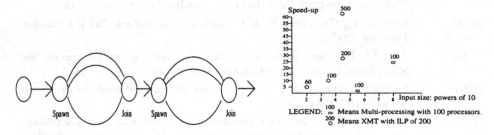

Fig. 3. Left figure: Parallel states from a Spawn to a Join and serial states. Right figure: Comparison to [Si97b].

N input size								P					
	10	20	30	40	50	60	80	100	200	500	1000	1500	2000
10^2	1.44	2.47	3.31	3.98	4.54	5.00	5.73	6.28	7.78	9.08	9.62	9.81	9.91
	alg 2	alg 4	alg 4	alg 4	alg 4	alg 4	alg 4	alg 4	alg 4	alg 4	alg 4	alg 4	alg 4
10^3	1.47	2.68	3.69	4.55	5.29	5.93	6.99	7.83	12.81	23.47	32.49	37.26	40.22
	alg 2	alg 2	alg 2	alg 2	alg 2	alg 2	alg 2	alg 2	alg 4	alg 4	alg 4	alg 4	alg 4
$5*10^3$	1.51	2.94	4.28	5.55	6.76	7.90	10.02	11.93	19.33	32.73	50.86	68.08	81.95
	alg 2	alg 2	alg 2	alg 2	alg 2	alg 2	alg 2	alg 2	alg 2	alg 3	alg 4	alg 4	alg 4
10^4	1.53	3.01	4.44	5.83	7.17	8.46	10.94	13.26	23.09	41.57	63.39	78.09	94.77
	alg 2	alg 2	alg 2	alg 2	alg 2	alg 2	alg 2	alg 2	alg 2	alg 2	alg 3	alg 3	alg 4
$5*10^4$	1.55	3.09	4.61	6.12	7.62	9.10	12.02	14.89	28.53	63.30	106.60	138.09	165.09
	alg 2	alg 2	alg 2	alg 2	alg 2	alg 2	alg 2	alg 2	alg 2	alg 2	alg 2	alg 2	alg 3
10^5	1.55	3.10	4.64	6.17	7.69	9.20	12.21	15.18	29.62	68.99	123.85	168.53	205.62
	alg 2	alg 2	alg 2	alg 2	alg 2	alg 2	alg 2	alg 2	alg 2	alg 2	alg 2	alg 2	alg 2
10^6	1.56	3.11	4.67	6.22	7.77	9.32	12.42	15.52	30.92	76.50	150.36	221.74	290.74
	alg 2	alg 2	alg 2	alg 2	alg 2	alg 2	alg 2	alg 2	alg 2	alg 2	alg 2	alg 2	alg 2
10^7	1.56	3.11	4.67	6.23	7.78	9.34	12.45	15.56	31.11	77.67	154.98	231.93	308.52
	alg 2	alg 2	alg 2	alg 2	alg 2	alg 2	alg 2	alg 2	alg 2	alg 2	alg 2	alg 2	alg 2

Table Above - Best Speedups Without RM's Algorithm These are the best speedups obtained when comparing the upper-bound on the parallel algorithms (not including RM's) with the lower-bound of the serial algorithm. *The*

algorithm used are: alg 1 - serial algorithm; alg 2 - work optimized coin tossing algorithm (No-Cut); alg 3 - time emphasized coin tossing algorithm (Cut-6); alg 4 - wyllie's (pointer jumping) algorithm. Horizontal and vertical lines indicate a different algorithm on each side of the line.

Table Below - Best Speedups With RM's Algorithm

Recall that the RM results (alg 5 below) are a bit optimistic, as explained in the paper.

N input size	10	20	30	40	50	60	80	100	200	500	1000	1500	2000
10^2	1.46	2.47	3.31	3.98	4.54	5.00	5.73	6.28	7.78	9.08	9.62	9.81	9.91
	alg 5	alg 4	alg 4	alg 4	alg 4	alg 4	alg 4	alg 4	alg 4	alg 4	alg 4	alg 4	alg 4
10^3	2.16	3.41	4.23	4.80	5.29	5.93	6.99	7.83	12.81	23.47	32.49	37.26	40.22
	alg 5	alg 5	alg 5	alg 5	alg 2	alg 2	alg 2	alg 2	alg 4	alg 4	alg 4	alg 4	alg 4
$5*10^3$	2.55	4.66	6.43	7.93	9.23	10.36	12.22	13.71	19.33	32.73	50.86	68.08	81.95
	alg 5	alg 5	alg 5	alg 5	alg 5	alg 5	alg 5	alg 5	alg 2	alg 3	alg 4	alg 4	alg 4
10^4	2.66	5.03	7.16	9.08	10.82	12.40	15.18	17.54	25.43	41.57	63.39	78.09	94.77
	alg 5	alg 5	alg 5	alg 5	alg 5	alg 5	alg 5	alg 5	alg 5	alg 2	alg 3	alg 3	alg 4
$5*10^4$	2.78	5.47	8.08	10.60	13.05	15.43	19.98	24.27	42.55	77.64	107.07	138.09	165.09
	alg 5	alg 5	alg 5	alg 5	alg 5	alg 5	alg 5	alg 5	alg 5	alg 5	alg 5	alg 2	alg 3
10^5	2.80	5.54	8.24	10.89	13.49	16.05	21.04	25.85	47.70	96.73	147.17	178.13	205.62
	alg 5	alg 5	alg 5	alg 5	alg 5	alg 5	alg 5	alg 5	alg 5	alg 5	alg 5	alg 5	alg 2
10^6	2.82	5.63	8.43	11.23	14.02	16.80	22.35	27.86	55.03	132.58	250.04	354.83	448.89
	alg 5	alg 5	alg 5	alg 5	alg 5	alg 5	alg 5	alg 5	alg 5	alg 5	alg 5	alg 5	alg 5
10^7	2.82	5.64	8.46	11.28	14.10	16.92	22.55	28.17	56.25	139.91	277.46	412.70	545.69
	alg 5	alg 5	alg 5	alg 5	alg 5	alg 5	alg 5	alg 5	alg 5	alg 5	alg 5	alg 5	alg 5

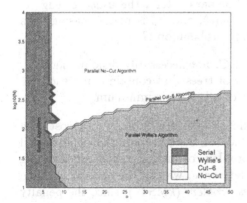
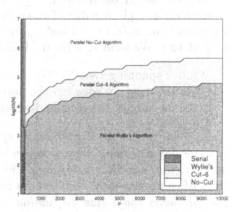

Fig. 4. The areas in which each of the four algorithms for list-ranking is dominant. Left figure: Parallelism up to 50. Right figure: Parallelism up to 10,000.

Finding Minimum Congestion Spanning Trees

Renato Fonseca F. Werneck, João Carlos Setubal, and Arlindo F. da Conceição

Institute of Computing, CP 6176, University of Campinas, SP, Brazil, 13083-970.
{werneck,setubal,973245}@dcc.unicamp.br

Abstract. Given a graph G and a positive integer k, we want to find k spanning trees on G, not necessarily disjoint, of minimum total weight, such that the weight of each edge is subject to a penalty function if it belongs to more than one tree. We present a polynomial time algorithm for this problem; the algorithm's complexity is quadratic in k. We also present two heuristics with complexity linear in k. In an experimental study we show that these heuristics are much faster than the exact algorithm also in practice, and that their solutions are around 1% of optimal for small values of k and much better for large k.

1 Introduction

Let $G = (V, E)$ be an undirected, weighted graph with n vertices and m edges. Let k be a positive integer. The edge weight function is denoted by w, and we assume all weights are positive integers. We extend the edge weight function to include an integer parameter i, $0 \le i \le k$. We call $w_p(e, i)$ the *penalized weight* of edge e for a given value of i; we call the parameter i the *usage* of edge e. By definition, $w_p(e, 0) = 0$ and $w_p(e, 1) = w(e)$; and w_p is nondecreasing with respect to i. We want to solve the following problem on G:

> Find k spanning trees $T_1, T_2, \ldots, T_k$ of G, not necessarily disjoint, such that the usage of edge e is the number of trees that contain e and such that the sum of the penalized weights of all edges is minimum.

In other words, we want to find k spanning trees of minimum total cost such that we pay a penalty every time we include an edge in more than one tree. If $w_p(e, i) = \infty$ for all $i > 1$, this is the problem of finding k *disjoint* spanning trees of minimum total cost. For future reference, we call the general problem the *minimum congestion k-spanning trees problem* (kMSTc), since the penalty function can be used to model congestion situations.

To our knowledge the general kMSTc problem has not been studied before. However, disjoint-trees versions of it are well-known. Nash-Williams [5] and Tutte [7] have studied the unweighted case. Roskind and Tarjan [6], building on work of Edmonds [3,4], presented a polynomial-time algorithm for the weighted case.

The kMSTc problem is interesting in its own right. However, our main interest in this paper is to explore it as an example of the following type of situation:

J.S. Vitter and C.D. Zaroliagis (Eds.): WAE'99, LNCS 1668, pp. 60–71, 1999.
© Springer-Verlag Berlin Heidelberg 1999

A problem for which a polynomial-time algorithm exists, but for which, *in practice*, it is better to use heuristics. Heuristics are commonly encountered and vastly studied in the realm of NP-complete problems. In the case of problems in class P, such results are much rarer.

Table 1. Summary of theoretical results

algorithm	running time
exact	$O(m \log m + k^2 n^2)$
A-Prim	$O(k(m + n \log n))$
A-Kruskal	$O(m \log m + k(m\alpha(2n, n) + n \log n))$
B	$O(m \log m + kn(\log m + \alpha(2n, n) \log n))$

The results we present are as follows. First we show that the kMSTc problem can be solved exactly in polynomial time. Then we present two algorithms for which we have no guarantees on solution quality, and hence are heuristics. We show that their theoretical running times are better than the running time of the exact algorithm. A summary of theoretical results is shown in Table 1. We then present an experimental study that shows that the heuristics are indeed much faster than the exact algorithm (for the graph instances tested), while finding solutions very close to the optimum.

2   An Exact Algorithm

The exact algorithm is a simple reduction of our problem to the weighted disjoint-trees problem (which we call the kMSTd problem). Our algorithm therefore is entirely based on Roskind and Tarjan's algorithm (hence-forward called RT). Our reduction assumes that the RT algorithm can handle parallel edges. This is an easy extension of the algorithm described in [6].

Here is how the reduction works. First note that a given edge e appearing in i trees of a given set contributes a total of $i w_p(e, i)$ to that set's weight; and that the total cost of a solution is $\sum_{e \in E} i_e w_p(e, i_e)$, where i_e is the usage of e. We can therefore define the *incremental cost* $c(e, i + 1)$ of edge e appearing in one more tree as

$$c(e, i + 1) = (i + 1)w_p(e, i + 1) - i w_p(e, i) . \tag{1}$$

The value of $c(e, i + 1)$ represents the increase in the solution's value by the inclusion of edge e in the $(i + 1)$-th tree.

We will now reduce an instance I of the kMSTc problem to an instance I' of the kMSTd problem. Given an instance of the kMSTc problem (defined above) create a graph $G' = (V, E')$ in the following way: for each edge $e = (u, v) \in E$, create k parallel edges $e_1, e_2, \ldots, e_k \in E'$ (all of them between vertices u and v)

such that $w'(e_i) = c(e, i)$, $1 \leq i \leq k$. Note that if G is connected, then G' will certainly contain at least k disjoint spanning trees. We remark that using parallel edges with incremental costs is a standard way of making such reductions; see for example [1, section 14.3].

We now claim that given a solution S' for I', we can obtain a solution S for I and vice versa. First note that given a pair of vertices (u, v) and the edges $e_1, e_2, \ldots, e_k$ between them in I', an edge e_i can belong to S' if and only if every edge e_j, $1 \leq j < i$, is also part of S' [to simplify the discussion, we assume $c(e, i)$ is strictly increasing w.r.t. i]. Furthermore, all these edges belong to different trees (because the trees must be disjoint and to avoid cycles). Based on this we can establish a one-to-one correspondence between a tree in S' and a tree in S. If in S', for a given pair (u, v), all edges up to e_l are used, this means that the usage of edge (u, v) in S will be l (and vice versa). The total cost of these edges in S or in S' is the same and that proves the claim.

We now very briefly describe the RT algorithm. It starts by sorting the edges in nondecreasing order by weight and creating a set F of k forests $F_1, F_2, \ldots, F_k$, each with n vertices and no edges. The algorithm executes the following augmenting step for each edge $e \in G$, in order: find an F_i such that e can be inserted in F_i without creating cycles (with the insertion possibly causing a rearrangement of edges in F). The algorithm stops when all forests become trees. The search for a valid F_i in each augmenting step is done by a labeling process.

Roskind and Tarjan show that this algorithm takes $O(m \log m + k^2 n^2)$ time. A direct application of the reduction above gives for our algorithm a running time of $O(km \log km + k^2 n^2)$ and a storage requirement of km edges. We now show that it is not necessary to store that many edges and that our algorithm can be made to run in the same time bound as the RT algorithm.

Given the input graph G and its m edges, the reduction gives a clear rule for determining all km edges of G'. This means that we may keep G' implicit, and generate each copy of an edge from G on the fly. In addition, note that as soon as the algorithm decides to discard an edge e_i, all other unexamined edges $e_j, j > i$ between the same pair of vertices can be discarded as well. This means that we don't even have to sort km values; we either discard an edge and all remaining parallel edges, or we use edge e_i and insert its next "copy" (e_{i+1}) into the data structure that contains the edges with the correct weight. The time bound depends on the data structure used to represent the sorted edge list. Since this list must also be updated, we need a priority queue. We assume that a binary heap is used.

Initialization of the algorithm consists of building the heap ($\Theta(m \log m)$ time) and creating the forests ($\Theta(kn)$). Selecting an edge from the heap costs $O(1)$. Once an edge is processed, we may need to discard it or replace it with the next copy. In terms of heap operations, this means deleting or updating an element, respectively. Both require $O(\log m)$ time.

As in the original algorithm, the labeling step is executed $O(kn)$ times. Each execution requires $O(kn)$ operations, in addition to the extra $O(\log m)$ time demanded by heap operations. There are up to $O(m)$ steps in which applying the labeling algorithm is not necessary, since the ends of the edge being examined belong to the same special subset ("clump", in RT's terminology) of vertices. Each of these steps requires $O(\log m)$ time to delete the edge from the heap. The overall time bound of the algorithm implemented in this way is $O(m \log m + k^2 n^2)$, which is an improvement over the direct application of the reduction and is the same time bound of the RT algorithm.

3   Heuristics

In this section we present two heuristics for the kMSTc problem. These heuristics were developed before we knew the complexity of the problem, and were thus contemplating the possibility that it might be NP-complete. Even though it turned out that the problem is polynomially solvable, the heuristics proposed are intuitive, asymptotically faster than the exact algorithm, and simple to implement.

3.1   Heuristic A: Tree-Greedy

This is a greedy heuristic, in that it computes each tree T_i based on all previously computed T_j trees, $1 \leq j < i$. We start with graph $G_1 = G$, and compute T_1 as its minimum spanning tree. We then update the weights obtaining graph G_2 to reflect that $n - 1$ edges were chosen. The updated weight of a chosen edge e will be $c(e, 2) = 2w_p(e, 2) - w_p(e, 1)$, its incremental cost as defined in (1). The next tree, T_2, will be the minimum spanning tree of G_2 and so on, until we have obtained k trees.

Any minimum spanning tree algorithm can be used to implement this heuristic. We analyze two possibilities: using Prim's algorithm and Kruskal's algorithm [2].

Implementing Prim's algorithm using a Fibonacci heap makes it run in $O(m + n \log n)$ time. As Heuristic A executes this algorithm k times, its complexity is $O(k(m + n \log n))$.

Kruskal's algorithm requires a preprocessing edge-sorting step that costs $O(m \log m)$. Using a union-find data structure with the usual path-compression and weighted-union techniques, the main loop costs $O(m\alpha(2n, n))$ time. The complexity of the algorithm is thus dominated by the complexity of the preprocessing step. As Heuristic A requires k applications of the algorithm, its overall complexity would be $O(km \log m)$. However, Heuristic A allows us to execute Kruskal's algorithm more efficiently. After the i-th ($1 \leq i < k$) execution of Kruskal's algorithm, the set of edges can be partitioned into two subsets, one (the first) containing the $n - 1$ edges of T_i, and another (the second) with the remaining edges. The latter is already sorted, since the incremental costs of its edges were not changed. For the next step, all we have to do is sort the first subset, merge it with the second, and apply Kruskal again. This takes $O(m\alpha(2n, n) + n \log n)$ time. This observation frees us from sorting the complete set of edges in every step. The overall complexity of the heuristic is hence $O(m \log m + k(m\alpha(2n, n) + n \log n))$.

3.2 Heuristic B: Edge-Greedy

Heuristic B is greedy too, but "grows" the trees of the desired solution all together, in a fashion reminiscent of the exact algorithm. We start with k forests, each forest F_i composed by the n vertices of G and no edges. We pick the edge with the smallest incremental cost (and hence need a priority queue) and check whether it can be inserted into some forest without creating a cycle. If we find such a *(valid)* forest, the edge is inserted and its incremental cost is updated in the queue; if a forest is not found, the edge can be discarded. This step is repeated until $k(n-1)$ edges are inserted into the forests, which by then are trees. There are at most m steps in which a valid forest is not found, and $k(n-1)$ steps in which an edge is actually inserted. In each step, we must perform two kinds of operations:

- *Edge selection*: We must either remove an edge from the queue or update its weight. We can perform both operations in $O(\log m)$ time if we represent the priority queue as a binary heap.
- *Cycle checking*: The most efficient way to manage cycle-related information is employing a union-find data structure. Since each forest may be checked in every step, all cycle-related operations (considering the entire execution of the algorithm) cost $O((kn + m)\alpha(2n, n))$ time.

All costs considered, the overall time bound of Heuristic B is $O((kn+m)(\log m + k\alpha(2n, n)))$. We will now show that this time bound can be substantially improved by using several different techniques.

Monitoring Clusters. The first idea is to use an extra union-find structure (in addition to the ones associated with forests) to represent *clusters*. Clusters are subsets of vertices that are in the same connected component in every forest F_i.

We start with n disjoint sets, each representing a vertex of G, and then proceed to the execution of the algorithm as previously described. If we test an edge $e = (u, v)$ against every relevant forest and find that none is valid for insertion, we conclude that vertices u and v belong to the same cluster, and execute $union(u, v)$ in the extra union-find structure. The performance gain will be achieved if we verify whether the endpoints of an edge belong to the same cluster before checking all forests in each step. If the vertices do belong to the same cluster, we simply discard the edge and proceed to the next one, avoiding a fruitless search.

The worst-case analysis of this approach is as follows: $k(n-1)$ successful searches, $O(n)$ unsuccessful searches (each resulting in a *union* operation), and $O(m)$ edges discarded without any search at all. Knowing that $O(\log m)$ operations are required to select an edge and that each search checks up to k forests, the overall time bound of this improved version of Heuristic B is $O(m\log m + kn(\log m + k\alpha(2n, n)))$.

Discarding Trees. Another idea is to test edges only against a relevant subset of the forests, those with more than one component (other forests can be discarded, since no edge can be added to them without creating cycles). To implement this, we keep the relevant forests (all of them, in the beginning of the algorithm) in a linked list. As soon the $(n-1)$-th edge is inserted into a forest F_i, we remove F_i from the list. This makes the algorithm faster, although its asymptotic time bound remains unchanged.

Ordering. For the running times presented so far, any forest scanning order can be used. One may even adopt different orders in each step. However, analyzing the forests in the exact same order in every step yields a significant improvement. It does not matter which order it is, as long as it is the same in every step.

To show the improvement we will need to compare connected components of the forests F_i. We will do so considering only the set of vertices in each component (not the edges). Suppose now that the fixed order in which forests are considered is $F_1, F_2, \ldots, F_k$. Then, the following is true:

Theorem 1. *Any connected component in a forest F_i is a subset (not necessarily proper) of some connected component of F_{i-1}, for $1 < i \le k$.*

Proof. If there were a component C in F_i whose vertices were not in the same component of F_{i-1}, then there would be an edge e joining some two vertices of C that could be inserted in F_{i-1}. This contradicts the fact that edges are inserted in the first valid forest.   □

The following corollary is immediate.

Corollary 1. *Let $e(u, v)$ be an edge of G and let F_i $(1 \le i < k)$ be the first forest in which the insertion of e does not create a cycle (i.e., in which u and v are in different connected components). Then, e does not create a cycle in F_j, for every j such that $i < j \le k$.*

As a consequence of this result, for each pair of vertices u, v, we can divide the set of forests into two subsets. In forests F_1 to F_{i-1}, u and v belong to the same connected component; in forests F_i to F_k, u and v are in different components. Our task is to find the first forest of the second subset. This can be done using binary search over the k forests. This reduces the overall complexity of the algorithm to $O(m \log m + kn(\log m + \alpha(2n, n) \log k))$. This complexity assumes that cluster monitoring (as described above) is used.

Applying the tree-discarding technique presented above is still possible. In fact, it becomes easier to implement when the forests are analyzed in a fixed order. It follows from Theorem 1 that a forest F_i $(1 < i \le k)$ can become a tree only after every forest F_j $(1 \le j < i)$ has already become a tree in previous steps. In other words, the subset of forests which are trees is either empty or can be expressed as $\{F_1, F_2, \ldots, F_s\}$, for some $s \le k$. Thus, when a forest F_i becomes a tree, all we have to do is set $s = i$ and restrict further searches to forests $F_{s+1}, F_{s+2}, \ldots, F_k$. Notice that the single variable s makes the linked list for the tree-discarding technique unnecessary.

Indexed search. Even with all the improvements mentioned so far, the running time of Heuristic B has a $k \log k$ factor, while Heuristic A is linear in k. As seen above, the extra $\log k$ factor is due to the cost of searching for a valid forest to insert a given edge. We now show that this search can be done in $O(\log n)$ time.

For this we need the concept of *component-equivalence*. Two forests $F_i(V, E_i)$ and $F_j(V, E_j)$ are component-equivalent if and only if their vertices are partitioned into the same connected components (recall that by "same component" we mean components with the same set of vertices, but not necessarily the same set of edges). With this definition we can now prove the following result.

Theorem 2. *In any step of the algorithm, there are no more than* $\min(n, k)$ *sets of component-equivalent forests.*

Proof. This is trivial if $k \leq n$, since k is the number of forests. Therefore, let us assume that $k > n$. An immediate consequence of Theorem 1 is that forests are always sorted in non-increasing order by number of edges. That is, given forests F_i and F_{i-1} $(1 < i \leq k)$, either $|E_i| < |E_{i-1}|$ or $|E_i| = |E_{i-1}|$. They cannot be component-equivalent if the former holds, since the number of edges is different. On the other hand, if the second statement is true, Theorem 1 guarantees the component-equivalence of F_{i-1} and F_i, which means they belong to the same set. Thus the number of sets is equal to the number of strict inequalities. Since a forest with n vertices may have no more than $n - 1$ edges, there are forests with at most n different number of edges in a given step of the algorithm (from 0 to $n - 1$). Hence, n is the maximum number of different sets of component-equivalent forests if $k > n$.   □

As described above, in every step of the algorithm we must look for a valid forest F_i for which i is minimum. The implementation discussed in the description of the ordering technique finds this forest by performing binary search on the list of forests. Using Theorem 2, we show that there is another and faster way of finding the same forest.

Let S_i $(0 \leq i \leq n - 1)$ be the set of component-equivalent forests with i edges. For a given edge e, we must find the largest i such that e can be inserted into a forest of S_i without creating a cycle. Although e could be inserted into any forest in S_i, we must choose the first forest in S_i according to the order in which forests are being considered. Therefore, we must keep, for each set S_i, a reference to such forest, making the list of component-equivalent sets act as an index to individual forests. To find the desired forest, we can perform a binary search on a vector of indices, which has size n, costing us $O(\log n)$. We call this index vector *cindex*. This is the general idea. However, there is a minor detail that we must consider. Since any set S_i may be empty in some steps of the algorithm, we must decide which forest should be referenced to by *cindex*[i] when this happens. We adopted the following solution: make *cindex*[i] represent the first forest with i edges *or fewer*. Hence, when there is no i-edge forest, we have *cindex*[i] = *cindex*[$i - 1$]. With this solution, *cindex* must be managed as follows:

- *Initialization:* Set $cindex[i] = 1$, for $0 \leq i \leq n - 1$.
- *Search:* Given an edge e, find, using binary search, the largest i such that e can be inserted in $cindex[i]$ and cannot be inserted in $cindex[i + 1]$.
- *Update:* Let F_i be the forest found in the previous step and $|E_i|$ be the number of edges in it (after e is inserted). (Since $|E_i|$ increased by one, we have to do something about $cindex[|E_i| - 1]$.) Set $cindex[|E_i| - 1] = i + 1$, which means that F_{i+1} replaces F_i as the first forest with $|E_i| - 1$ edges or fewer.

Notice that the second phase (search) can be unsuccessful, because there may be, in some step of the algorithm, no i such that e can be inserted in F_i. When this happens, e can be removed from the queue.

We can now present the final analysis of Heuristic B. The only difference we have introduced is the number of comparisons made when searching for a forest to insert an edge. When $cindex$ is used, $O(\log n)$ comparisons are required per search, as opposed to $O(\log k)$ without that auxiliary vector. In both cases, all comparisons related to a certain forest take at most $O(n\alpha(2n, n))$ total time. Therefore, Heuristic B with cluster control and indexed search will run in $O(m \log m + kn(\log m + \alpha(2n, n) \log n))$ time.

4 An Experimental Study

We have implemented and tested the exact algorithm, Heuristic A with Kruskal's algorithm, Heuristic B, and a random algorithm (see next paragraph). Heuristic A with Prim's algorithm and a binary heap was implemented, but preliminary testing showed that it was considerably slower than A-Kruskal, and hence it was not part of further testing.

The random algorithm builds k trees, and for each tree it selects $(n-1)$ edges randomly, using a union-find data structure to avoid cycle-creating edges. This algorithm was implemented so that the proposed heuristics could be judged not only on how close they get to the optimal solution but also how far they are from a randomly found solution.

The implementation of the exact algorithm contains the following practical improvement in algorithm RT motivated by our experience in developing Heuristic B. When trying to insert an edge into a forest, we try all forests (in the same fixed order) rather than invoke the labeling procedure after failing to insert the edge into forest F_1.

All programs were implemented in C++ and compiled with the GNU C Compiler using the -O3 optimization flag. All tests were done on a DEC Alpha 600au with 1 GB of RAM. Every instance fit in memory, thus limiting I/O operations to reading the input graph. The penalty function used was $w_p(e, i) = iw(e)$. This results in a quadratic objective function.

The implementations were mostly tested on program-generated families of instances. The generators were written by the authors. The families are as follows:

- complu, a complete graph, with uniform distribution of distinct weights.
- hyperb, a hypercubic lattice with a biased distribution of distinct weights; the bias favors small weights; these graphs are sparse ($m = 4n$).
- random, a random connected graph with uniform distribution of weights.

In reporting results we use the concept of solution *quality*. The quality of solution S given by algorithm X is simply the ratio between the cost of S and the optimal cost.

4.1 Results

Table 2 compares the three implementations with respect to running time, and the two heuristics and the random algorithm with respect to solution quality, for family complu and $n = 100$ with varying k. Table 3 is similar, but the instances come from family hyperb, $k = 100$ and n varies. In both cases, running times in each row are averages over three different instances (different seeds); and the qualities reported are maxima (i.e. worst), except for the random algorithm. The random algorithm was run three times on each instance, and the value shown in each row is the best solution found in the nine instances thus solved.

Both tables make it clear how much faster the heuristics are with respect to the exact algorithm. In all these cases heuristic solution quality is within 0.05% of optimal. The random algorithm is significantly worse than the heuristics. The difference in the quality values obtained by the random algorithm in families complu and hyperb shows that this algorithm is sensitive to the weight distribution, as expected.

To get an idea of how quality changes with different values of k we ran Heuristic A on a graph from TSPLIB (called brazil58). This is a complete graph with $n = 58$. The results are shown in Figure 1. Results for Heuristic B are essentially the same. We see that the worst case occurs for $k = 5$, and that for large values of k quality tends to 1. For this instance and $k = 1000$, Heuristic A took 0.52 secs, Heuristic B took 0.21 secs, while the exact algorithm took 22 secs. Similar quality behavior was observed for other inputs. The random algorithm in this experiment had the following behavior: for $k = 1$, its quality was 6.99; for larger values of k, the quality improved, reaching 2.02 for $k = 1000$.

The last experiment compares the performance of heuristics A and B with respect to running time, on family random, $n = 500$, $k = 1000$. We varied the number of edges from $m = 1200$ (very sparse) to $m = 124750$ (complete), on a total of 10 m values (twelve instances for each value of m). Results are shown in Figure 2.

We can see there that the km factor in the theoretical running time of Heuristic A does have a considerable influence in its performance in practice. We note that running time variance in this experiment was large, which may explain the odd behavior of B's running time towards the right end of the graph.

Table 2. Results for family complu, $n = 100$

	time (secs)			quality		
k	exact	A	B	A	B	random
100	2.32	0.08	0.04	1.000492	1.000496	4.92
200	6.97	0.13	0.08	1.000150	1.000149	4.51
300	12.37	0.20	0.12	1.000297	1.000302	4.29
400	17.67	0.28	0.16	1.000198	1.000197	4.13
500	21.99	0.36	0.20	1.000130	1.000130	4.06

Table 3. Results for family hyperb, $k = 100$

	time (secs)			quality		
n	exact	A	B	A	B	random
81	0.81	0.01	0.02	1.000461	1.000403	52.29
256	10.75	0.05	0.10	1.000254	1.000258	54.48
625	76.93	0.17	0.32	1.000310	1.000310	29.14
1296	364.21	0.41	0.85	1.000325	1.000325	15.69

5   Final Remarks

The parallel edge technique is a standard way of dealing with certain nonlinear optimization problems, such as the kMSTc problem, studied here. We have shown that greedy heuristics, well implemented, represent an interesting alternative to solve this problem, if optimality is not crucial but running time is. In addition these heuristics obtain results that are far better than random solutions.

We have also investigated the possibility of adding a lower bound computation to our heuristics. A simple lower bound for the kMSTc problem is a set of $k(n-1)$ edges (not necessarily distinct) of G such that the sum of their penalized weights is minimum and such that no more than k copies of any given edge are selected. The computation of this lower bound can be done as follows. Initially, we must insert every edge $e \in E$ into a priority queue and set $counter_e = 0$. In each step of the algorithm, we remove the first edge (e) from the queue and increment $counter_e$. If the new value of $counter_e$ is k, e can be discarded; if $counter_e < k$, we must update the incremental cost of e and reinsert it into the queue. We stop after $k(n-1)$ steps. If the priority queue is implemented as a binary heap, each step will require $O(\log m)$ time. This yields a running time of $O(kn \log m)$, asymptotically better than any of the algorithms discussed so far.

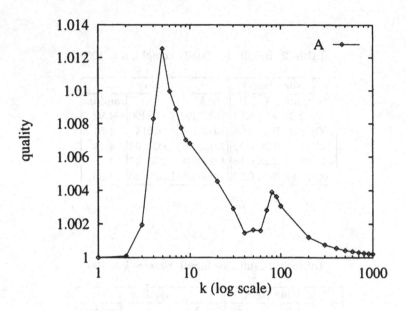

Fig. 1. Quality values of Heuristic A on instance brazil58.

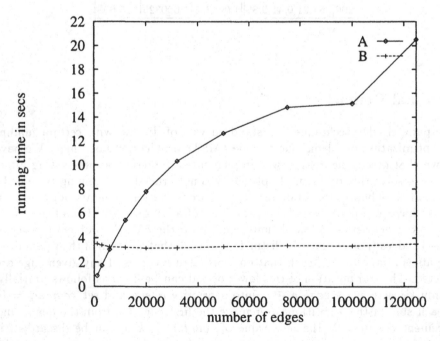

Fig. 2. Comparison of Heuristics A and B on instances from family random; $n = 500$ and $k = 1000$.

This lower bound is so simple that its computation can be incorporated into any of the heuristics, thereby creating a program that will output a quality measure of the solution obtained, while still being much faster than the exact algorithm. However, preliminary experiments have shown that this lower bound is not consistently strong for this purpose (there are cases in which it attains only 50% of the optimum value). We are currently trying to improve this bound and at the same time keep it fast to compute.

Several other lines of future research suggest themselves: is there a faster exact algorithm for kMSTc? Are the heuristics in fact approximation algorithms? What is the behavior of the heuristics for other penalty functions? In what other congestion problems can these greedy techniques be applied?

Acknowledgments

We thank Jorge Stolfi for fruitful conversations and the anonymous referees for helpful suggestions. RFFW was supported by FAPESP undergraduate fellowship 98/13423-5; JCS was supported in part by CNPq fellowship 301169/97-7 and by the Pronex project 107/97; and AFC was supported by CNPq MSc fellowship 131945/97-1.

References

1. R. Ahuja, T. Magnanti, and J. Orlin. *Network flows*. Prentice-Hall, 1993.
2. T. Cormen, C. Leiserson, and R. Rivest. *Introduction to Algorithms*. MIT Press/McGraw Hill, 1990.
3. J. Edmonds. Minimum partition of a matroid into independent subsets. *J. Res. Nat. Bur. Standards*, 65B:67–72, 1965.
4. J. Edmonds. Lehman's switching game and a theorem of Nash-Williams. *J. Res. Nat. Bur. Standards*, 65B:73–77, 1965.
5. C. St. J. A. Nash-Williams. Edge-disjoint trees of finite graphs. *J. London Math. Soc.*, 36:445–450, 1961.
6. J. Roskind and R. Tarjan. A note on finding minimum-cost edge-disjoint spanning trees. *Mathematics of Operations Research*, 10(4):701–708, 1985.
7. W. Tutte. On the problem of decomposing a graph into n connected factors. *J. London Math. Soc.*, 36:221–230, 1961.

Evaluation of an Algorithm for the Transversal Hypergraph Problem

Dimitris J. Kavvadias[1] and Elias C. Stavropoulos[2*]

[1] University of Patras, Department of Mathematics, GR-265 00 Patras, Greece
djk@math.upatras.gr
[2] University of Patras, Computer Engineering & Informatics Department
GR-265 00 Patras, Greece
estavrop@ceid.upatras.gr

Abstract. The Transversal Hypergraph Problem is the problem of computing, given a hypergraph, the set of its minimal transversals, i.e. the hypergraph whose hyperedges are all minimal hitting sets of the given one. This problem turns out to be central in various fields of Computer Science. We present and experimentally evaluate a heuristic algorithm for the problem, which seems able to handle large instances and also possesses some nice features especially desirable in problems with large output such as the Transversal Hypergraph Problem.

1 Introduction

Hypergraph theory [2] is one of the most important areas of discrete mathematics with significant applications in many fields of Computer Science. A hypergraph $\mathcal{H}$ is a generalized graph defined on a finite set $\mathcal{V}$ of nodes, with every hyperedge $\mathcal{E}$ of $\mathcal{H}$ being a subset of $\mathcal{V}$. A hypergraph is a convenient mathematical structure for modeling numerous problems in both theoretical and applied Computer Science and discrete mathematics. One of the most intriguing problems on hypergraphs is the problem of computing the *transversal hypergraph* of $\mathcal{H}$, denoted $\mathrm{tr}(\mathcal{H})$. The transversal hypergraph is the family of all minimal hitting sets (*transversals*) of $\mathcal{H}$, that is, all sets of nodes T such that (a) T intersects all hyperedges of $\mathcal{H}$, and (b) no proper subset of T does. TRANSVERSAL HYPERGRAPH is the problem of generating $\mathrm{tr}(\mathcal{H})$ given a hypergraph $\mathcal{H}$, and is an important common subproblem in many practical applications. Its importance arises from the fact that problems referring to notions like minimality or maximality are quite common in various areas of Computer Science.

For example, concepts of propositional circumscription [18, 3] and minimal diagnosis [5] restrict interest to the set of models of an expression that are minimal. In circumscription, model checking for circumscriptive expressions reduces to determining the minimality of a model, whereas in model-based system diagnosis, finding a minimal diagnosis is equivalent to finding the prime implicants of an expression and, next, finding the minimal cover of them. Moreover, finding a

---

* Research supported by the University of Patras Research Committee (Project Caratheodory under contract no.1939).

maximal model is essential task in model-preference default inference [20], since, given a set of default rules, our aim is to find a maximal (most preferred) model of these rules. It can also be seen that the problem of computing the transversal hypergraph is an alternative view of the problem of generating all maximal models of a Boolean expression in CNF, having all its variables negated. Such an expression can be seen as a hypergraph whose hyperedges are the clauses of the expression and each node of a hyperedge corresponds to a negative variable of the corresponding clause. The maximal models of the expression then correspond to the transversals of the hypergraph. A symmetric situation holds for the minimal models of an expression having all its variables positive. Complexity questions related to minimal or maximal models have been discussed in [3, 4, 1]; more recent results can be found in [13].

An encyclopedic exposition of the applications of the TRANSVERSAL HYPER-GRAPH problem can be found in [6, 7]. We briefly state from there certain problems in the design of relational databases [15, 16], in distributed databases [8], and in model-based diagnosis [5]. Another interesting connection was pointed out between the TRANSVERSAL HYPERGRAPH problem and the rapidly growing field of knowledge discovery in databases, or data mining [9, 17].

In this paper we present and experimentally evaluate a heuristic algorithm for solving the TRANSVERSAL HYPERGRAPH problem. The algorithm was implemented and tested on a number of randomly generated problem instances. The experimental results show that the algorithm computes all transversals of a given hypergraph correctly and efficiently. This fact makes our heuristic suitable for solving problems in all areas mentioned above.

One has to be careful in defining efficiency of algorithms for problems like the TRANSVERSAL HYPERGRAPH. It is not hard to see that the transversal hypergraph tr($\mathcal{H}$) of a hypergraph $\mathcal{H}$ may have exponentially many hyperedges with respect to the number of nodes and the number of hyperedges of $\mathcal{H}$. It will therefore require exponential amount of time to compute tr($\mathcal{H}$) in the worst case. Therefore, the usual distinction between tractable and intractable problems based on the existence or not of a polynomial-time algorithm, clearly does not apply here. Instead, more elaborate complexity measures have to be defined, that will take into account the *size of output*, too. It is natural to consider as tractable a problem with large output if it can be solved by an algorithm that is polynomial in *both* the input and the output. Such algorithms are called *output-polynomial*. A slightly stronger requirement is that the algorithm generates a new output bit in time polynomial in the input *and the output so far*. These latter algorithms are called *incrementally output-polynomial*. An even stronger requirement is that the algorithm generates two consecutive output bits in time bounded by a polynomial in the input size. These are called *polynomial delay* algorithms. There has been a recent surge of interest in such algorithms. For discussions of algorithms with output and performance criteria see [11, 12, 19].

The precise complexity of the TRANSVERSAL HYPERGRAPH problem is still unknown. The brute force algorithm given by Berge [2] needs time exponential in both the input and the output. However, several special cases can be solved in polynomial time [6]. Recently, an output-subexponential algorithm was given by Fredman and Khachiyan in [7]. There, it was shown that the duality of two monotone Boolean expressions in DNF can be checked in time $O(n^{\log n})$, where n is the combined size of the input and the output. It is not hard to see that this problem is another disguised form of the TRANSVERSAL HYPERGRAPH problem (see [7, 10] for further details on these issues).

Our algorithm presents in practice a remarkable uniformity in its output rate; averaging over relatively small parts of the output (e.g. 100 transversals out of a total output of tenths of thousands), we get delays deviating by at most 3 mean values. (We mention however, that at present we can prove no bound for the delay between consecutive outputs.) This happens partially because our algorithm operates in a *generate-and-forget* fashion i.e., no previous transversal is required for the generation of the next ones. In contrast, both the brute force algorithm and the Fredman–Khachiyan algorithm require all previous transversals to be stored. Moreover, the former will output the first transversal after exponentially long delay. Our approach also greatly reduces the memory requirements, since previously generated transversals need not be stored. In a different situation, the memory requirements could be devastating as the total number of transversals can be enormous. In addition, absolute time delays are very small, allowing the successful handling of large problems.

The rest of this paper is organized as follows: In Section 2 we describe our heuristic algorithm for generating all transversals of a hypergraph. In the next section we compare our algorithm with some other approaches while in Section 4 we give some implementation details. Experimental results on the performance of the algorithm are summarized in Section 5. Finally, in Section 6 some conclusions are given.

2 Description of the Algorithm

The proposed algorithm is based on the following simple algorithm of Berge (see [2, 6]): Consider a hypergraph $\mathcal{H} = \{\mathcal{E}_1, \ldots, \mathcal{E}_m\}$. Assume that we have already computed the transversal hypergraph $\mathrm{tr}(\mathcal{G})$ of $\mathcal{G} = \{\mathcal{E}_1, \ldots, \mathcal{E}_k\}$, for some $k < m$. It is easy to see that $\mathrm{tr}(\mathcal{G} \cup \{\mathcal{E}_{k+1}\}) = \{\min\{t \cup \{v\}\} : t \in \mathrm{tr}(\mathcal{G})$ and $v \in \mathcal{E}_{k+1}\}$, where by $\min\{t \cup \{v\}\}$ we denote the set of minimal subsets of $t \cup \{v\}$ that are hitting sets of $\mathcal{G} \cup \{\mathcal{E}_{k+1}\}$. Based on this observation we may find all transversals of G by starting from the transversals of $\mathcal{E}_1$ (note that the transversals of a hypergraph with a single hyperedge are all its nodes) and adding one-by-one the rest of the hyperedges, computing at each step the set of transversals of the new hypergraph. The algorithm terminates after the addition of $\mathcal{E}_m$.

There are several drawbacks in the above scheme regarding its efficiency which are explained in detail in the next section. We only mention here the most severe one, in view of the complexity measures for this problem: The computation of the first *final* transversal (a transversal of the input hypergraph $\mathcal{H}$) is accomplished after all transversals of the graph $\{\mathcal{E}_1, \ldots, \mathcal{E}_{m-1}\}$ have been computed. This means that it may take an exponentially long time before the first transversal is output. No less important are the memory requirements that also emerge from the above: all intermediate transversals have to be stored and kept until used for the computation of the new transversal set.

We now explain our algorithm. Consider a hypergraph $\mathcal{H} = \{\mathcal{E}_1, \ldots, \mathcal{E}_m\}$ defined on a set V of n nodes. We call a set of nodes $\mathcal{X} \subseteq V$ a *generalized node* if all nodes in $\mathcal{X}$ belong in exactly the same hyperedges of $\mathcal{H}$. Assume that the hypergraph $\mathcal{H}$ has a generalized node $\mathcal{X}$ with cardinality $|\mathcal{X}| \geq 2$. Consider the hypergraph $\mathcal{H}'$ which follows from $\mathcal{H}$ by replacing all nodes in $\mathcal{X}$ in all hyperedges that they appear by a new node $v_{\mathcal{X}}$. Let now $\text{tr}(\mathcal{H}')$ be the set of transversals of $\mathcal{H}'$. The importance of the concept of generalized node follows from the observation that $\text{tr}(\mathcal{H}) = (t \setminus v_{\mathcal{X}}) \times \mathcal{X}$, for all $t \in \text{tr}(\mathcal{H}')$ such that $v_{\mathcal{X}} \in t$. In other words, the transversals of $\mathcal{H}$ follow by taking one by one the transversals of $\mathcal{H}'$ that include the node $v_{\mathcal{X}}$ and replacing $v_{\mathcal{X}}$ by each node in $\mathcal{X}$ in turn.

It is obvious that the number of transversals of $\mathcal{H}$ produced from a single transversal of $\mathcal{H}'$ is $|\mathcal{X}|$. The transversals of $\mathcal{H}'$ that do not include $v_{\mathcal{X}}$ remain as they are, since they hit $\mathcal{H}$. Going one step further, if a hypergraph $\mathcal{H}$ has two generalized nodes, $\mathcal{X}_1$ and $\mathcal{X}_2$, we can compute the transversals of the hypergraph $\mathcal{H}'$ that follows from $\mathcal{H}$ by removing the nodes of $\mathcal{X}_1$ and $\mathcal{X}_2$ and replacing them with two new nodes, $v_{\mathcal{X}_1}$ and $v_{\mathcal{X}_2}$, respectively. We compute now the transversals of $\mathcal{H}$ by taking the transversals of $\mathcal{H}'$ and substituting the generalized nodes $v_{\mathcal{X}_1}$ and $v_{\mathcal{X}_2}$ (where they appear) by all possible combinations of (simple) nodes in $\mathcal{X}_1$ and $\mathcal{X}_2$, respectively. Clearly, this procedure can be generalized to any number of generalized nodes.

We exploit the concept of the generalized vertex in such a way that, during all intermediate steps, we keep only the generalized transversals which, in turn, are split after the addition of the new hyperedge. This dramatically reduces the number of intermediate transversals, especially at the early stages (where the generalized nodes are few but large) and greatly improves the time performance and the memory requirements.

Example 1. Assume that the first two hyperedges have 100 nodes each: $\mathcal{E}_1 = \{v_1, \ldots, v_{100}\}$ and $\mathcal{E}_2 = \{v_{51}, \ldots, v_{150}\}$. The partial hypergraph $\{\mathcal{E}_1, \mathcal{E}_2\}$ has 2550 transversals (2500 with two nodes and 50 with one) which must be kept for the subsequent stage if we use the straightforward scheme. Using the generalized node approach, we have only 2 transversals to store, namely the set $\{\{v_{51}, \ldots, v_{100}\}\}$ and the set $\{\{v_1, \ldots, v_{50}\}, \{v_{101}, \ldots, v_{150}\}\}$.

The second improvement to the simple scheme of Berge was especially designed having in mind the rate of output of the algorithm. Recall that the simple scheme must make a lot of computation before outputting the first transversal, after which all the rest follow almost with zero delay one from the other. This occurs because the simple scheme is based on a sort of *breadth-first* computation of the transversals (all transversals are computed after a new hyperedge is added). Instead, our implementation computes the transversals in a *depth-first* manner: At a certain level, we compute a transversal of the partial hypergraph and then

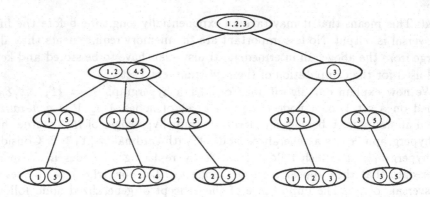

Fig. 1. Transversal tree of the hypergraph $\mathcal{H} = \{\{1, 2, 3\}, \{3, 4, 5\}, \{1, 5\}, \{2, 5\}\}$. The tree is visited in preorder.

add to it the next hyperedge. From this transversal several others follow, as we described above. However, instead of computing them all, we pick one, add the next hyperedge and continue until all hyperedges have been added; in this case we output the final transversal. We then backtrack to the previous level, pick the next transversal, etc. The whole scheme resembles a preorder visit of a tree of transversals with root the single (generalized) transversal of the first hyperedge, and internal nodes at some level, the generalized transversal of the partial hypergraph at that level. The descendants of a transversal are the transversals of the next hypergraph which include this transversal. Finally, the leaves of the tree at level m are the transversals of the original hypergraph.

Example 2. Consider the hypergraph with 5 nodes and 4 hyperedges $\mathcal{H} = \{\{1, 2, 3\}, \{3, 4, 5\}, \{1, 5\}, \{2, 5\}\}$. The tree of transversals which corresponds to the addition of the hyperedges according to the giver order (top to bottom) is shown in Fig. 1. Generalized nodes are denoted by circles with thin lines. For instance, a partial transversal of the hypergraph consisting of the first two hyperedges is $\{\{1, 2\}, \{4, 5\}\}$.

The efficiency of the above is further improved by a selective way of producing new transversals at an intermediate level. This idea results in ruling out regeneration of transversals at the cost that some intermediate nodes may have no descendants. The advantage of this approach is that search in some subtrees stops at higher levels instead of exhaustively generating everything that would subsequently need to be compared to previous transversals and, possibly, discarded. The method and its correctness is described and proved in [14].

3   Comparison to Other Approaches

We already mentioned that the algorithm of Berge outputs its first transversal near the end of the total computation. Thus, it may take exponentially long time for the first transversal to output. In contrast our algorithm, as the experiments show, delivers its output in quite a uniform way from the beginning to the end of the computation. From the implementation point of view the algorithm of Berge requires all intermediate transversals to be stored and kept until used for the computation of the new transversal set. Therefore the memory requirements are proportional to the size of the transversal tree. Since the number of transversals can be exponential, we conclude that the memory requirements can become devastating. Our algorithm instead, operates in a generate-and-forget fashion by visiting the transversal tree in preorder and so its memory requirements are proportional to the depth of the transversal tree rather than to its size. There are also other points of improvement in our algorithm compared to the algorithm of Berge; one has to do with the concept of generalized variables. In addition to the compact form of storing intermediate transversals, our algorithm also excludes the possibility of generating a transversal more than once which is a possibility in Berge's algorithm. The additional copies of a transversal must be identified and removed, otherwise they will result in a blow-up of the total number of partial transversals at the next steps. Additionally, there is an unnecessarily large number of intermediate transversals (especially in problem instances with many nodes) that have to participate in the computation of the transversal set of the new hypergraph. This is something that further adds to the blow-up of the transversals of the new hypergraph. In contrast, by using the concept of generalized variables, our algorithm greatly reduces the number of intermediate transversals, as illustrated in Example 1. Regarding the total running time of Begre's algorithm, as an example we mention that we ran Berge's algorithm for hypergraphs with 30 nodes and 30 hyperedges. The running time was more than 60 seconds while, as illustrated in Table 1, our algorithm requires only 1 second for hypergraphs with this size.

The algorithm of Fredman and Khachiyan has the best provable bound on its time performance. The algorithm actually solves the decision problem, namely the problem of deciding, given hypergraphs $\mathcal{H}$ and $\mathcal{G}$, whether $tr(\mathcal{H}) = \mathcal{G}$ and, if not, it returns a minimal hitting set of one of the two hypergraphs that is not a hyperedge of the other. The algorithm runs in time $O(n^{\log n})$ where n is the combined size of $\mathcal{H}$ and $\mathcal{G}$. By using repeatedly this algorithm as a subroutine, the generation problem can be solved in incremental time $O(n^{\log n})$ [10]. We are not aware of any implementation of the Fredman-Khachiyan algorithm. However, as in the previous case, both the input and the output so far have to be stored and consequently the same problem regarding the memory requirements exists here as well. Regarding its time performance, this algorithm operates in each decision step on both the input and the output so far and thus, the delay increases after each output step. In contrast, our algorithm seems to remain relatively stable in this regard. In any case, a careful implementation of the Fredman-Khachiyan algorithm would be of great interest.

Finally, for the sake of completeness, we mention the possibility of computing the transversal hypergraph by generating all possible hitting sets of the hypergraph (which are of the order of 2^n) and subsequently checking each if it is a minimal hitting set. Naturally, this simple algorithm overcomes the problem of storing all transversals, but its time performance is unacceptable (it can be used for hypergraphs having no more than 15 nodes). Yet, we have implemented it, since it is a reliable test for verifying the correctness of our algorithm.

4   Implementation Issues

The main body of the program consists basically of two procedures. (We only mention here the more important formal parameters.)

```
procedure add_next_hyperedge(t,h) {
    Update the generalized nodes;
    while generate_next_transversal(t,t') {
        if h is last, then output t
        else {
            Let h' be the next hyperedge;
            add_next_transversal(t',h');
        };
    };
};
```

and the boolean **function generate_next_transversal(t,t')**.

The first adds to the partial transversal t the next hyperedge h and repeatedly calls the second one which returns the next partial transversal t' of the new hypergraph. **generate_next_transversal** is called until no more transversals follow from t after the addition of h, in which case **generate_next_transversal** becomes false. After a new transversal t' is returned, **add_next_hyperedge** is called recursively for t' and the next hyperedge. The recursive implementation was chosen as a fast solution to the task of developing a fairly complex code. Since **add_next_hyperedge** is called once for each hyperedge, the depth of recursive calls equals the number of hyperedges. Hence, for some problem instances this can be quite large, resulting in poor memory usage and even in a code not as fast as it could be. We plan however to remove recursion from future versions of the program thus solving the above problems. This is why we have chosen to count primarily as a measure of performance the number of recursive calls between consecutive generations and not absolute time. Note however that, as shown in the next section, even the absolute time performance is quite satisfactory even for large problems. We stress that this can be improved further.

The performance of the program is very much affected by the data types used for storing generalized nodes. This is because the use of generalized nodes resulted in a code that does intensive set manipulation operations. These sets represent sets of nodes (actually integers from 1 to the number of nodes of the hypergraph). Several different methods were tried in different versions of the code. The fastest implementation represents a set as a bit vector that spans as many computer words as required depending on its size (our machine has 32-bit word). This choice may slightly limit the portability of the code but it is by far the fastest. Set operations (union, intersection, complementation, etc.) are then accomplished as low level bit operations (or, and, etc.).

5   Experimental Results

In this section we present the experimental evaluation of the performance of the algorithm on a set of test cases. Experiments were carried out on a Sun Enterprise 450, using GNU C++ 2.8.1 with compiler setting -O3.

Since real data were not available, the program was evaluated using a number of randomly generated hypergraphs. A random hypergraph generator was implemented and used for this task. Given the number of nodes, n, and the desired number of hyperedges, m, the random hypergraph generator uniformly and independently generates m sets of nodes, each of them corresponding to a hyperedge of the instance. The cardinality of each set lies between 1 and $n - 1$, and is randomly chosen, too. The outcoming hypergraph is simple, that is, there is no hyperedge that appears twice or is fully included in another one. Reports are averages over 50 different runs for each instance size, using a different initial seed for the random hypergraph generator in every run.

The first few experiments aimed at verifying the correctness of the code. This was done by implementing the simple algorithm that computes all transversals of a particular hypergraph using exhaustive search and then checking for minimality. Since this simple algorithm needs exponentially many steps to compute all minimal transversals, it could only be used for small instances (hypergraphs with at most 15–16 nodes). The resulting transversal hypergraph was then compared to the output of the program.

Testing the correctness of the code for larger instances was accomplished by applying the duality theorem of transversal hypergraph: $\mathrm{tr}(\mathrm{tr}(\mathcal{H})) = \mathcal{H}$ (see [2, 6] for more theoretical issues on hypergraphs). We run our algorithm to the output for a particular instance and verified that the new output was the original input. Notice that the above theorem also offers the possibility for generating non-random problem instances with specific properties of the output that would otherwise be impossible to generate by any conventional random instance generator; we were therefore able to test our algorithm for problems with very large number of hyperedges but very few transversals.

The results of the experiments are summarized in the tables that follow. Problem sizes are identified by the number of nodes, n, and the number of hyperedges, m, of the hypergraph. In Table 1, the first three columns (after column m) give the size of the output (the minimum, the maximum, and the average), in thousands of transversals, over the 50 experiments that were conducted for each pair n and m. It is important to notice that these numbers also characterize the problem size. Larger problem instances with respect to n and m resulted in a number of transversals too large to be generated within some moderate time. We next report the performance of the algorithm in terms of its rate of output. Specifically, the next three columns give the delay time (in number of calls of procedure add_next_hyperedge) between consecutive outputs. The first one reports the maximum delay that was observed in the whole process of generating the transversals. This number characterizes the worst-case performance of the algorithm. The unexpectedly good behavior of this parameter is what we believe it deserves further theoretical investigation. The next column presents the average delay while the last one reports the total running time for generating all transversals. As before, every entry in these columns is the average (over the 50 runs) of the corresponding parameters. The last column of Table 1 reports the total CPU time (in seconds) required, on the average, for each run.

The experiments summarized in Table 2 aimed in better studying the rate of the output. To this end, the standard deviation σ of the delays was calculated. Each row in Table 2 corresponds to a single problem (as opposed to results from 50 runs that were reported in Table 1) since averaging standard deviations is absurd. We have chosen however, from the 50 different experiments with the same n and m, to report the one with the worst behavior with respect to σ. Even in this case, the values of σ remain relatively small (3 times the average delay at most). The size of the output (transversals) as well as the total time (number of calls of procedure add_next_hyperedge for generating all transversals) and the CPU time required are also reported.

In problems reported in Tables 1 and 2 all randomly generated hypergraphs have a small number of hyperedges and a very large number of transversals. Problems of this kind have relatively small delays, since the output is very large. The duality property of the transversal hypergraph mentioned above provides a way of testing the algorithm in non-random instances with very large number of hyperedges and a small (known a priori) number of transversals. As a result, the delays are now large, comparable to the total running time. The technique for doing this was by running the algorithm with input the output of a particular randomly generated instance. This also serves as a test of the correctness of the algorithm as already explained. The results are summarized in Table 3. Notice that the problem size is now defined by the number of nodes, n, and the number of transversals, while the number of edges, m, is now the size of the output.

6 Conclusion

In this paper we describe an implementation of a new algorithm for solving the TRANSVERSAL HYPERGRAPH problem. This problem has certain peculiarities regarding its efficiency, in that both the total running time and the rate of the output (delay between consecutive outputs) are of equal importance. Our experiments show a surprisingly even rate of the output, something which calls for further theoretical justification. Moreover, both the delays between consecutive outputs and the total running time suggest that quite large instances (with respect to both their input and output sizes) can be solved efficiently. Future work will include the removal of recursion, since this will further speed up the code and more in-depth study of the performance measures of the program.

Acknowledgments

The authors wish thank the referees and the program committee for their helpful comments.

Table 1. Delay time and CPU time for various problem instances. Reports are averages over 50 runs. Each problem instance is defined by the number of nodes n and the number of hyperedges m. *Transversals* corresponds to the number of generated transversals while *Delay Time* to the number of calls of procedure add_next_hyperedge.

	m	Transversals $(\times 10^3)$			Delay Time $(\times 10^3)$			CPU Time
		min	*max*	*average*	*max*	*average*	*total*	*(in seconds)*
$n = 15$	10	0.03	0.1	0.1	0.01	0.004	0.3	0
	30	0.08	0.3	0.2	0.1	0.02	3	0
	50	0.1	0.4	0.3	0.2	0.03	9	0.1
	100	0.2	0.6	0.5	0.5	0.1	30	0.2
	200	0.3	1	0.6	1	0.1	90	1
	400	0.4	1.2	0.9	2	0.3	250	2
	600	0.7	1.5	1.1	4	0.5	500	4
	800	1	1.6	1.4	5	0.6	800	7
	1000	1.4	2	1.7	6	0.7	1200	10
$n = 20$	10	0.1	0.4	0.2	0.02	0.004	0.6	0
	30	0.3	1	0.7	0.1	0.02	10	0.1
	50	0.4	2	1.2	0.3	0.03	30	0.2
	100	0.5	3	2	0.7	0.1	100	1
	200	2	6	3	2	0.1	500	5
	400	1	9	6	4	0.3	1800	20
	600	2	11	8	7	0.5	3800	50
	800	2	13	9	9	0.6	5800	70
	1000	2	14	9	12	0.9	8200	110
$n = 30$	10	0.2	2	0.6	0.01	0.003	2	0
	30	2	12	5	0.2	0.01	70	1
	50	2	20	10	0.4	0.02	200	2
	70	5	40	20	1	0.03	700	10
	100	5	80	30	1	0.05	1800	20
	200	10	200	80	3	0.1	9100	150
	400	30	350	200	9	0.2	46000	900
	600	15	500	300	12	0.4	11000	7700
$n = 40$	10	0.1	3	1.3	0.01	0.003	4	0
	30	5	60	25	0.4	0.01	300	4
	50	11	300	90	1	0.02	1600	20
	70	13	400	200	2	0.03	4400	80
	100	13	1300	400	3	0.04	1600	360
$n = 50$	10	0.3	8	3	0.02	0.003	7	0.1
	30	5	400	100	0.5	0.01	900	10
	50	40	2000	470	1	0.02	7200	140
	70	20	4100	1200	4	0.02	28000	2300
$n = 60$	10	1	15	5	0.02	0.002	12	0.3
	30	20	1500	300	1	0.01	2600	50
	50	30	3700	1300	5	0.02	21000	450

Table 2. Statistical measures (*Average Delay* and standard deviation σ) for individual problem instances. Entries at the 1st and 2nd column (*Transversals* and *Total Time*, respectively) correspond to the number of generated transversals and the number of calls of procedure add_next_hyperedge, respectively.

	m	Transversals $(\times 10^3)$	Total Time $(\times 10^3)$	Average Delay $(\times 10^3)$	$\sigma(\times 10^3)$	CPU Time (in seconds)
$n = 15$	10	0.1	0.4	0.01	0.002	0
	30	0.1	2	0.02	0.022	0
	50	0.3	10	0.04	0.040	0.1
	100	0.2	30	0.1	0.1	0.2
	200	0.4	70	0.2	0.2	1
	400	0.5	200	0.4	0.5	2
	600	0.7	450	0.6	0.8	4
	800	1	820	0.7	0.8	7
	1000	2	1200	0.8	1	10
$n = 20$	10	0.2	0.8	0.01	0.01	0
	30	0.3	6	0.02	0.03	0.1
	50	1	40	0.04	0.06	0.4
	100	2	100	0.1	0.1	2
	200	3	650	0.3	0.4	10
	400	6	250	0.4	0.6	40
	600	6	5000	0.7	0.9	70
	800	6	6000	1	1.4	100
	1000	5	8000	1.6	2	130
$n = 30$	10	0.6	3	0.004	0.01	0
	30	4	80	0.02	0.03	1
	50	10	500	0.04	0.08	10
	70	20	100	0.07	0.1	20
	100	30	3000	0.1	0.2	70
	200	80	2000	0.2	0.4	400
	400	70	33000	0.5	1	1000
	600	300	180000	0.7	1.2	6000
$n = 40$	10	2	6	0.003	0.003	0
	30	40	600	0.01	0.03	10
	50	70	2000	0.03	0.06	40
	70	300	11000	0.04	0.13	300
	100	700	40000	0.06	0.14	1100
$n = 50$	10	2	10	0.005	0.01	0.1
	30	80	1000	0.01	0.03	10
	50	600	13000	0.02	0.05	300
	70	4000	120000	0.03	0.1	6000
$n = 60$	10	2	8	0.005	0.01	0.1
	30	400	4200	0.01	0.04	100
	50	3400	63000	0.02	0.06	1600

Table 3. Total time, CPU time, and statistical measures for individual non-random (dual) problem instances. The size of the output, m, is known a priori. Each row corresponds to individual runs. *Total Time* is the number of calls of procedure add_next_hyperedge for the generation of all transversals.

	Transversals	m	Total Time $(\times 10^3)$	Average Delay $(\times 10^3)$	$\sigma(\times 10^3)$	CPU Time (in seconds)
$n = 15$	80	10	0.7	0.07	0.03	0
	200	30	10	0.4	0.3	0.1
	200	50	10	0.2	0.1	0.1
	500	100	90	1	0.8	1
	800	200	200	1	0.6	2
	1000	400	450	1	0.8	4
	1100	600	500	0.8	0.6	4
	1200	800	700	1	0.6	5
	1700	1000	1500	1.5	1.1	10
$n = 20$	250	10	4	0.4	0.2	0.1
	500	30	40	1.5	1	1
	1100	50	20	3	2	4
	1400	100	400	4	4	10
	3000	200	1900	10	8	50
	8000	400	25000	60	50	500
	8000	600	29000	5	40	500
	9000	800	20000	20	20	500
	12000	1000	44000	40	40	1000
$n = 30$	600	10	20	2	2	1
	4200	30	2700	90	130	300
	20000	50	103000	2100	2700	12000
$n = 40$	1300	10	50	5	5	20
	4300	20	3000	150	190	500
	14000	30	51000	1700	1900	12000
$n = 50$	2600	10	100	10	10	70
	32000	20	27000	1400	1600	45000

References

1. R. Ben-Eliyahu and R. Dechter. On computing minimal models. *Annals of Mathematics and Artificial Inteligence*, 18:3–27, 1996.
2. C. Berge. *Hypergraphs*, volume 45 of *North Holland Mathematical Library*. Elsevier Science Publishers B.V., Amsterdam, 1989.
3. M. Cadoli. The complexity of model checking for circumscriptive formulae. *Information Processing Letters*, 42:113–118, 1992.
4. Z. Chen and S. Toda. The complexity of selecting maximal solutions. *Information and Computation*, 119:231–239, 1995.
5. J. de Kleer, A. K. Mackworth, and R. Reiter. Characterising diagnosis and systems. *Artificial Intelligence*, 56:197–222, 1992.
6. T. Eiter and G. Gottlob. Identifying the minimal transversals of a hypergraph and related problems. *SIAM J. Computing*, 24(6):1278–1304, December, 1995.

7. M. L. Fredman and L. Khachiyan. On the complexity of dualization of monotone disjunctive normal forms. *Journal of Algorithms*, 21:618–628, 1996.
8. H. Garcia-Molina and D. Barbara. How to assign votes in a distributed system. *Journal of ACM*, 32(4):841–860, 1985.
9. D. Gunopulos, R. Khardon, H. Mannila, and H. Toinonen. Data mining, hypergraph transversals, and machine learning. In *Proc. of Sixteenth ACM SIGACT-SIGMOD-SIGART Symposium on Principles of Database Systems*, pages 209–216, Tucson, Arizona, USA, May 12–14, 1997.
10. V. Gurvich and L. Khachiyan. Generating the irredundent conjunctive and disjunctive normal forms of monotone Boolean functions. Technical Report LCSR-TR–251, Department of Computer Science, Rutgers University, New Brunswick, NJ 08903, 1995.
11. D. S. Johnson, M. Yannakakis, and C. H. Papadimitriou. On generating all maximal independent sets. *Information Processing Letters*, 27:119–123, 1988.
12. D. Kavvadias and M. Sideri. The inverse satisfiability problem. *SIAM J. Computing*, 28(1):152–163, 1999.
13. D. J. Kavvadias, M. Sideri, and E. C. Stavropoulos. Generating all maximal models of a Boolean expression. Submitted.
14. D. J. Kavvadias and E. C. Stavropoulos. A new algorithm for the transversal hypergraph problem. Technical Report CTI TR 99.03.03, Computer Technology Institute, Patras, Greece, March 1999.
15. H. Mannila and K. J. Räihä. Design by example: An application of Armstrong relations. *Journal of Computer and System Sciences*, 32(2):126–141,1986.
16. H. Mannila and K. J. Räihä. Algorithms for inferring functional dependencies. *Data & Knowledge Engineering*, 12(1):83–99, February, 1994.
17. H. Mannila and H. Toivonen. Levelwise search and borders of theories in knowledge discovery. Technical Report C-1997-8, Department of Computer Science, University of Helsinki, Finland, 1997.
18. J. McCarthy. Cirmumscription-a form of nonmonotonic reasoning. *Artificial Inteligence*, 13:27–39, 1980.
19. C. H. Papadimitriou. NP-completeness: A retrospective. In *Proc. of ICALP 98*, Bologna, Italy, 1998.
20. B. Selman and H. K. Kautz. Model preference default theories. *Artificial Inteligence*, 45:287–322, 1990.

Construction Heuristics and Domination Analysis for the Asymmetric TSP

Fred Glover[1], Gregory Gutin[2], Anders Yeo[3], and Alexey Zverovich[2]

[1] School of Business, University of Colorado, Boulder
CO 80309-0419, USA
Fred.Glover@Colorado.EDU
[2] Department of Mathematics and Statistics
Brunel University, Uxbridge, Middlesex, UB8 3PH, U.K.
fax: +44 (0) 1895 203303
{Z.G.Gutin,Alexey.Zverovich}@brunel.ac.uk
[3] Department of Mathematics and Statistics
University of Victoria, P.O. Box 3045, Victoria B.C.
Canada V8W 3P4
yeo@Math.UVic.CA

Abstract. Non-Euclidean TSP construction heuristics, and especially asymmetric TSP construction heuristics, have been neglected in the literature by comparison with the extensive efforts devoted to studying Euclidean TSP construction heuristics. Motivation for remedying this gap in the study of construction approaches is increased by the fact that such methods are a great deal faster than other TSP heuristics, which can be important for real time problems requiring continuously updated response. The purpose of this paper is to describe two new construction heuristics for the asymmetric TSP and a third heuristic based on combining the other two. Extensive computational experiments are performed for several different families of TSP instances, disclosing that our combined heuristic clearly outperforms well-known TSP construction methods and proves significantly more robust in obtaining high quality solutions over a wide range of problems. We also provide a short overview of recent results in domination analysis of TSP construction heuristics.

1 Introduction

A construction heuristic for the traveling salesman problem (TSP) builds a tour without an attempt to improve the tour once it is constructed. Most of the construction heuristics for the TSP [13, 18] are very fast; they can be used to produce approximate solutions for the TSP when the time is restricted, to provide good initial solutions for tour improvement heuristics, to obtain upper bounds for exact branch-and-bound algorithms, etc.

Extensive research has been devoted to construction heuristics for the Euclidean TSP (see, e.g., [18]). Construction heuristics for the non-Euclidean TSP are much less investigated. Quite often, the greedy algorithm is chosen as a construction heuristic for the non-Euclidean TSP (see, e.g., [13]). Our computational experiments show that this heuristic is far from being the best choice in terms

J.S. Vitter and C.D. Zaroliagis (Eds.): WAE'99, LNCS 1668, pp. 85–94, 1999.
© Springer-Verlag Berlin Heidelberg 1999

of quality and robustness. Various insertion algorithms [18] which perform very well for the Euclidean TSP produce poor quality solutions for non-Euclidean instances.

Hence, it is important to study construction heuristics for the asymmetric TSP (we understand by the asymmetric TSP the general TSP which includes both asymmetric and symmetric instances). Our aim is to describe two new construction heuristics for the asymmetric TSP as well as a combined algorithm based on those heuristics. In this paper we also present results of our computational experiments obtained for several different families of TSP instances. These results show that the combined algorithm clearly outperforms well-known construction heuristics for the TSP. While other heuristics produce good quality tours for some families of TSP instances and fail for some other families of instances, the combined algorithm appears much more robust. Being a heuristic the combined algorithm is not always a winner among various heuristics. However, we show that it obtains (relatively) poor quality solutions rather seldom.

We also discuss recent results in the new area of domination analysis of TSP construction heuristics. Domination analysis is aimed to provide some theoretical foundations to construction heuristics.

The vertex set of a weighted complete digraph K is denoted by $V(K)$; the weight of an arc xy of K is denoted by $c_K(xy)$. (We say that K is *complete* if the existence of an arc xy in K, $x \neq y \in V(K)$, implies that the arc yx is also in K.) The *asymmetric traveling salesman problem* is defined as follows: given a weighted complete digraph K on n vertices, find a Hamiltonian cycle (*tour*) H of K of minimum weight. A cycle factor of K is a collection of vertex-disjoint cycles in K covering all vertices of K. A *cycle factor* of K of minimum (total) weight can be found in time $O(n^3)$ using assignment problem (AP) algorithms (for the corresponding weighted complete bipartite graph) [3, 14, 15]. Clearly, the weight of the lightest cycle factor of K provides a lower bound to the solution of the TSP (AP lower bound).

We will use the operation of *contraction* of a (directed) path $P = v_1 v_2 ... v_s$ of K. The result of this operation is a weighted complete digraph K/P with vertex set $V(K/P) = V(K) \cup \{p\} - \{v_1, v_2, ..., v_s\}$, where p is a new vertex. The weight of a arc xy of K/P is

$$c_{K/P}(xy) = \begin{cases} c_K(xy) & \text{if } x \neq p \text{ and } y \neq p \\ c_K(v_s y) & \text{if } x = p \text{ and } y \neq p \\ c_K(xv_1) & \text{if } x \neq p \text{ and } y = p \end{cases} . \qquad (1)$$

Sometimes, we contract an arc a considering a as a path (of length one).

2   Greedy and Random Insertion Heuristics

These two heuristics were used in order to compare our algorithms with well-known ones.

The greedy algorithm finds the lightest arc a in K and contracts it (updating the weights according to (1)). The same procedure is recursively applied to the contracted digraph $K := K/a$ till K consists of a pair of arcs. The contracted arcs and the pair of remaining arcs form the "greedy" tour in K.

The random insertion heuristic chooses randomly two initial vertices i_1 and i_2 in K and forms the cycle $i_1 i_2 i_1$. Then, in every iteration, it chooses randomly a vertex ℓ of K which is not in the current cycle $i_1 i_2 ... i_s i_1$ and inserts ℓ in the cycle (i.e., replaces an arc $i_m i_{m+1}$ of the cycle with the path $i_m \ell i_{m+1}$) such that the weight of the cycle increases as little as possible. The heuristic stops when all vertices have been included in the current cycle.

3 Modified Karp-Steele Patching Heuristic

Our first heuristic (denoted by GKS) is based on the well-known Karp-Steele patching (KSP) heuristic [14, 15]. The algorithm can be outlined as follows:

1. Construct a cycle factor F of minimum weight.
2. Choose a pair of arcs taken from different cycles in F, such that by patching (i.e. removing the chosen arcs and adding two other arcs that join both cycles together) we obtain a cycle factor (with one less cycle) of minimum weight (within the framework of patching).
3. Repeat step 2 until the current cycle factor is reduced to a single cycle. Use this cycle as an approximate solution for the TSP.

The difference from the original Karp-Steele algorithm is that instead of joining two shortest cycles together it tries all possible pairs, using the best one. (The length of a cycle C is the number of vertices in C.)

Unfortunately, a straightforward implementation of this algorithm would be very inefficient in terms of execution time. To partly overcome this problem we introduced a pre-calculated $n \times n$ matrix C of patching costs for all possible pairs of arcs. On every iteration we find a smallest element of C and perform corresponding patching; also, the matrix is updated to reflect the patching operation that took place. Having observed that only a relatively small part of C needs to be re-calculated during an iteration, we cache row minima of C in a separate vector B, incrementally updating it whenever possible. If it is impossible to update an element of B incrementally (this happens when the smallest item in a row of C has been changed to a greater value), we re-calculate this element of B by scanning the corresponding row of C in the beginning of the next iteration. Finally, instead of scanning all n^2 elements of C in order to find its minimum, we just scan n elements of B to achieve the same goal.

Although the improved version has the same $O(n^3)$ worst-case complexity as the original algorithm, our experiments show that the aforementioned improvements yield significant reduction of execution time. A detailed description of the approach outlined earlier follows.

Pseudo-code
$BestCost$ is C, $BestNode$ is B

Arguments:
N - number of nodes
$W = (w_{i,j})$ - $N \times N$ matrix of distances

(* $Next[i] = j$ if the current cycle factor contains the arc (i, j) *)
$Next$: **array**[1..N] **of integer**;
(* $Cost[i, j]$ is cost of removing arcs $(i, Next[i])$ and $(j, Next[j])$, and
    adding $(i, Next[j])$ and $(j, Next[i])$ instead. *)
$Cost$: **array**[1..N, 1..N] **of integer**;
(* $BestCost[i]$ contains a smallest value found in the i-th row of $Cost$ *)
$BestCost$: **array**[1..N] **of integer**;
(* $BestNode[i]$ is column index of corresponding $BestCost[i]$ in $Cost$ *)
$BestNode$: **array**[1..N] **of integer**;

(* Whenever possible, caches row minimum in $BestNode$ and $BestCost$ *)
procedure UpdateCost(r, c, $newCost$: **integer**);
begin
    if ($r < c$) **then** Swap(r, c); (* Exchange r and c values *)

    $Cost[r, c] := newCost$;

    if $BestNode[r] \neq -1$ **and** $newCost < BestCost[r]$ **then**
        (* New value is smaller than current row minimum *)
        $BestNode[r] := c$;
        $BestCost[r] := newCost$;
    else if $BestNode[r] = c$ **and** $newCost > BestCost[r]$ **then**
        (* Current row minimum has been updated to a greater value *)
        $BestNode[r] := -1$; (* stands for "unknown" *)
        $BestCost[r] := +\infty$;
    end if;
end;

function GetPatchingCost(i, j: **integer**): **integer**;
begin
    if vertices i and j belong to the same cycle **then**
        return $+\infty$; (* patching not allowed *)
    else
        return $W[i, Next[j]] + W[j, Next[i]] - W[i, Next[i]] - W[j, Next[j]]$;
    end if;
end;

BEGIN

Build a cycle factor by solving LAP on the distances matrix, store result in *Next* and number of cycles obtained in M;

(* Initialize *Cost*, *BestCost* and *BestNode* *)
for $i := 1$ **to** N **do**
  $bn := 1$;

  for $j := 1$ **to** i **do**
    $Cost[i, j] := GetPatchingCost(i, j)$;
    if $Cost[i, j] < Cost[i, bn]$ **then** $bn := j$;
  end for;

  $BestNode[i] := bn$;
  $BestCost[i] := Cost[i, bn]$;
end do;

repeat $M - 1$ **times**

  Find the smallest value in *BestCost* and store its index in i;
  $j := BestNode[i]$;

  update *Cost*:
    1) for each pair of nodes k and l, such as k belongs to
      the same cycle as i, and l belongs to the same cycle as j:
        $UpdateCost(k, l, +\infty)$;
    2) for each node m, which does not belong to the same cycle as
      either i or j:
        $UpdateCost(i, m, GetPatchingCost(i, m))$;
        $UpdateCost(m, j, GetPatchingCost(m, j))$;

  Patch two cycles by removing arcs $(i, Next[i])$, $(j, Next[j])$, and
  adding $(i, Next[j]) and (j, Next[i])$; update *Next* to reflect the patching
  operation;

  (* re-calculate *BestNode* and *BestCost* if necessary *)
  for $i := 1$ **to** N **do**
    if $BestNode[i] = -1$ **then**
    (* Needs re-calculating *)
    $bn := 1$;

    for $j := 1$ **to** i **do**
      if $Cost[i, j] < Cost[i, bn]$ **then** $bn := j$;
    end for;

$$BestNode[i] := bn;$$
$$BestCost[i] := Cost[i, bn];$$
      end if;
    end for;

  end repeat;

END.

4 Recursive Path Contraction Algorithm

The second heuristic originates from [19]. One of the main features of this heuristic is the fact that its solution has a large domination number. We discuss some recent results on domination analysis of TSP algorithms in Section 7.

The algorithm (denoted by RPC) proceeds as follows:

1. Find a minimum weight cycle factor F.
2. Delete a heaviest arc of each cycle of F and contract the obtained paths one by one.
3. If the number of cycles is greater than one, apply this procedure recursively.
4. Finally, we obtain a single cycle C. Replace all vertices of C with the corresponding contracted paths and return the tour obtained as a result of this procedure.

5 Contract-or-Patch Heuristic

The third heuristic (denoted by COP - contract or patch) is a combination of the previous two algorithms. It proceeds as follows:

1. Fix a threshold t.
2. Find a minimum weight cycle factor F.
3. If there is a cycle in F of length (= number of vertices) at most t, delete a heaviest arc in every short cycle (i.e. of length at most t) and contract the obtained paths (the vertices of the long cycles are not involved into the contraction) and repeat the above procedure. Otherwise, patch all cycles (they are all long) using GKS.

Our computational experiments (see the next section) showed that $t = 5$ yields a quite robust choice of the threshold t. Therefore, this value of t has been used while comparing COP with other heuristics.

6   Computational Results

We have implemented all three heuristics along with KSP, the greedy algorithm (GR), and the random insertion algorithm (RI), and tested them on the following seven families of instances of the TSP:

1. all asymmetric TSP instances from TSPLib;
2. all Euclidean TSP instances from TSPLib with the number of nodes not exceeding 3038;
3. asymmetric TSP instances with cost matrix $C = (c_{i,j})$, with $c_{i,j}$ independently and uniformly chosen random numbers from $\{0, 1, 2, ..., 10^5\}$;
4. asymmetric TSP instances with cost matrix $C = (c_{i,j})$, with $c_{i,j}$ independently and uniformly chosen random numbers from $\{0, 1, 2, ..., i \times j\}$;
5. symmetric TSP instances with cost matrix $C = (c_{i,j})$, with $c_{i,j}$ independently and uniformly chosen random numbers from $\{0, 1, 2, ..., 10^5\}$ ($i < j$);
6. symmetric TSP instances with cost matrix $C = (c_{i,j})$, with $c_{i,j}$ independently and uniformly chosen random numbers from $\{0, 1, 2, ..., i \times j\}$ ($i < j$);
7. sloped plane instances ([12]), which are defined as follows: for a given pair of nodes $p_i = (x_i, y_i)$ and $p_j = (x_j, y_j)$ the distance is

$$c(i,j) = \sqrt{(x_i - x_j)^2 + (y_i - y_j)^2} - max(0, y_i - y_j) + 2 \times max(0, y_j - y_i) . \quad (2)$$

We have tested the algorithms on sloped plane instances with independently and uniformly chosen random coordinates from $\{0, 1, 2, ..., 10^5\}$.

For the families 3-7, the number of cities n was varied from 100 to 3000, in increments of 100 cities. For $100 \leq n \leq 1000$, all results are average over 10 trials each, and for $1000 < n \leq 3000$, the results are average over 3 trials each.

Our implementations of RPC and COP algorithms make use of a shortest augmenting path algorithm, described in [20], for solving the assignment problem. This algorithm is of high performance in practice and provides a partial explanation for the relatively small execution times for RPC and COP seen in Table 2.

All tests were executed on a Pentium II 333 MHz machine with 128MB of RAM. All results for TSPLib instances are compared to optima, results for other instances are compared to a lower bound obtained by solving the corresponding assignment problem.

Tables 1 and 2 show an overview of the results we have obtained.

Observe that COP is the only heuristic from the above six that performs well on all the tested families. The others fail on at least one of the families. Note also that for symmetric TSP instances as well as for the instances close to symmetric (the Random Sloped Plane instances) the lower bound that we used is far from being sharp. Apart from being effective, COP is also comparable with respect to the execution time with (often used for the asymmetric TSP) the greedy algorithm.

Table 1. Average excess over optimum or AP lower bound

Family	Instances	GR	RI	KSP	GKS	RPC	COP
Family 1	26	30.62%	17.36%	4.29%	3.36%	18.02%	4.77%
Family 2	69	18.29%	11.61%	15.00%	17.26%	36.72%	17.52%
Family 3	160	320.13%	1467.38%	3.11%	3.09%	106.65%	1.88%
Family 4	160	515.10%	1369.13%	2.06%	2.02%	146.73%	1.11%
Family 5	160	246.65%	1489.95%	744.22%	586.92%	183.57%	79.87%
Family 6	160	405.77%	1386.32%	562.79%	195.06%	229.80%	83.77%
Family 7	160	2201.19%	41.78%	44.20%	46.33%	72.17%	47.29%

Table 2. Average execution time (sec)

Family	Instances	GR	RI	KSP	GKS	RPC	COP
Family 1	26	0.0917	0.0100	0.0547	0.0597	0.0508	0.1949
Family 2	69	3.2688	0.3611	1.2474	2.3377	1.1672	3.9341
Family 3	160	6.2109	0.3016	1.4624	1.5602	1.3112	2.6764
Family 4	160	7.1253	0.3027	2.7324	2.8338	2.5634	4.9129
Family 5	160	2.9795	0.2565	1.7093	2.4085	1.7268	2.4354
Family 6	160	3.3299	0.2568	2.8662	3.8108	3.0338	3.8821
Family 7	160	7.0776	0.3016	6.9904	7.8782	7.9787	12.0041

7   Algorithms of Factorial Domination Number

An equivalent of the following notion of domination number of an algorithm was introduced by Punnen [16] and Glover and Punnen [5]. The *domination number*, $\mathrm{domn}(\mathcal{A}, n)$, of an approximation algorithm $\mathcal{A}$ for the TSP is the maximum integer $d = d(n)$ such that, for every instance $\mathcal{I}$ of the TSP on n cities, $\mathcal{A}$ produces a tour T which is not worse than at least d tours in $\mathcal{I}$ including T itself. Clearly, every exact TSP algorithm is of domination number $(n - 1)!$. Thus, the domination number of an algorithm close to $(n - 1)!$ may indicate that the algorithm is of high quality.

As we pointed out in Section 4, the heuristic RPC has a large domination number. Let c_i be the number of cycles in the i-th cycle factor (the first one is F) and let m be the number of cycle factors derived in RPC. Then one can show that the domination number of the tour constructed by RPC is at least $(n - c_1 - 1)!(c_1 - c_2 - 1)!...(c_{m-1} - c_m - 1)!$ [19]. This number is quite large when the number of cycles in cycle factors is small (which is often the case for 'pure' asymmetric instances of the TSP). Still, when most of the cycles are of length two, the above product becomes not so 'big'. It was proved [19] that $\mathrm{domn}(RPC, n) = \Omega(n!/r^n)$ for every $r > 3.15$.

This result was improved in [8] to $\mathrm{domn}(\mathcal{B}, n) = \Omega(n!/r^n)$ for every $r > 1.5$, where $\mathcal{B}$ is a polynomial approximation TSP algorithm introduced in [8]. In [9], Gutin and Yeo introduced a polynomial time heuristic GEA and proved that $\mathrm{domn}(GEA, n) \geq (n - 2)!$, resolving a conjecture of Glover and Punnen [5]. Using one of the main results in [9], Punnen and Kabadi [17] showed that some well-known heuristics, including RI and KSP, are of domination number at least $(n-2)!$.

Gutin and Yeo [10] introduced another polynomial time TSP algorithm. The algorithm is demonstrated to be of domination number $(n-1)!/2$. Note that the proof is based on a reported yet unpublished theorem by R. Häggkvist [11] on Hamiltonian decompositions of regular digraphs. Also, the algorithm is somewhat slow and thus impractical. Still, we believe that this result in [10] indicates a possibility for improvement on currently known construction heuristics for the TSP.

Heuristics yielding tours with exponential yet much smaller domination numbers were introduced in the literature on so-called exponential neighbourhoods for the TSP (for a comprehensive survey of the topic, see [4]). Exponential neighbourhood local search [5-7] has already shown its high computational potential for the TSP (see, e.g., [1, 2]).

8 Conclusions

The results of our computational experiments show clearly that our combined algorithm COP can be used for wide variety of the TSP instances as a fast heuristic of good quality. It also demonstrates that theoretical investigation of algorithms that produce solutions of exponential domination number can be used in practice to design effective and efficient construction heuristics for the TSP.

Acknowledgments

The research of GG was partially supported by a grant from the Nuffield Foundation. The research of AY was partially supported by a grant from DNSRC (Denmark).

References

1. E. Balas and N. Simonetti, Linear time dynamic programming algorithms for some new classes of restricted TSP's. *Proc. IPCO V*, LNCS **1084**, Springer Verlag, 1996, 316-329.
2. J. Carlier and P. Villon, A new heuristic for the traveling salesman problem. *RAIRO* **24**, 245-253 (1990).
3. W.J. Cook, W.H. Cunninghan, W.R. Pulleyblank and A. Schrijver, Combinatorial Optimization, Wiley, New York, 1998.
4. V. Deineko and G.J. Woeginger, A study of exponential neighbourhoods for the travelling salesman problem and for the quadratic assignment problem. TR Woe-05, TU of Graz, Graz, Austria, 1997.
5. F. Glover and A.P. Punnen, The travelling salesman problem: new solvable cases and linkages with the development of approximation algorithms, J. Oper. Res. Soc., 48 (1997) 502-510.
6. G. Gutin, Exponential neighbourhood local search for the traveling salesman problem. *Computers & Operations Research* **26** (1999) 313-320.

7. G. Gutin and A. Yeo, Small diameter neighbourhood graphs for the traveling salesman problem: at most four moves from tour to tour. *Computers & Operations Research* **26** (1999) 321-327.
8. G. Gutin and A. Yeo, TSP heuristics with large domination number. Manuscript, 1998.
9. G. Gutin and A. Yeo, Polynomial approximation algorithms for the TSP and the QAP with factorial domination number (submitted).
10. G. Gutin and A. Yeo, TSP tour domination and hamiltonian cycle decomposition of regular digraphs (submitted).
11. R. Häggkvist, Series of lectures on Hamilton decomposition, Seminar Orsey, France, 1986 and Hindsgavl's Seminar, Denmark, 1994.
12. D.S. Johnson, private communication, 1998.
13. D.S. Johnson and L.A. McGeoch, The traveling salesman problem: a case study in local optimization. *Local Search in Combinatorial Optimization*, E.H.L. Aarts and J.K. Lenstra (eds.), Wiley, N.Y., 215-310 (1997).
14. R.M. Karp, A patching algorithm for the nonsymmetric Traveling Salesman Problem. *SIAM J. Comput.* **8** (1979) 561-73.
15. R.M. Karp and J.M. Steele, Probabilistic analysis of heuristics, in *The Traveling Salesman Problem*, E.L.Lawler, et al. (eds.), Wiley, N.Y., 1985, pp.181-205.
16. A.P. Punnen, The traveling salesman problem: new polynomial approximation algorithms and domination analysis. Manuscript, December (1996).
17. A.P. Punnan and S.N. Kabadi, Domination analysis of some heuristics for the asymmetric traveling salesman problem (submitted).
18. G. Reinelt, *The traveling salesman problem: Computational Solutions for TSP Applications. Springer Lecture Notes in Computer Sci.* **840**, Springer-Verlag, Berlin (1994).
19. A. Yeo, Large exponential neighbourhoods for the TSP, preprint, Dept of Maths and CS, Odense University, Odense, Denmark, 1997.
20. R. Jonker and A. Volgenant, A shortest augmenting path algorithm for dense and sparse linear assignment problems, *Computing* **38** (1987) 325-340.

Counting in Mobile Networks: Theory and Experimentation*

K. Hatzis[1,2], G. Pentaris[1,2], P. Spirakis[1,2], and B. Tampakas[1,3]

[1] Computer Technology Institute Kolokotroni 3, 26221 Patras, Greece
{hatzis,pentaris,spirakis,tampakas}@cti.gr
[2] Computer Engineering and Informatics Department, Patras University
26500 Rion, Patras, Greece
[3] Technological Educational Institute (TEI) of Patras
M Alexandrou 1, 26334, Koukouli, Patras, Greece

Abstract. In this work we study the problem of counting the number of mobile hosts in mobile networks. Mobile networks aim to provide continuous network connectivity to users regardless of their location. Host mobility introduces a number of new features and requirements for the distributed algorithms. In this case, the use of conventional distributed algorithms from mobile hosts results in a number of serious drawbacks. The *two tier* principle has been proposed (see [2]) to overcome these problems. The use of this principle for structuring distributed algorithms for mobile hosts means that the computation and communications requirements of an algorithm is borne by the static hosts to the maximum extend possible.

The Distributed Systems Platform (DSP) is a software platform that has been designed for the implementation, simulation and testing of distributed protocols. It offers a set of subtools which permit the researcher and the protocol designer to work under a familiar graphical and algorithmic environment. The use of DSP gave us considerable input and permitted us to experimentally test the two tier principle for the counting problem of mobile hosts. Moreover it helped us to design new distributed algorithms for this problem, improve them and experimentally test them, validating their performance under various conditions.

1 Introduction

Generally there is a considerable gap between the theoretical results of Distributed Computing and the implemented protocols, especially in the case of networks of thousands of nodes. On the other hand, well-designed tools would possibly offer to the researchers a more practical view of the existing problems in this area, and this, in turn, could give better (in the content of flexibility and efficiency) protocol design. Our work shows that a platform, suitably designed, can become a flexible tool for the researcher and offer a valuable help both in the verification and the extension of theoretical results (see also [19]).

* This work was partially supported by the EU ESPRIT LTR ALCOM-IT. (contract No. 20244).

The Distributed System Platform (DSP) is a software tool designed and developed during the sequel of ALCOM [1] projects and took its current form as a platform during the ALCOM-IT project. It provides an integrated environment for the implementation, simulation and testing of distributed systems and protocols. The DSP offers an integrated graphical environment for the design and implementation of simulation experiments of various ranges. It can provide visualization (animation) for networks of restricted number of nodes, or support experiments with networks of hundreds or thousands of nodes. It provides a set of simple, algorithmic languages which can describe the topology and the behaviour of distributed systems and it can support the testing process (on line simulation management, selective tracing and presentation of results) during the execution of specific and complex simulation scenarios. The DSP can support the hierarchical simulation of more than one type of protocols at the same execution. The latter is suggested in the case of pipelined protocols (the protocols of the upper level use the final output of the protocols of the lower level, e.g. leader election and counting protocols) or layered protocols (the protocols of the upper level call and use in every step the protocols of the lower level, e.g. synchronizers). Moreover, in its last version DSP supports the simulation of mobile protocols. The reader can find more about DSP in Section 3 and in [8], [10] and [24].

In this work we use this platform for the design, testing and verification of distributed protocols related with mobile computing. The problem of *process or node counting* (size of network) is extensively studied in the case of networks with static hosts. Various solutions have been proposed in the past (i.e. see [17],[18], depending on the model and the assumptions of the fixed network (i.e. timing conditions, network topology, existence of a network leader, dynamic or static network, distinguishable processes or not). The problem of process counting is one of the most fundamental problems in network control. Also note that the mobility of the hosts introduces new technical difficulties which, at first level, seem to invalidate the known solutions to the problem.

In the case of mobile networks the problem seems to preserve its significance. Indeed, the knowledge of how many mobile users are currently connected in a mobile networks is generally valuable and can be used both by the control (i.e. routing, data management) and the application (i.e. sales and inventory applications, see [11]) level of the network. Furthermore, this fundamental problem gives a good insight on the methodology of applying several distributed protocols to mobile networks and their performance. We believe that solutions of this problem in the mobile network settings will provide basic building blocks for solutions to more complicated problems (e.g. election of a leader etc.).

To facilitate continuous network coverage for mobile hosts, a static network is augmented with mobile support stations or MSSs that are capable of directly communicating with Mobile Hosts (MHs) within a limited geographical area ("cell"). In effect, MSSs serve as access points for an MH to connect to the static network and the cell, from which a MH connects to the static network, represents

---

[1] The ALCOM Projects are basic research projects funded by the European Union.

its current "location". MHs are thereby able to connect to the static segment of the network from different locations at different time. Consequently, the overall network topology changes dynamically as MHs move from one cell to another. This implies that distributed algorithms for a mobile computing environment cannot assume that a host maintains a fixed and universally known location in the network at all time; a mobile host must be first located("searched") before a message can be delivered to it. Furthermore, as hosts change their locations, the physical connectivity of the network changes. Hence, any logical structure, which many distributed algorithm exploit, cannot be statically mapped to a set of physical connections within the network. Second, bandwidth of the wireless link connecting an MH to an MSS is significantly lower than that of ("wired") links between static hosts [20],[23]. Third, mobile hosts have tight constraints on power consumption relative to desktop machines [20],[21],[22], since they usually operate on stand-alone sources such as battery cell. Consequently, they often operate in a "doze mode" or voluntarily disconnect from the network. Lastly, transmission and reception of messages over the wireless link also consumes power at an MH, and so distributed algorithms need to minimize communication over the wireless links. These aspects are characteristic of mobile computing and need to be considered in the design of distributed algorithms.

The main result of this work is a new correct and efficient (in number of messages) distributed protocol for the problem of counting mobile hosts. A significant part of the work is a demonstration of some principles of *Distributed Algorithmic Engineering* (see also [19]). Specifically, starting from the *two tier* principle we show how to successfully modify classical solutions aided by the use of the DSP and by experiments conducted on this platform.

The remainder of this work is organized as follows. In Section 2 we give the system model for mobile networks with fixed base stations. A brief description of the DSP tool is presented in Section 3. Section 4 contains the presentation of a counting algorithm executed distributedly by the mobile hosts. In Section 5 a counting protocol based in the two tier principle is presented. Finally, Section 6 contains simulation results and enhancements for the two tier protocol.

2   The System Model

A host that can move while retaining its network connections is a *mobile host* ([4]). The geographical area that is served by the fixed base station network is divided into smaller regions called *cells*. Each cell has a base station also referred to as the *Mobile Service Station (MSS)* of the cell. All mobile hosts that have identified themselves with a particular MSS are considered to be *local* to the cell of this MSS. At any instance of time, a mobile host belong to *only one* cell. When a mobile host enters new cell it sends a $< join(mh - id) >$ message to the new MSS. Therefore it is added in the list of local mobile hosts of this MSS.

Each MSS is connected to the service stations of neighbouring cells by a fixed high bandwidth network. The communication between a mobile host and its MSS is based on the use of low-bandwidth wireless channels. If a mobile host

h_1 wants to send a message to another mobile host h_2 it first sends the message to the local MSS over a wireless channel. The MSS then forwards through the fixed network the message to the local MSS of h_2 which forwards it to h_2 over its local wireless channels. Since the location of a mobile host within the network is neither fixed nor universally known at the whole network (its "current" cell may change with every move), the local MSS of h_1 needs first to determine the MSS that currently serves h_2. This means that for each message transmission between two mobile hosts incurs also a *search cost*. The following notation has been proposed in [1] for the description of the cost of messages exchanged in the network:

- C_{fixed}: The cost of sending a point-to-point message between any two fixed hosts.
- $C_{wireless}$: The cost of sending a message from a mobile host to its local MSS over a wireless channel (and vice versa). An extra cost may incur in order to allocate the wireless channels ([15], [16]).
- C_{search}: The cost (messages exchanged among the MSSs of the fixed network) to locate a mobile host and forward a message to its current local MSS. We consider that $C_{search} = aC_{fixed}$ where a a constant depending on the location management strategy used (e.g. [3]).

Let $G(V, E)$, where $|V| = n$ and $|E| = O(n^2)$, be the graph which describes the fixed part of the network (the network of the MSSs). Each vertex models an MSS. There exist an edge between two vertices if and only if the corresponding MSSs communicate directly (point-to-point) in the fixed network. Let m be the number of mobile hosts (we suppose that usually $m >> n$). We define by D the diameter of G.

Based on the above notation, the search cost C_{search} is approximately $O(D)$. A message sent from a mobile host to another mobile host incurs a cost $2C_{wireless}$ $+C_{search}$. This means that any algorithm based on the communication between mobile hosts requires a large number of messages to be exchanged over the fixed network and the wireless channels.

3   A Brief Description of the DSP Tool

DSP is a software tool that provides an integrated environment for the simulation and testing of distributed protocols. It follows the principles proposed by the books of G. Tel ([14])and N. Lynch ([13]) and aims in describing precisely and concisely the relevant aspects of a whole class of distributed computing systems. A distributed computation is considered to be a collection of discrete events, each event being an atomic change in the state of the whole system. This notion is captured by the definition of transmission systems. What makes such a system distributed is that each transmission is only influenced by, and only influences, part of the state, basically the local state of a single process. In DSP, each process is represented by a finite state machine with a local transmission table.

Events affecting transmissions include the arrival of a message at a node, time-outs, node (and link) failures and mobile process movement. The power of the processes is not limited in any other way since they are allowed to have local memory and (unbounded) local registers.

The platform allows in addition the modeling of mobile processes, the calling of a DSP library protocol from another user protocol and user control of local (virtual) clocks. The DSP platform thus differs from all existing "simulators" or "languages" of distributed systems because of its generality, algorithmic simplicity and clarity of semantics of its supported features. It aims in providing to the distributed algorithms designer an ideal environment of what a general distributed system "is expected to be".

The basic components of the platform include: I. A set of algorithmic languages that allow the description of a distributed protocol and the specification of a distributed system. II. A discrete event simulator that simulates the execution of a specified distributed protocol on a specified distributed system. III. A data base for distributed protocols that can also be used as a distributed protocol library. IV. Graphical user interface for the protocol and topology specification and the interaction during the simulation.

The DSP can support simulation and testing of protocols for fixed base station mobile networks in the following way: The user can describe MSSs and Mobile Hosts as static and mobile DSP processes respectively. Given a network topology the nodes can be specified to execute the MSS static process protocol in order to form the fixed base station network. Static and mobile processes may communicate by the use of radio messages. When a mobile process transmits a radio message this message is received from the static process on the node where the mobile process currently resides. On the other hand, a static process like an MSS can transmit radio messages which are received from all mobile processes in its node. By this way the DSP simulates the structure of cells. Any number of mobile processes (and thus Mobile Hosts) may be created in the network. The mobile processes may move from one node to another without restrictions representing the move of a mobile host from one cell to another.

4   The *Virtual Topology Algorithm*: A Counting Algorithm for Fixed Networks Executed Distributedly by the Mobile Hosts

Suppose that m Mobile Hosts are moving throughout a fixed base station mobile network. One of the mobile hosts (the initiator of the algorithm) wants to find the size of the mobile network (the number of the Mobile Hosts). We assume that the communication between the MSSs is based on the *asynchronous timing* model. We also assume that the Mobile Hosts are willing to control by themselves the execution by avoiding the participation of the MSSs. Note that, sometimes, this would be the case if the protocol requires computational power that will increase the overhead of the MSSs (since their basic activity is to control the communications in the network).

```
int routing_table[]; the routing table of the MSS
set local_MH; the set of local MHs of the MSS
do forever
    on receive < join, mh − id >
        insert mh-id to local_MH;
        execute the routing protocol with the other MSSs to update their routing tables;
    on receive < message, receiver − mh >
        if (receiver-mh is in local_MH) then
            transmit < message > to the cell;
        else
            look up the routing_table to find the next MSS i for this receiver-mh
            send < message, receiver − mh > to i;
end;
```

Fig. 1. The protocol executed by the MSSs

A fundamental algorithm for the solution of the counting problem in fixed networks can be found in [14] (p.190). It is basically a wave algorithm which can be used if a unique starter process (an initiator) is available. The application of this algorithm in a mobile network implies the definition of a "virtual" topology on the mobile hosts (each mobile host should be somehow assigned a set of "neighbours" at the beginning of the execution). By assuming that the virtual topology has $O(m)$ "edges", the total cost of the protocol would approximately be $O(m)C_{wireless} + O(Dm)C_{fixed}$. The appearance of the D factor is due to the search cost in the fixed network.

Besides the high cost of message transmissions in the fixed network, another drawback of this approach is that it requires the participation of every mobile host in order to maintain the connectivity of the virtual topology and cannot therefore permit to any one of them to disconnect during the execution (which is very usual in the case of mobile hosts). Thus, the total protocol execution time is a very crucial parameter since it increases the consumption of battery power in the mobile hosts.

The Mobile Service Stations act mainly as routers by forwarding messages from one Mobile Host to another and updating their routing tables when a mobile host joins their cell. The protocol executed by the MSSs is presented in Figure 1.

4.1 Test Case Scenarios and Measurement Parameters

An important factor for the performance of the protocols executed in a mobile network is the speed and the type of movement of the mobile hosts. Obviously when a mobile host is moving fast it will change many cells during the execution of the protocol and thus the overhead of keeping the routing information updated will grow.

Since we could not describe the speed of the mobile hosts in terms of physics (e.g. in kilometers or miles per hour) our approach was to associate the speed of the hosts with the propagation delay of messages in the fixed network. We consider a slow mobile host as a host which does not change its cell for $O(D)$

time units, where D is the diameter of the fixed network. Practically this means that the host keeps its position during the time interval required for a message to propagate from its host MSS to any other MSS. A fast mobile host is assumed to be a host which moves from a cell to another in time much less than $O(D)$.

In order to conduct the experiments we prepared five different topologies with different node cardinality ranging from 20 to 100 nodes (MSSs). The transmission delay in all links was unary on order to avoid the overhead of message delay in the protocol execution time. In all cases the protocol was initialized with ten mobile hosts on each node. The experiment parameters are presented in Figure 2.

Topology	Nodes	Links	Diameter	Mobile Hosts
1	20	50	5	200
2	40	100	8	400
3	60	180	11	600
4	80	260	15	800
5	100	310	21	1000

Fig. 2. The parameters used in the experiments

For all the experiments the measurements concern the following parameters:

- The number of messages exchanged in the fixed network in order to deliver messages between the mobile hosts. We did not take into account messages used from the routing protocol since they are irrelevant to the execution of the basic counting protocol itself.
- The number of radio messages transmitted by the mobile hosts. In this case we did not count the $< join >$ messages since they are also used for network control and routing purposes.
- The execution time of the protocol.

The consumption of battery power of a Mobile Host during the execution of the protocol is expressed by the last two parameters. It depends on the number of transmissions made by the host and the time that the host remains active (protocol execution time).

4.2 The DSP Simulation Settings of the Protocol

We will omit the details of this implementation due to the lack of space. An example of the methodology used for protocol specification in DSP is presented in the Appendix. In order to simplify the protocol, we did not implement a separate distributed routing protocol since the behavior of the routing part and the messages exchanged for the update of routing tables were not included in our measurements. Instead, we used a global routing table shared among all MSSs. This table was updated by a separate "daemon" routing process.

At the beginning of the simulation the mobile hosts were left to move randomly in the network in order to take random positions before the protocol was started. After the beginning of the execution, the virtual topology was constructed in the following way. Each Mobile Host with identity i considered as

```
int size=0; the final size of the network
boolean counted=false; a flag indicating if the host has been counted
the initiator mobile host:
begin
    broadcast < count >;
    on receive < size, s >
      size=s;
end
the other mobile hosts:
begin
    on receive < count >
      if (not counted)
        begin
          broadcast < count_me >;
          counted=true;
        end
    on receive < size, s >
      size=s;
end
```

Fig. 3. The counting protocol executed by the mobile hosts
"neighbours" the Mobile Hosts with identities $i-1$ and $i+1$ constructing a "virtual" line. This is an efficient [2] virtual topology for this case since it contains $m-1$ edges.

5   A New Counting Protocol Based on the *Two Tier* Principle

If the protocol execution is not very complex to require much of the MSSs computational power, more efficient solutions can be provided. A guiding principle for this case was presented in ([2]) and is called the *two tier* principle:

Computation and communication costs of an algorithm should be based on the static portion of the network. This leads to avoid locating a mobile participant and lowers the total search cost of the algorithm. Additionally, the number of operations performed at the mobile hosts is decreased and thereby consumption of battery power (which is critical resource for mobile hosts) is kept to a minimum.

The application of this simple principle on the design of distributed algorithms for mobile hosts has been studied in [2] in the case of a classical algorithm for mutual exclusion in distributed systems (Le Lann's token ring, [12]). In this section we propose and we study the behaviour of a protocol based on this principle that solves the counting problem in a mobile network with fixed base stations.

The proposed counting scheme is based on the execution of the *Echo* protocol. The protocol executed by the mobile hosts is presented in Figure 3. As can be

---

[2] The "star" topology would seem to be more appropriate but in this case one Mobile Host (the one in the middle of the "star") would already know the size of the network, since all the other hosts would be its neighbours.

seen, the execution is started by the *initiator* Mobile Host which broadcasts a $< count >$ message. Afterwards, the initiator itself does not respond to any $< count >$ message. The execution on the fixed part of the network (the MSSs) is described as follows:

1. In the first phase, the initiator MSS (the MSS serving the initiator) broadcasts a $< count >$ message in its cell and then spreads along the base station network the request for counting by using the *Echo* algorithm. The algorithm starts by sending $< count\_tok >$ messages to all neighbouring MSSs.
2. Upon receiving such a message ($< count\_tok >$), an MSS broadcasts a $< count >$ message to its cell and waits to collect answers from mobile hosts in this cell (we assume that an operational mobile host responds to the $< count >$ message with a $< count\_me >$ message immediately). The MSS also forwards a $< count\_tok >$ message to its neighbours in order to continue the execution of the *Echo*. When the MSS receives a $< count\_me >$ message it increases $size_p$, the number of counted mobile hosts in its cell. The MSSs that become "leaves" in the execution of the *Echo* respond with a message $< size, size_p >$ to their "parents". When a "parent" receives a $< size, s >$ message from a "child", it adds s to $< size_p >$. Upon collecting answers from all its "children" it reports a $< size, size_p >$ message to its own "parent".
3. After the completion of the *Echo* the initiator MSS knows the total number of mobile hosts in the network (its $< size_p >$ variable).
4. The initiator base station broadcasts a $< size, size_p >$ message in its cell and then forwards an $< inform\_tok, size_p >$ message to its "children". Upon receiving such a message an MSS broadcasts a $< size, size_p >$ message to its cell and forwards a $< inform\_tok, size_p >$ message to its "children" in order to continue the execution. After the propagation of the *inform_tok* messages in the base station network, all the MSSs have broadcasted the size of the mobile network in their cells and the mobile hosts have been informed about it.

It is very easy to see that a mobile host cannot be counted twice even if it moves throughout the network during the execution of the protocol (it will respond only to one $< count >$ message). The above protocol requires m broadcasts by the mobile hosts. The protocol needs $3E = O(n^2)$ messages to be exchanged in the fixed network yielding a total cost of $mC_{wireless} + 3|E|C_{fixed}$ which approximately is $O(m)C_{wireless} + O(n^2)C_{fixed}$. The protocol is completed in time $3D$. The implementation of the protocol is presented in detail in the Appendix.

6 Simulation Results and Algorithm Enhancements

6.1 The Case of Slow Mobile Hosts

Both protocols were executed 100 times for each one of the topologies described in Figure 2. The simulations proved the correctness of both algorithms, i.e. the total number of Mobile Hosts was reported correctly by the initiator host in all cases. As expected, the number of messages exchanged by the two tier algorithm *(tta)* remained almost unchanged in all simulations, while in the case of the virtual topology algorithm *(vta)* the number of messages exchanged varied highly, depending on the random "virtual topology" of the mobile hosts. Furthermore, the simulation results show the remarkable advantage of the two tier algorithm over the virtual topology algorithm in both, battery consumption (yielding from the much smaller execution time and the fifty percent lower radio message transmissions) and fixed network load. In particular, the two tier algorithm showed a remarkable decrease of total messages and battery consumption in all cases.

The simulations lead to the conclusion that in the case of slow mobile hosts, a simple two tier echo algorithm suffices to count the Mobile Hosts correctly and efficiently. The simulation results are presented in Figure 4.

Topology	Fixed Network Messages		Radio Messages		Execution Time	
	vta	tta	vta	tta	vta	tta
Topology1	1266	150	400	200	941	17
Topology2	4497	300	800	400	2729	26
Topology3	8562	540	1200	600	4154	35
Topology4	18352	780	1600	800	10904	48
Topology5	29350	930	2000	1000	18512	63

Fig. 4. Simulation results for slow mobile hosts (mean values over 100 experiments)

6.2 The Case of Fast Mobile Hosts: The Discovery of an Error

As in the previous case, both protocols were executed 100 times for each one of the topologies of Figure 2. The virtual topology algorithm was proven to be correct, i.e. the total number of Mobile Hosts was computed correctly in all simulations. The number of messages exchanged in the fixed network showed the same behaviour as in the first case. On the other hand, the two tier algorithm, despite being faster in all simulations, did not compute the total number of Mobile Hosts correctly in all simulations. In fact, the calculated number was in many cases slightly less than the total number of Mobile Hosts in the network. This can also be observed by examining the total number of radio message transmissions which was lower than the number of Mobile Hosts, revealing that some Mobile Hosts did not participate in the execution. The results of some of these runs are shown in Figure 5.

By using the DSP debugging facilities, the problem of the two tier algorithm could be traced. In fact, the situation can be described as follows: Suppose that MSSs S_1 and S_2 complete the execution of the MSS protocol steps 1 and 2 at time t_1 and t_2 respectively, where $t_2 > t_1$. It is possible that a mobile host starts moving from the cell of S_2 towards the cell of S_1 at time $t'_1 < t_1$ and appears at the cell of S_1 at time t'_2 where $t_1 < t'_2 < t_2$. This may happen if the path from the root to S_1 is shorter than the one to S_2. In this case the mobile host does not participate in the algorithm even if it is willing to do. This situation does

Topology	Radio Messages
Topology1	195
Topology2	392
Topology3	584
Topology4	782
Topology5	975

Fig. 5. Radio transmissions of the two tier protocol in the case of fast mobile hosts)

not arise in the case of slow Mobile Hosts because practically they do not move during this time interval.

6.3   A Corrected *Two Tier* Algorithm

The solution to the problem was to maintain each MSS active (transmit $< count >$ and receive $< count\_me >$ messages) until all other MSSs are informed for the protocol execution (have received $< count\_tok >$ messages). In order to eliminate the faulty behaviour, the protocol was modified in the following way:

1. After a base station has been informed about the execution of a counting protocol (by receiving a $< count\_tok >$ message) in the network it behaves as follows: By receiving a $< count\_me >$ message it increases the number of counted mobile hosts in its cell. If a base station receives a $< join(mh-id) >$ message (from a mobile host joining its cell) it broadcasts again a $< count >$.
2. After the completion of the first *Echo*, all of the MSSs have been informed about the execution of a counting algorithm in the network and have broadcasted a $< count >$ message in their cell.
3. The initiator base station starts a second execution of the *Echo* algorithm by sending a $< size\_tok, 0 >$ message to its neighbours. This execution aims to collect the $size_p$ variables from all MSSs to the initiator. After completing the execution of the second *Echo* (by receiving answers from all children and sending its $size_p$ variable to its parent as in step 2 of the previous version), an MSS stops to broadcast $< count >$ messages when a new mobile host joins its cell.
4. In order to avoid the appearance of the previous problem , the informing phase is also implemented by using a double execution of the *Echo*. The initiator base station broadcasts a $< size, size_p >$ message in its cell and then starts the third execution of the *Echo* by sending a $< inform\_tok, size_p >$ message to its neighbours. Upon receiving such a message, an MSS broadcasts a $< size, size_p >$ message to its cell and forwards an $< inform\_tok, size_p >$ message to its neighbours in order to continue the execution. If a base station receives a $< join(mh - id) >$ message, it broadcasts the $< size, size_p >$ again. After the third completion of the *Echo*, all the MSSs have broadcasted the size of the mobile network in their cells. The initiator starts a fourth execution of the *Echo* to inform the MSSs about the completion of the counting. After the completion of the fourth *Echo* an MSSs stops to broadcast $< size >$ messages when a new mobile host joins its cell.

Lemma 1. *The modified two tier protocol counts the number of Mobile Hosts correctly.*

Proof: We consider an MSS as *notified* if it has received a $< count\_tok >$ message and therefore collects answers and broadcasts $< count >$ messages in its cell. If a mobile host travels from an MSS S_2 which is *not notified* to an MSS S_1 which is *notified*, S_1 still waits to collect answers (the execution of the first *Echo* has not been completed yet since S_2 is *not notified*) and therefore the mobile host will receive a $< count >$ from S_1 and reply to S_1 with a $< count\_me >$ message. ∎

By the nature of the *Echo* algorithm which is executed four times the protocol needs $8E = O(n^2)$ messages to be exchanged in the fixed network yielding a total cost of $mC_{wireless} + 8|E|C_{fixed}$ which approximately is still $O(m)C_{wireless} + O(n^2)C_{fixed}$. The protocol is completed in time $8D$.

The simulation results on this version of the protocol are presented in Figure 6. The number of radio message transmissions verifies that the Mobile Hosts are counted correctly, with a linear increase of the fixed network messages and the execution time. However, both performance parameters show a good advantage over the virtual topology algorithm, suggesting the application of the two tier principle even in the case of fast Mobile Hosts.

Topology	Fixed Network Messages	Radio Messages	Execution Time
Topology1	400	200	44
Topology2	800	400	68
Topology3	1440	600	88
Topology4	2080	800	126
Topology5	2480	1000	171

Fig. 6. Simulation results of the corrected two tier protocol

7   Conclusions and Future Work

In this work we studied how to use *distributed algorithmic engineering* paradigms (such as the *two tier* principle) supported by the Distributed Systems Platform and suitable experiments in order to arrive at a new and more efficient Mobile Host counting protocol. Current work of ours extends this methodology to the problem of counting the Mobile Hosts in ad-hoc networks, where the MSSs are missing.

Acknowledgments

We wish to thank Richard Tan for inspiring discussions about the issue.

References

1. "Impact of mobility on Distributed Computations", A. Acharya, B. R. Badrinath, T. Imielinski, Operating Systems Review, April 1993.
2. "Structuring distributed algorithms for Mobile Hosts", A. Acharya, B. R. Badrinath, T. Imielinski, 14th International Conference on Distributed Computing systems, June 1994.
3. "Concurrent online tracking of mobile users", B. Awerbuch, D. Peleg.
4. "IP-based protocols for mobile internetworking", D. Duchamp, G. Q. Maquire, J. Ioannidis, In Proc. of ACM SIGCOM, September 1991.
5. "Brief Announcement: Fundamental Distributed Protocols in Mobile Networks", K. Hatzis, G. Pentaris, P. Spirakis, V. Tampakas, R. Tan, Eighteenth ACM Symposium on Principles of Distributed Computing (PODC '99), Atlanta, GA, USA.
6. "Fundamental Control Algorithms in Mobile Networks", K. Hatzis, G. Pentaris, P. Spirakis, V. Tampakas, R. Tan, Eleventh ACM Symposium on Parallel Algorithms and Architectures (SPAA 99), June 27-30 1999, Saint-Malo, France.
7. *"The description of a distributed algorithm under the DSP tool: The BNF notation"*, ALCOM-IT Technical Report, 1996.
8. *"The design of the DSP tool"*, ALCOM-IT Technical Report, 1996.
9. *"DSP: Programming manual"*, 1998.
10. *"The specifications of the DSP tool*, ALCOM-IT Technical Report, 1996.
11. "Mobile Computing", T. Imielinski, H. F. Korth, Kluwer Academic Publishers, 1996.
12. "Distributed Systems towards a formal approach", G. Le Lann, IFIP Congress, 1977.
13. Nancy Lynch, *"Distributed algorithms"*, Morgan Kaufmann Publishers, 1996.
14. Gerard Tel, *"Introduction to distributed algorithms"*, Cambridge University Press, 1994.
15. "Distributed Dynamic Channel Allocation for Mobile Computing", R. Prakash, N. Shivaratri, M. Sighal, In Proc. of ACM PODC 1995.
16. "Competitive Call Control in Mobile Networks", G. Pantziou, G. Pentaris, P. Spirakis, 8th International Symposium on Algorithms and Computation, ISAAC 97, December 1997.
17. "Symmetry Breaking in Distributive Networks", A. Itai and M. Rodeh, In proceedings of 22nd FOCS, 1981, pp 150-158.
18. "Approximating the Size of a Dynamically Growing Asynchronous Distributed Network", B. Awerbuch and S A. Plotkin, Technical Report MIT/LCS/TM-328, April 1987.
19. "Implementation and testing of eavesdropper protocols using the DSP tool", K. Hatzis, G. Pentaris, P. Spirakis and V. Tampakas, In Workshop on Algotithmic Engineering WAE 98, Saarbruecken, Aug. 1998. Postcript version in http://helios.cti.gr/alcom-it/foundation
20. "Agent -Mediated Message Passing for constraint Environment", A. Athas and D. Duchamp, In USENIX Symposium on Mobile and Location Independent Computing, Aug. 1993.
21. "Unix for Nomads: Making Unix support mobile computing", M. Bender et al., In USENIX Symposium on Mobile and Location Independent Computing, Aug. 1993.
22. "Power efficient filtering on data on the air" , T. Imielinski et al. , In EDBT'94, 1994.
23. " System Issues in Mobile Computing" B. Marsh, F. Douglis, and R. Caceres, T.R. MITL-TR-50-93, MITL, 1993.
24. DSP Web Site, http://helios.cti.gr/alcom-it/dsp

Appendix: The Implementation of the *Two-Tier* Protocol Under the DSP Environment

Protocol specification

The implementation is based on the protocol description language of the DSP. This language provides the ability to describe a protocol in an algorithmic form, similar to the one met in the literature for distributed systems. It includes usual statements met in programming languages like C and Pascal, and special structures for the description of distributed protocols (e.g. states, timers) and communication primitives (e.g. shared variables, messages). The statements of the language support the use of these structures during the specification of a distributed protocol in the areas of data modelling, communication, queuing, process and resource management (e.g. send a message through a link, execute a function as an effect of a process state transition). Some of the language statements support the interface of the specified protocol with the DSP graphical environment (e.g. show a node or a link with different color). The BNF notation of this language is presented in [7] and a more detailed description is given in the user manual ([9]).

The basic object of the DSP protocol description language is the *process*. The processes can be static (residing permanently on a node) or mobile (moving throughout the network). Our protocol involves two types of processes, a static (the *Mobile Service Station*) and a mobile one (the *Mobile Host*). Each one of them is described separately in the protocol file as a distinct object.

The protocol starts with the *TITLE* definition. It is followed by the definition of the messages used which are *count*, *count_me*, *count_tok*, *size* and *inform_tok*. The *Mobile_Host* is defined to be a *mobile* process. The *INIT* procedure is executed during the initialization of the process and includes the initialization of the process variables by assigning them constant values, or values created by specific language statements. The *Mobile_Host* intializes its *counted* flag to *FALSE* and stores its identity in local memory. The procedure *PROTOCOL* includes the core algorithm executed by the process. It is an event-driven procedure, which means that for every specific event that the process is expected to handle, a corresponding set of statements is executed. In this case the *Mobile_Host* process handles the following events:

- INITIALIZE which indicates the awakening of the process. Only one mobile host will process this event (the initiator) and will transmit a < *count* > message.
- MOBILE_RECEIVE_RADIO_MESSAGE which means that the process receives a message broadcasted by an MSS. The logical structure of the statements executed in this case is similar to the one presented in Figure 3.
- MOBILE_PROCESS_ARRIVAL meaning that a mobile process arrived on a node. If this event concerns the mobile process which currently executes the protocol it

starts an internal timer *timer_to_stay* to remain on its current node for *stay_period* time units.
- TIME_OUT which indicates the time out of the internal timer. In this case the process moves to a (randomly chosen) neighbour. In the case of slow mobile hosts the timeout period was arranged to $2D$, while in the case of fast mobile hosts this period was adjusted to one time unit.

The *Mobile_Service_Station* process is a static process and it is implemented in a similar way. The interested reader may find the correspondence between the DSP specification and the description of a similar algorithm presented in [14] (page 190). A *STATIC_RECEIVE_RADIO_MESSAGE* event concerns a radio message which has been transmitted by a *Mobile_Host* in the cell while a *RECEIVE_MESSAGE* event indicates a message received from the fixed network. The initiator MSS calls the procedure *wake_up()* in order to start the execution of the *Echo*. On the second phase of the execution (while *size* messages are propagating from the leaves towards the root) each MSS identifies the set of its children which is then used for the propagation of the *inform_tok* messages.

```
TITLE "A counting protocol for mobile net-
works with fixed base stations";

MESSAGE count END;
MESSAGE count_me END;
MESSAGE count_tok END;
MESSAGE size
    INT sizep; END;
MESSAGE inform_tok
    INT info; END;
```

MOBILE PROCESS Mobile_Host

```
BEGIN
  STATE dummy;
  TIMER timer_to_stay;
  CONST
     stay_period=100;
  VAR
     INT size, my_id, arr_id, neigh;
     BOOLEAN counted;

  PROCEDURE wake_up();
     BEGIN
        TRANSMIT_NEW_MESSAGE count
        BEGIN END;
     END;

  PROCEDURE move();
     BEGIN
        PUT_RANDOM_NEIGHBOUR TO neigh;
        MOVE_TO_NODE neigh;
     END;

  INIT();
     BEGIN
        size=0;
        PUT_MY_PROCESS_ID TO my_id;
        counted=FALSE;
     END;
```

```
PROTOCOL();
  BEGIN
  ON EVENT INITIALIZE DO CALL wake_up();
  ON EVENT MOBILE_RECEIVE_RADIO_MESSAGE DO
    BEGIN
    IF (CURRENT_MESSAGE_TYPE==count) THEN
      IF (counted==FALSE) THEN
        BEGIN
        TRANSMIT_NEW_MESSAGE count_me
          BEGIN END;
        counted=TRUE;
        END;
      IF (CURRENT_MESSAGE_TYPE==size) THEN
        size=CURRENT_MESSAGE_DATA.sizep;
    END;
  ON EVENT MOBILE_PROCESS_ARRIVAL DO
    BEGIN
    PUT_ID_OF_ARRIVING_PROCESS TO arr_id;
    IF (arr_id==my_id) THEN
      START timer_to_stay
      TIMEOUT stay_period;
    END;
  ON EVENT TIME_OUT DO CALL move();
  END;
END;
```

STATIC PROCESS Mobile_Service_Station

```
BEGIN
  STATE dummy;
  CONST
    undefined=-1;
  VAR
    INT receivep,fatherp,sizep,q,k,
        messages_send;
    SET neighbours, children;

  PROCEDURE wake_up();
    BEGIN
    SEND_NEW_MESSAGE count_tok TO
                      ALL  NEIGHBOURS
      BEGIN END;
    END;

  INIT();
    BEGIN
    receivep=0;fatherp=undefined;sizep=0;
    messages_send=0;
    PUT_ALL_NEIGHBOURS TO neighbours;
    END;

  PROTOCOL();
    BEGIN
    ON EVENT STATIC_RECEIVE_RADIO_MESSAGE DO
      BEGIN
      IF (CURRENT_MESSAGE_TYPE==count) THEN
        CALL wake_up();
      IF (CURRENT_MESSAGE_TYPE==count_me) THEN
        sizep=sizep+1;
      END;
    ON EVENT RECEIVE_MESSAGE DO
      BEGIN
      IF (CURRENT_MESSAGE_TYPE==count_tok) THEN
        BEGIN
        PUT_SENDER_OF_CURRENT_MESSAGE TO q;
        fatherp= q;
        DELETE q FROM neighbours;
        FOR ALL k IN neighbours DO
          BEGIN
          SEND_NEW_MESSAGE count_tok TO k
            BEGIN END;
          messages_send=messages_send+1;
```

```
        END;
      IF (messages_send==0) THEN
        SEND_NEW_MESSAGE size TO fatherp;
          BEGIN size=sizep; END;
      END;
    IF (CURRENT_MESSAGE_TYPE==size) THEN
      BEGIN
      PUT_SENDER_OF_CURRENT_MESSAGE TO q;
      INSERT q IN children;
      receivep=receivep+1;
      sizep=sizep+CURRENT_MESSAGE_DATA.size;
      IF (receivep==messages_send) THEN
        BEGIN
        IF (fatherp!=undefined) THEN
          SEND_NEW_MESSAGE size TO fatherp
            BEGIN size=sizep; END;
        ELSE
          BEGIN
          TRANSMIT_NEW_MESSAGE size
            BEGIN size=sizep END;
          FOR ALL k IN children DO
            SEND_NEW_MESSAGE inform_tok TO k
              BEGIN
              info=sizep;
              END;
          END;
        END;
      END;
    IF (CURRENT_MESSAGE_TYPE==inform_tok)
      THEN
      TRANSMIT_NEW_MESSAGE size
        BEGIN size=CURRENT_MESSAGE_DATA.info;
      END;
    END;
  END;
```

Protocol initialization
The initialization language of DSP provides statements that assigns process types (as specified in the protocol) to nodes of the topology and create and place mobile processes on nodes. An instance of an initialization file used during a simulation is presented below. It is used to assign the *Mobile_Service_Station* process type to all nodes and create five mobile processes of type *Mobile_Host* on each node of a network of 20 nodes. In this example the *Mobile_Host* 0 is the initiator and is initiated at an arbitrary time instance between 5 and 20.

INITIALIZATION FILE "init file1" FOR PROTOCOL "Counting protocol based on the two tier principle"

```
SET ALL NODES TO Mobile_Service_Station
PUT MOBILE PROCESS Mobile_Host ON NODE 0-20
PUT MOBILE PROCESS Mobile_Host ON NODE 0-20
PUT MOBILE PROCESS Mobile_Host ON NODE 0-20
PUT MOBILE PROCESS Mobile_Host ON NODE 0-20
PUT MOBILE PROCESS Mobile_Host ON NODE 0-20
INIT MOBILE PROCESS 0 RANDOMLY FROM 5 TO 20
```

Dijkstra's Algorithm On–Line: An Empirical Case Study from Public Railroad Transport

Frank Schulz, Dorothea Wagner, and Karsten Weihe

Universität Konstanz, Fakultät für Mathematik und Informatik
Fach D188, 78457 Konstanz, Germany
{schulz,wagner,weihe}@fmi.uni-konstanz.de
http://www.fmi.uni-konstanz.de/~{schulz,wagner,weihe}

Abstract. Traffic information systems are among the most prominent real–world applications of Dijkstra's algorithm for shortest paths. We consider the scenario of a central information server in the realm of public railroad transport on wide–area networks. Such a system has to process a large number of on–line queries in real time. In practice, this problem is usually solved by heuristical variations of Dijkstra's algorithm, which do not guarantee optimality. We report results from a pilot study, in which we focused on the travel time as the only optimization criterion. In this study, various optimality–preserving speed–up techniques for Dijkstra's algorithm were analyzed empirically. This analysis was based on the timetable data of all German trains and on a "snapshot" of half a million customer queries.[1]

1 Introduction

Problem. From a theoretical viewpoint, the problem of finding a shortest path from one node to another one in a graph with edge lengths is satisfactorily solved. In fact, the Fibonacci–heap implementation of Dijkstra's algorithm requires $\mathcal{O}(m + n \log n)$ time, where n is the number of nodes and m the number of edges [8]. However, various practical application scenarios impose restrictions that make this algorithm impractical. For instance, many scenarios impose a strict limitation on space consumption.[2]

In this paper, we consider a different scenario: space consumption is not an issue, but the system has to answer a potentially infinite number of customer queries on–line. The real–time restrictions are soft, which basically means that the average response time is more important than the maximum response time. The concrete scenario we have in mind is a central server for public railroad

---

[1] With special courtesy of the *TLC Transport-, Informatik- und Logistik–Consulting GmbH/EVA–Fahrplanzentrum*, a subsidiary of the *Deutsche Bahn AG*.

[2] To give a concrete example: if a traffic information system is to be distributed on CD–Rom or to be run on an embedded system, a naive implementation of Dijkstra's algorithm would typically exceed the available space.

J.S. Vitter and C.D. Zaroliagis (Eds.): WAE'99, LNCS 1668, pp. 110–123, 1999.
© Springer-Verlag Berlin Heidelberg 1999

transport, which has to process a large number of queries (*e.g.* a server that is directly accessible by customers through terminals in the train stations or through a WWW interface).

Algorithmic problems of this kind are usually approached heuristically in practice, because the average response time of optimal algorithms seems to be inacceptable. In a new long-term project, we investigate the question to what extent optimality–preserving variants of Dijkstra's algorithm have become competitive on contemporary computer technology. Here we give an experience report from a pilot study, in which we focused on the most fundamental kind of queries: find the fastest connection from some station A to some station B subject to a given *earliest departure time*.

This scenario is an example of a general problem in the design of practical algorithms, which we discussed in [14]: computational studies based on artificial (*e.g.* random) data do not make much sense, because the characteristics of the real–world data are crucial for the success or failure of an algorithmic approach in a concrete use scenario. Hence, experiments on real–world data are the method of choice.

Related work. Various textbooks address speed–up techniques for Dijkstra's algorithm but have no concrete applications in mind, notably [4] and [10]. In [13], Chapter 4, a brief, introductory survey of selected techniques is given (with a strong bias towards the use scenario discussed here). Most work from the scientific side addresses the single–source variant, where a spanning tree of shortest paths from a designated root to all other nodes is to be found. Moreover, the main aspect addressed in work like [6] is the choice of the data structure for the priority queue. In Section 2 below (paragraph on the "search horizon"), we will see that the scenario considered in this paper requires algorithmic approaches that are fundamentally different, and Section 3 will show that the choice of the priority queue is a marginal aspect here.

On the other hand, most application–oriented work in this field is commercial, not scientific, and there is only a small number of publications. In fact, we are not aware of any publication especially about algorithms for wide–area railroad traffic information systems.

Some scientific work has been done on *local* public transport. For example, [12] gives some insights into the state of the art. However, local public transport is quite different from wide–area public transport, because the timetables are very regular, and the most powerful speed–up techniques are based on the strict periodicity of the trains, busses, ferries, etc. In contrast, our experience is that the timetables of the national European train companies are not regular enough to gain a significant profit from these techniques.[3]

On the other hand, *private* transport has been extensively investigated in view of wide–area networks. Roughly speaking, this means "routing planners" for cars on city and country maps [2, 3, 5, 7, 9, 11]. This problem is different to

---

[3] This experience is supported by personal communications with people from the industry.

ours in that it is two–dimensional, whereas train timetables induce the time as a third dimension: due to the lack of periodicity, the earliest departure time is significant in our scenario. In contrast, temporal aspects do not play any role in the work quoted above.[4] So it is not surprising that the research has focused on purely geometric techniques.

For completeness, we mention the work on variants of Dijkstra's algorithm that are intended to efficiently cope with large data in secondary memory. Chapter 9 of [1] gives an introduction to theoretical and practical aspects. As mentioned above, the problems caused by the slow access to secondary memory are beyond the scope of our paper.

Contribution of the paper. We implemented and tested various optimality–preserving speed–up techniques for Dijkstra's algorithm. The study is based on all train data (winter period 1996/97) of the Deutsche Bahn AG, the national railroad and train company of Germany. The processed queries are a "snapshot" of the central *Hafas*[5] server of the Deutsche Bahn AG, in which all queries of customers were recorded over several hours. The result of this snapshot comprises more than half a million queries, which might suffice for a representative analysis (assuming that the typical query profile of customers does not vary dramatically from day to day).

Due to the above–mentioned insight that the periodicity of the timetables is not a promising base for algorithmic approaches, the question is particularly interesting whether geometric techniques like those in routing planners are successful, although the scenario has geometric *and* temporal characteristics. We will see that this question can indeed be answered in the affirmative.

2   Algorithms

Train graph. The arrival or departure of a train at a station will be called an *event*. The train graph contains one node for every event. Two events v and w are connected by a directed edge $v \to w$ if v represents the departure of a train at some station and w represents the very next arrival of this train at some other station. On the other hand, two successive events at the same station are connected by an edge (in positive time direction), which means that every station is represented in the graph by a cycle through all of its events (the cycle is closed by a turn–around edge at midnight).

In each case, the length of an edge is the time difference of the events represented by its endnodes. Obviously, a query then amounts to finding a shortest path from the earliest event at the start station not before the earliest departure time to an arbitrary arrival event at the destination.

The data contains $933,066$ events on $6,961$ stations. Consequently, there are $933,066 \cdot 3/2 = 1,399,599$ edges in the graph.

---

[4] In principle, temporal aspects would also be relevant for the private transport, for example, the distinction between "rush hours" and other times of the day.

[5] Hafas is a trademark of the Hacon Ingenieurgesellschaft mbH, Hannover, Germany.

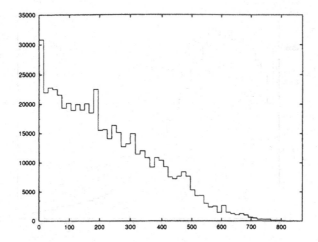

Fig. 1. The frequency distribution histogram of the queries from the "snapshot" according to the Euclidean distance between the start station and the destination (granularity: 15 kilometers).

Priority queue. Dijkstra's algorithm relies on a priority queue, which manages the nodes on the current "frontier line" of the search. As mentioned above, the best general worst–case bound, $\mathcal{O}(m+n \log n)$, is obtained from Fibonacci heaps, where n is again the number of nodes and m the number of edges. We do not use a Fibonacci heap but a normal heap (also often called a *2–heap*), which yields an $\mathcal{O}((n + m) \log n)$ bound [8]. Since $m \in \mathcal{O}(n)$ in train graphs, both bounds reduce to $\mathcal{O}(n \log n)$.

As an alternative to heaps, we also implemented the *dial variant* as described in [4]. Basically, this means that the priority queue is realized by an array of buckets with a cyclically moving array index. The nodes of the frontier line are distributed among the buckets, and it is guaranteed that the very next non-empty bucket after the current array index always contains the candidates to be processed next.

Search horizon. Of course, we deviate from the "textbook version" of Dijkstra's algorithm in that we do not compute the distance of every node from the start node but terminate the algorithm immediately once the first (and thus optimal) event at the destination is processed. The most fundamental optimality-preserving speed–up technique for our scenario is then a reduction of the search to a (hopefully) small part of the graph, which contains all relevant events. Figs. 1–3 demonstrate that such a reduction is crucial. In fact, they reveal that for the majority of all queries only small fractions of the total area and time horizon are relevant.

To our knowledge, some commercial implementations remove nodes and edges from the graph (more or less heuristically, *i.e.* losing optimality) before the search itself takes place. In contrast, we aim at an evaluation of optimality-preserving strategies, so our approach is quite different. First of all, we apply

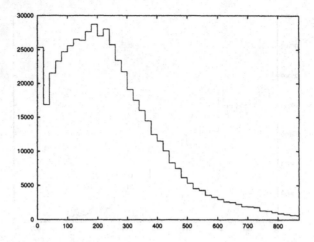

Fig. 2. Like Fig. 1 except that the abscissa now denotes the minimal travel time in minutes (granularity: 20 minutes).

the amortization technique discussed in [13] to obtain a sublinear expected run time per query. The only obstacle to sublinearity is the initialization of all nodes with infinite distance labels in the beginning of the textbook algorithm; in fact, Figs. 1 and 2 strongly suggest that on average the main loop of the algorithm only processes a very small fraction of the graph until the destination is seen and the algorithm terminates.

As described in [13], every node is given an additional *time stamp*, which stores the number of the query in which it was reached in the main loop. Whenever a node is reached, its time stamp is updated accordingly. If this update properly increases the time stamp, the distance label is regarded as infinite, otherwise the value of the distance label is taken as is. Consequently, there is no need for an expensive initialization phase, and no event outside the "search horizon" of the main loop is hit at all.

The following two additional techniques rely on this general outline. They are independent of each other in the sense that one of them may be applied alone, or both of them may be applied simultaneously.

Angle restriction. This technique additionally relies on the coordinates associated with the individual stations. In a preprocessing step, we apply Dijkstra's algorithm to each event to compute shortest paths from this event to all other stations.[6] The results are not stored (this would require too much space) but only used to compute two values α and β for each edge. These $2 \cdot m$ values are stored and then used in the on–line system. More specifically, these values are to be interpreted as angles in the plane. Let $v \to w$ be an edge, and let s be the

---

[6] This preprocessing takes several hours, which is absolutely acceptable in practice. Hence, there is no need to optimize the preprocessing.

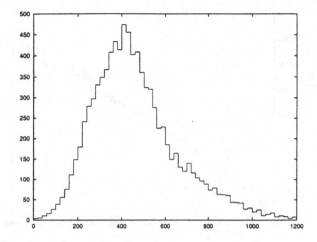

Fig. 3. For each minimal travel time (granularity: 20 minutes) the total CPU times of all queries yielding this value of the travel time are summed up to reveal which range of travel times contributes most of the total CPU time.

station of event v. Then the values α and β stored for this edge span a circle sector with center s. The meaning is this: if the shortest path from event v to some station s' contains $v \to w$, then s' is in this circle sector. Clearly, the brute–force application of Dijkstra's algorithm in the preprocessing allows one to compute the narrowest possible circle sector for each edge subject to this constraint.

Consequently, edge $v \to w$ may be ignored by the search if the destination is *not* in the circle sector of this edge. The restriction of the search to edges whose circle sectors contain the destination is the strategy that we will call "angle restriction" in the following.

Selection of stations. The basic idea behind this technique is similarly implemented in various routing planners for the private transport [2, 3, 5, 9]. A certain small set of nodes is selected. For two selected nodes v and w, there is an edge $v \to w$ if, and only if, there is a path from v to w in the train graph such that no internal node of this path belongs to the selected ones. In other words, every connected component of non–selected stations (plus the neighboring selected stations) is replaced by a directed graph defined on the neighboring selected stations. This constitutes an additional, *auxiliary* graph.

The length $\ell(v, w)$ of an edge $v \to w$ in this auxiliary graph is defined as the minimum length of a path from v to w in the train graph that contains no selected nodes besides v and w. The auxiliary graph and these edge weights are also constructed once and for all in a preprocessing step (which only takes a few minutes). Each query is then answered by the computation of a shortest path in the auxiliary graph, and this path corresponds to a shortest path in the train graph.

We implement this general approach as follows. First of all, note that it is not necessary to reconstruct the path in the train graph that corresponds to

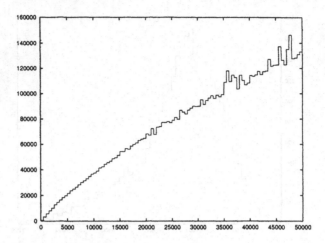

Fig. 4. The relation of the number of edges hit without (abscissa) and with (ordinate) strategy angle selection (granularity: 500 edges).

the shortest path computed in the auxiliary graph. In fact, what we really want to have from the computation is a sequence of trains and the stations where to change train. This data can be attached to the edges of the auxiliary graph in an even more compact (less redundant) and thus more efficiently evaluable fashion than to the edges of the train graph.

We select a set of stations, and the events of these stations are the selected nodes. Clearly, there is a trade-off. Roughly speaking, the smaller the number of selected stations is, the larger the resulting connected components are and, even worse, the larger the number of selected stations neighboring to a component. Since the number of edges depends on the latter number in a quadratic fashion, an improvement of performance due to a rigorous reduction of stations is soon outweighed by the tremendous increase in the number of edges.

It has turned out that in our setting, a minor refinement of this strategy is necessary and sufficient to overcome this trade-off. For this, let u, v, and w be three selected events such that edges $u \rightarrow v$, $v \rightarrow w$, and $u \rightarrow w$ exist in the auxiliary graph. If $\ell(u, v) + \ell(v, w) \leq \ell(u, w)$, then edge $u \rightarrow w$ is dropped in the auxiliary graph. Again, optimality is preserved. The number of edges grows only moderately after this modification, so a quite small set of selected stations becomes feasible.

In the data available to us, every station is assigned an "importance number," which is intended to rank its degree of "centrality" in the railroad network. The computational study is based on the 225 stations in the highest categories (see Fig. 7). These stations induce 95, 423 events in total, which means that the number of events is approximately reduced by a factor of 10, and the number of stations is reduced by a factor of 31. This discrepancy between these two factors is not surprising, because central stations are typically met by more trains than marginal stations.

Combination of both strategies. In principle, these two strategies can be combined in two ways, namely the angle restrictions can be computed for the auxiliary graph, or they can be computed for the train graph and simply taken over for the auxiliary graph. Not surprising, we will see in the next section that the former strategy outperforms the latter one.

3   Analysis of the Algorithmic Performance

The experiments were performed on a SUN Sparc Enterprise 5000, and the code was written in C++ using the GNU compiler.

Table 1 presents a summarizing comparison of all combinations of strategies. Note that the total number of algorithmic steps is asymptotically dominated by the number of operations inside the priority queue. In other words, these operations are representative operation counts in the sense of [4], Sect. 18. More specifically, for a heap the number of exchange operations is representative, whereas for the dial variant the number of cyclic increments of the moving array index is representative. The average total number per query of these operations are listed in the last part of Table 1.

For both implementations of the priority queue, the CPU times imply the same strong ranking of strategies. Figs. 7 and 8 might give a visual impression why this ranking is so unambiguous. Moreover, the discrepancy between the heap and the dial implementation also decreases roughly from row to row. This is not surprising: the overhead of the heap should be positively correlated with the size of the heap, which is significantly reduced by both strategies.

Our experience with several versions of the code is that the exact CPU times are strongly sensitive to the details of the implementation, but the general tendency is maintained and seems to be reliable. In particular, the main question raised in this paper (whether optimality–preserving techniques are competitive) can be safely answered in the affirmative at least for the restriction of the problem to the total travel time as the only optimization criterion.

However, a detailed look at the results is more insightful. Fig. 4 shows that there is a very strong linear correlation between the number of edges hit with/without strategy "angle restriction." On the other hand, Fig. 5 shows a detailed analysis of one particular, exemplary combination of strategies. A comparison of these diagrams reveals an interesting effect, which is also found in the analogous diagrams of the other combinations: the CPU times of both the heap and the dial implementation are linear in the number of nodes hit by the search. This correlation is so strong and the variance is so small that corresponding diagrams in Fig. 5 look almost identical. Figure 6 reveals the cause.

In other words, in both cases the operations on the priority queue take constant time on average, even when the average is taken over each query separately! This is in great contrast to the asymptotic worst–case complexity of these data structures.

Selection of stations	Angle restriction	CPU heap	CPU dial
no	no	0.310	0.103
no	yes	0.036	0.018
yes	no	0.027	0.012
yes	train	0.007	0.005
yes	auxiliary	0.005	0.003

Selection of stations	Angle restriction	Nodes	Edges
no	no	17576	31820
no	yes	4744	10026
yes	no	2114	3684
yes	train	1140	2737
yes	auxiliary	993	2378

Selection of stations	Angle restriction	Ops. heap	Ops. dial
no	no	246255	23866
no	yes	24526	3334
yes	no	26304	3660
yes	train	4973	1197
yes	auxiliary	3191	932

Table 1. A summary of all computational results for the individual combinations of techniques. The entries "train" and "auxiliary" in column #2 refer to the graph in which the angles were computed (see the last paragraph in Section 2). The columns #3–#4 give the average over all queries of the "snapshot." More specifically, the first table gives the average raw CPU times, the second table the average number of nodes and edges hit by the search, and the third table the average operation counts.

4   Conclusion and Outlook

The outcome of this study suggests that geometric speed–up techniques are a good basis for the computation of provably optimal connections in railroad traffic information systems. The question raised in this paper is answered for the total travel time: the best combinations of strategies are by far faster than is currently needed in practice. This success is a bit surprising, because the underlying data is not purely geometric in nature.

Another surprising outcome of the study is that both the normal heap and the dial data structure only require an (amortized) constant time per operation, whereas the worst–case bound is logarithmic for heaps and even linear for dials. Note that no amortization over a set of queries must be applied to obtain a constant time per operation; the variance is small enough that the average run time per operation within a single query can essentially be regarded as bounded by a constant. Due to the fact that the variance is negligible, a "classical" statistical analysis would not make any sense.

The minimal travel time is certainly an empirical research topic in its own right, not only because it is the most fundamental objective in practice. However, a practical algorithm must consider further criteria and restrictions. For example, the ticket costs and the number of train changes are also important objectives. Moreover, certain trains do not operate every day, and certain kinds of tickets are not valid for all trains, so it should be possible to exclude train connections in a query. A satisfactory compromise must be found between the speed of the algorithm and the quality of the result. Thus, the problem is not purely technical anymore but also involves "business rules," which are usually *very* informal.

In the future, an extensive requirements analysis will be necessary, which means that the work will be no longer purely "algorithmical" in nature. Such an analysis must be very detailed because otherwise there is no hope to match the *real* problem. Unfortunately, there is high evidence that the general problems addressed in [14] will become virulent here: a sufficiently simple formal model that captures all relevant details does not seem to be in our reach, and many details are "volatile" in the sense that they may change time and again in unforeseen ways. Future research will show whether these conflicting criteria can be simultaneously fulfilled satisfactorily.

References

1. S. Albers, A. Crauser, and K. Mehlhorn: *Algorithmen für sehr große Datenmengen.* MPI Saarbrücken (1997).[7]
2. K. Ishikawa, M. Ogawa, S. Azume, and T. Ito: *Map Navigation Software of the Electro Multivision of the '91 Toyota Soarer.* IEEE Int. Conf. Vehicle Navig. Inform. Syst. (VNIS '91), 463–473.
3. R. Agrawal and H. Jagadish: *Algorithms for Searching Massive Graphs.* IEEE Transact. Knowledge and Data Eng. 6 (1994), 225–238.
4. R.K. Ahuja, T.L. Magnanti, and J.B. Orlin: *Network Flows.* Prentice–Hall, 1993.
5. A. Car and A. Frank: *Modelling a Hierarchy of Space Applied to Large Road Networks.* Proc. Int. Worksh. Adv. Research Geogr. Inform. Syst. (IGIS '94), 15–24.
6. B.V. Cherkassky, A.V. Goldberg und T. Radzik: *Shortest Paths Algorithms: Theory and Experimental Evaluation.* Mathematical Programming 73 (1996), 129–174.
7. S. Shekhar, A. Kohli, and M. Coyle: *Path Computation Algorithms for Advanced Traveler Information System (ATIS).* Proc. 9th IEEE Int. Conf. Data Eng. (1993), 31–39.
8. T.H. Cormen, C.E. Leiserson, and R.L. Rivest: *Introduction to Algorithms.* MIT Press and McGraw–Hill, 1994.
9. S. Jung and S. Pramanik: *HiTi Graph Model of Topographical Road Maps in Navigation Systems.* Proc. 12th IEEE Int. Conf. Data Eng. (1996), 76–84.
10. T. Lengauer: *Combinatorial Algorithms for Integrated Circuit Layout.* Wiley, 1990.

---

[7] Available from http://www.mpi-sb.mpg.de/units/ag1/extcomp.html. The lecture notes are in German, however, the chapter on shortest paths (Chapter 9) is in English.

11. J. Shapiro, J. Waxman, and D. Nir: *Level Graphs and Approximate Shortest Path Algorithms*. Network 22 (1992), 691–717.
12. T. Preuss and J.-H. Syrbe: *An Integrated Traffic Information System*. Proc. 6th Int. Conf. Appl. Computer Networking in Architecture, Construction, Design, Civil Eng., and Urban Planning (europIA '97).
13. K. Weihe: *Reuse of Algorithms — Still a Challenge to Object–Oriented Programming*. Proc. 12th ACM Symp. Object–Oriented Programming, Systems, Languages, and Applications (OOPSLA '97), 34–48.
14. K. Weihe, U. Brandes, A. Liebers, M. Müller–Hannemann, D. Wagner, and T. Willhalm: *Empirical Design of Geometric Algorithms*. To appear in the Proc. 15th ACM Symp. Comp. Geometry (SCG '99).

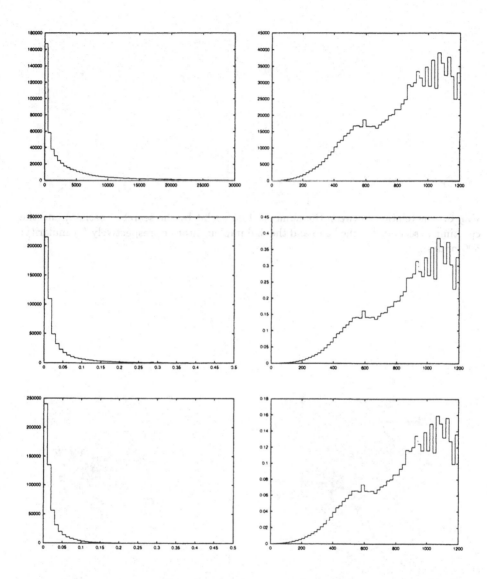

Fig. 5. An exemplary sequence of diagrams for one particular combination of strategies: "angle restriction" is applied, but "selection of stations" is not. First column: the frequency distribution histogram of all queries in the "snapshot" according to (a) the number of nodes met by the search (granularity: 500 nodes), (b) the CPU time for the heap implementation and (c) for the dial implementation (granularity: 10 milliseconds). Second column: the average of (a) the number of nodes met and (b/c) the CPU times for the heap/dial variant taken over all queries with roughly the same resulting total travel times (granularity: 20 minutes). The strong resemblance of the diagrams in each row nicely demonstrates the linear behavior of both priority–queue implementations.

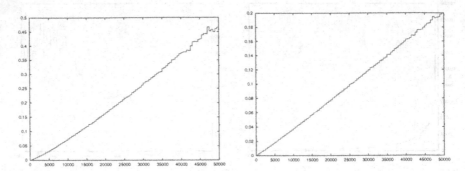

Fig. 6. The relation between the number of nodes hit by the search (abscissa) and the cpu time in seconds for the heap and the dial implementation, respectively (granularity: 500 nodes).

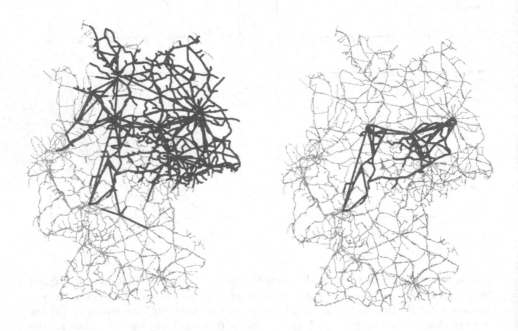

Fig. 7. The left picture shows the edges hit by Dijkstra's algorithm from Berlin Main East Station until the destination Frankfurt/Main Main Station is reached. In the right picture, the strategy "angle selection" was applied to the same query.

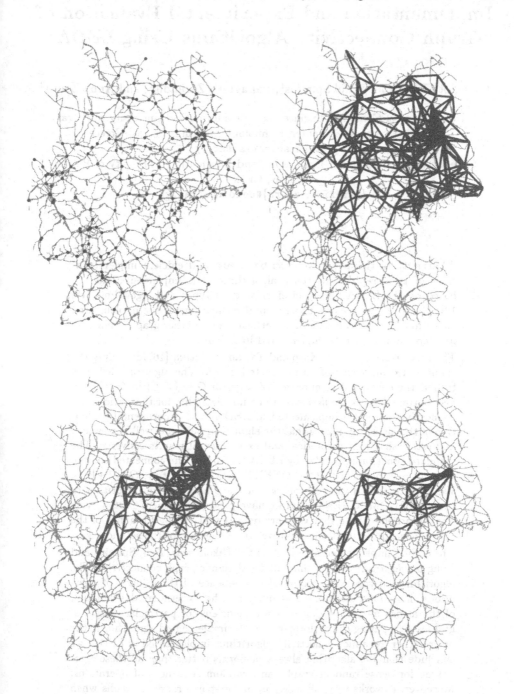

Fig. 8. The first picture shows the 225 stations selected for the study on strategy "selection of stations." The remaining three pictures refer to the same query as in Fig. 7. However, now the strategy "selection of stations" is applied with no angle restriction (upper right), with angles computed from the train graph (lower left), and with angles computed from the auxiliary graph itself. The train graph is shown in the background. The highlighted edges are the edges of the auxiliary graph hit by the search.

Implementation and Experimental Evaluation of Graph Connectivity Algorithms Using LEDA[*]

Panagiota Fatourou[1], Paul Spirakis[1], Panagiotis Zarafidis[2], and Anna Zoura[2]

[1] Department of Computer Engineering and Informatics, University of Patras, Patras, Greece & Computer Technology Institute, Patras, Greece
{faturu,spirakis}@cti.gr
[2] Department of Computer Engineering and Informatics, University of Patras, Patras, Greece
{zarafidi, zoura}@ceid.upatras.gr

Abstract. In this paper we describe robust and efficient implementations of two graph connectivity algorithms. The implementations are based on the LEDA library of efficient data types and algorithms [18, 19]. Moreover, we provide experimental evaluations of the implemented algorithms and we compare their performance to other graph connectivity algorithms currently implemented in LEDA.

The first algorithm is the Karp and Tarjan algorithm [16] for finding the connected components of an undirected graph. The algorithm achieves to find the connected components of a graph $G = \langle V, E \rangle$ in $O(|V|)$ expected time. This is the *first* expected-time algorithm for the static graph connectivity problem implemented in LEDA. The experimental evaluation of the algorithm proves that the algorithm performs well in practice, and establishes that theoretical and experimental results converge. The standard procedure provided by LEDA for finding the connected components of a graph, called COMPONENTS has running time $O(|V| + |E|)$. We have compared the performance of Karp and Tarjan's algorithm to the one of COMPONENTS and we have proved that there exists a wide class of graphs (those that they are dense) that the performance of the first algorithm dramatically improves upon the one of the second.

The second implemented algorithm is the Nikoletseas, Reif, Spirakis and Yung *polylogarithmic* algorithm [20] for dynamic graph connectivity. The algorithm can cope with any random sequence of three kinds of operations: insertions, deletions and queries. The experimental evaluation of the algorithm proves that it is very efficient for particular classes of graphs. Comparing the performance of the implemented algorithm to the one of other dynamic connectivity algorithms implemented in LEDA, we conclude that the algorithm always performs better than all these algorithms for dense random graphs and random sequences of operations. Moreover, it works very efficiently even for sparse random graphs when the sequence of operations is long.

* This work has been supported in part by the European Union's ESPRIT Long Term Research Project ALCOM-IT (contract # 20244).

J.S. Vitter and C.D. Zaroliagis (Eds.): WAE'99, LNCS 1668, pp. 124–138, 1999.

1   Introduction

In many areas of computer science, graph algorithms play an important role: problems modeled as graphs are solved by computing a property of the graph. If the underlying problem instance changes incrementally, algorithms are needed that quickly compute the property in the modified graph. A problem is called fully dynamic when both insertions and deletions of edges take place in the underlying graph. The goal of a dynamic graph algorithm is to update the solution after a change, doing so more efficiently than re-computing it at that point from scratch. To be precise, a graph dynamic algorithm is a data structure that supports the following operations: (1) insert an edge e; (2) delete an edge e; (3) test if the graph fulfills a certain property, e.g. are two given vertices connected? The adaptivity requirements (insertions/deletions of edges) usually make dynamic algorithms more difficult to design and analyze than their static counterparts.

Graph connectivity is one of the most basic problems with numerous applications and various algorithms in different settings. The area of dynamic graph algorithms has been a blossoming field of research in the last years, and it has produced a large body of algorithmic techniques [4, 8, 11, 12, 15, 20]. Most of the results [8, 11, 12, 14] on efficient fully dynamic structures for general graphs were based on clustering techniques. This has led to deterministic solutions of an inherent time bound of $O(n^\epsilon)$, for some $\epsilon < 1$, since the key problem encountered by these techniques is that the algorithm must somehow balance the work investing in maintaining the component of the cluster structure, and the work on the cluster structure (connecting the components).

The use of probabilistic techniques has been also considered for solving dynamic connectivity problems. The theory of random graphs [5] provides a rich variety of techniques that can be helpful in analyzing or improving the performance of dynamic graph connectivity algorithms. This theory has been extensively used for the theoretical evaluation of several interesting dynamic problems. Besides, the use of random inputs of updates on random graphs for the experimental evaluation of the performance of dynamic graph algorithms is quite common.

Nikoletseas, Reif, Spirakis and Yung [20] have presented a fully dynamic graph connectivity algorithm with *polylogarithmic* execution time per update for random graphs. Their algorithm can cope with any random sequence of three kinds of operations:

Property-Query(parameter): Returns true if and only if the property holds (or returns a subgraph as a witness to the property). For a connectivity query (u,v), a true answer means that the vertices u, v are in the same connected component.

Insert(x,y): Inserts a new edge joining x to y (assuming that $\{x, y\} \notin E$).

Delete(x,y): Deletes the edge $\{x, y\}$ (assuming $\{x, y\} \in E$).

In the previous operations it is assumed that random updates (insertions and deletions) have equal probability 1/2 and the edges to be deleted/inserted are chosen randomly. The algorithm admits an amortized expected cost of $O(\log^3 n)$

time per update operation with high probability and an amortized expected cost
of $O(1)$ per query.

Another interesting work, published at about the same period, is by M. Henzinger and V. King [15]. Their work presents a technique for designing dynamic
algorithms with *polylogarithmic* time per operation and applies this technique
to the dynamic connectivity, bipartiteness, and cycle equivalence problem. The
resulting algorithms are Las-Vegas type randomized algorithms. The connectivity algorithm achieves $O(\log^3 n)$ update time and $O(\log n)$ query time for both
random and non-random sequences of updates.

The area of dynamic graph connectivity algorithms is reach of theoretical
results. However, it seems that the implementation and experimental study of
this kind of algorithms are on the opposite poor. It is only recently that some of
these algorithms have been implemented in C++ and tested under the LEDA
Extension Package Dynamic Graph Algorithms (LEPDGA) [3, 17]. To the best
of our knowledge, there are only three works towards this direction. In the first
of these works [1], G. Amato *et al.* implemented and tested Frederickson's algorithms [12], and compared them to other dynamic algorithms. In the second
work [2], the dynamic minimum spanning tree based on sparsification by Eppstein, Galil, Italiano and Spencer [7] and the dynamic connectivity algorithm
presented by M. Henzinger and V. King [15] were implemented and experimentally evaluated. For random inputs the algorithm by Henzinger and King is the
fastest. For non-random inputs sparsification was the fastest algorithm for small
sequences, while for medium and large sequences of updates, the Henzinger and
King's algorithm was faster. The third work [13], by D. Frigioni *et al.* concentrates on the study of practical properties of many dynamic algorithms for
directed acyclic graphs. More specifically, the practical behaviour of several dynamic algorithms for transitive closure, depth first search and topological sorting
for such graphs have been studied.

In this paper, we implement and experimentally evaluate two graph connectivity algorithms. The first one, namely GCOMPONENTS, is a linear expected-
time algorithm for the static connectivity problem. The algorithm, proposed by
R. Karp and R. Tarjan [16], achieves to find the connected components of a
graph $G = (V, E)$ in $O(n)$ expected time, where $n = |V|$. It is worth pointing
out that all deterministic algorithms for the same problem work in $O(n + m)$
execution time in the worst case, where $m = |E|$. The only algorithm for this
problem provided by LEDA is such a deterministic algorithm and thus, it runs
in $O(n + m)$ execution time in the worst case. Apparently, the theoretical results
imply that GCOMPONENTS improves over COMPONENTS by an additive
factor of m. Our work proves that this theoretical result can be verified experimentally. We prove through a sequence of experiments that the performance
time of GCOMPONENTS on dense random graphs is extremely faster than the
one of COMPONENTS.

The second work, namely NRSY, is the Nikoletseas, Reif, Spirakis and Yung
algorithm for dynamic connectivity. We implement the algorithm in C++ using
LEDA and LEPDGA and evaluate its performance. More specifically, our results

imply that the algorithm behaves efficiently in case of dense random graphs, as stated by theory. Comparing the performance of the implemented algorithm to the performance of other dynamic connectivity algorithms [1] implemented in LEDA, we found that the algorithm performs better than those algorithms for random graphs with at least $O(n \ln n)$ edges, where n is the number of nodes in the graph. For such graphs and for random sequences of operations, our implementation is proved the fastest; it is faster even than the Henzinger and King algorithm, which has been considered the fastest implemented algorithm for such inputs thus far (see e.g., [1]). For sparser graphs, our implementation is also very efficient in case the sequence of updates is not too short.

All our implementations are written in C++ and use LEDA/LEPDGA. The source codes are available at the following URL:

http://www.ceid.upatras.gr/~faturu/projects.htm

The rest of our paper is organized as follows. Section 2 describes how Karp and Tarjan's algorithm for finding the connected components of a graph has been implemented, and provides experimental results on its performance. Section 3 concentrates on the implementation and performance evaluation of the NRSY dynamic connectivity algorithm. We conclude with a brief discussion of our results in Section 4.

2   The Karp and Tarjan's Algorithm

2.1   Algorithm Description

In this section, we present the main ideas of the randomized algorithm of Karp and Tarjan, described in [16], for finding the connected components of a graph.

Consider any graph $G = \langle V, E \rangle$ and let $n = |V|$ and $m = |E|$. The algorithm finds the connected components of G in $O(n)$ expected time. The algorithm proceeds in two stages. The first stage, called sampling, finds a *giant* connected sub-graph of G as follows:

(1) Let $E_0 = \emptyset$.
(2) If $|E \setminus E_0| < n$, let $D = E \setminus E_0$. Otherwise, let D be a set of n distinct edges randomly selected from $E \setminus E_0$.
(3) Let $E_0 = E_0 \cup D$.
(4) Use DFS (Depth First Search, see e.g. [6]) to find the connected components of $G_0 = (V, E_0)$. If $E = E_0$ or G_0 has a connected component of at least θn vertices, where $\theta > 1/2$ is a constant, stop. Otherwise, return to Step (2).

The second stage of the algorithm, called cleanup, finishes the grouping of vertices into components and is executed if a giant component is found in the sampling stage . This stage uses the standard depth-first search procedure with a few small changes. Initially, all vertices in the giant component of G_0 are marked giant, while the remaining vertices are unmarked. An unmarked vertex is selected as a start vertex and a depth first search is initiated from this vertex.

If an edge leading to the giant component is found, the search is immediately aborted and all vertices reached during the search are marked giant. If the search finishes without reaching the giant component, all vertices reached are marked as being in a new component. The process of carrying out such a search from an unmarked vertex is repeated until no unmarked vertices remain.

The design of the algorithm, as well as its good performance is based upon the following three observations: (1) there is a well-known depth-first search procedure [6] that finds the connected components of an undirected graph in $O(m)$ worst-case time; (2) Erdos and Renyi [9] have shown that there exists a constant $\theta > 1/2$ such that with probability tending to one as n tends to infinity, a random graph with n vertices and n edges has a giant connected component with at least θn vertices; (3) there is no need to examine all the edges of a graph in order to find its connected components; it suffices to exhaust the adjacency lists of the vertices in all but one of the components and to look at enough edges in the remaining component to verify that it is connected.

Karp and Tarjan have proved that their algorithm computes the connected components of any graph in $O(|V|)$ expected time; moreover, the probability that the running time exceeds $c|V|$, where $c > 0$ is a constant, tends to zero exponentially fast as $|V|$ tends to infinity.

2.2   Implementation

For the implementation of the algorithm, the LEDA library has been used. More specifically, the graph data structure provided by LEDA, as well as useful operations implemented on graphs (e.g., node or edge arrays, iterators, etc.) have been particularly helpful for the implementation of the algorithm.

A function, called GCOMPONENTS, has been implemented. It takes as arguments a graph $G = \langle V, E \rangle$ and a float θ, and computes the connected components of the underlying undirected graph, i.e., for every node $v \in V$, an integer compnum[v] from $[0 \ldots \text{num\_of\_components} - 1]$ is returned, where the value of the variable num_of_components expresses the number of connected components of G and v belongs to the i-th connected component if and only if compnum[v] = i. GCOMPONENTS returns a boolean, indicating the existence or not of a giant component with θn vertices in G.

The implementation of GCOMPONENTS is implied directly by the algorithm of Karp and Tarjan [16]. The two stages of the algorithm are indicated by the call of two procedures, called sampling and cleanup. Procedure sampling is called first and if a giant component of the graph G exists, cleanup is executed. The main idea of sampling is that a new graph $G_0 = (V, E_0)$ is incrementally created, by adding edges of G. Initially, E_0 is empty, while as many nodes as in G are inserted in V_0. A node array is used for direct mapping of G_0 nodes to G nodes. Procedure sampling executes a while loop. During each iteration of the loop, a set of at most $|V|$ distinct edges are inserted in E_0 and the steps of the first stage, as described in Section 2.1 are executed. The existing procedure COMPONENTS of LEDA is used for finding the connected components of G_0. Since the running time of COMPONENTS is $O(n + m)$ and G_0 has $O(n)$ edges, the running time

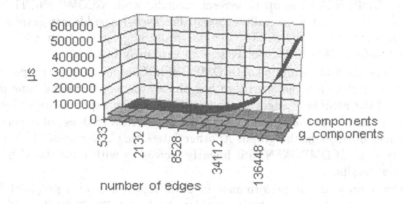

Fig. 1. The execution times of GCOMPONENTS and COMPONENTS for a graph with 1600 nodes.

of COMPONENTS on G_0 is $O(n)$. Procedure `cleanup` simply executes a depth-first search procedure starting from every UNMARKED node in order to share all remaining nodes in the proper connected components.

The code described above has been compiled and tested under the C++ compiler g++. In order to testify its correctness a flexible main function has been created and a lot of executions have been performed for graphs with different set of nodes and edges. The connected components calculated by GCOMPONENTS have been compared with those calculated by COMPONENTS. For graphs with small node/edge sets this comparison was straightforward. However, such comparisons can not be easily performed when the number of nodes becomes large. In this case, we used proper UNIX scripts for comparing the results produced by the two algorithms.

2.3 Performance Evaluation

The time performance of GCOMPONENTS has been compared to the one of COMPONENTS. COMPONENTS performs slightly better for sparse graphs, that is, for graphs with a small set of edges. Apparently, this is natural since if $|E| \in O(|V|)$, the $O(|V| + |E|)$ running time of COMPONENTS becomes $O(|V|)$, while additionally the implementation of COMPONENTS is simpler than the one of GCOMPONENTS. However, GCOMPONENTS performs dramatically better than COMPONENTS for dense graphs, that is, graphs with large sets of edges (e.g., $|E| \in O(|V| \log |V|)$ or $|E| \in O(|V|^2)$).

For example, on a graph of 200 nodes, COMPONENTS performs slightly better than GCOMPONENTS if the number of edges is less than 5000. For denser graphs GCOMPONENTS becomes better than COMPONENTS. For large graphs, e.g., graphs with 5000 nodes and at least 500000 edges, there exists a tremendous difference on the performance of the two algorithms (for instance, the execution time of COMPONENTS is up to several seconds, while GCOMPONENTS terminates after at most a few milliseconds). For even larger/denser graphs, the execution time of COMPONENTS becomes really huge (more than a few minutes), while GCOMPONENTS behaves in an extremely faster way.

The experimental evaluation of GCOMPONENTS implies that for graphs with fixed number of nodes, the algorithm has almost the same running time independent of the number of edges in the graph. Thus, the experiments verify the theoretical result that the running time of GCOMPONENTS does not depend on the number of edges in the graph. Another interesting observation is that the performance of GCOMPONENTS is linearly increasing with n, as stated by the theoretical results.

Figure 1 presents the performance of both algorithms for a graph of 1600 nodes[1]. All our experiments took place on an ULTRA ENTERPRISE 3000 SUN SPARC station with two processors working at 167MHz with 256 Mbytes of memory.

3   The NRSY Algorithm

3.1   Description

In this section, we describe the dynamic connectivity algorithm of Nikoletseas, Reif, Spirakis and Yung (NRSY), presented in [20].

The algorithm alternates between two epochs, namely the *activation* epoch and the *retirement* epoch. As long as the algorithm is in the activation epoch, each operation is performed in at most polylogarithmic time. Queries are performed in constant time. However, in order to perform these operations fast, the algorithm may examine many edges per delete operation, which means that after a small number of delete operations the algorithm has probably looked at the entire graph in which case it cannot use the fact that the graph is random. If this happens, the algorithm switches to the retirement epoch, where any dynamic connectivity algorithm presented in the literature, with expected update time at most $O(n)$, should be used for performing a specific number of operations, probably in a higher cost than a polylogarithmic one. However, the number of operations performed during a retirement epoch is not too large, so that the amortized cost per operation of the algorithm has been proved to be polylogarithmic.

Initially, the algorithm uses a slightly modified version of the linear expected-time algorithm for finding connected components by Karp and Tarjan [16] to compute a forest of spanning trees for the original graph. We call the above procedure of computing the forest of spanning trees a *total reconstruction*. We say that a total reconstruction is *successful* if it achieves to find a giant component

---

[1] More diagrams demonstrating the above observations are provided in [10].

of the graph. It is worth pointing out that a total reconstruction is an expensive operation, since all data structures of the algorithm should be re/initiated.

A graph activation epoch is entered after a successful total reconstruction. The activation epoch lasts at most $c_1 n^2 \log n$ operations, where $c_1 > 1$ is a constant. However, the epoch may end before all these operations are performed. This happens when a deletion operation disconnects the spanning tree of the giant component and the attempted fast reconnection fails. A graph retirement epoch starts when the previous graph activation epoch ends. However, the algorithm may start by a retirement epoch if the initial total reconstruction is not successful. During the course of an execution of a retirement epoch, the algorithm does not maintain any data structure. It simply uses some other algorithm for dynamic connectivity (any algorithm presented in the literature with update time at most $O(n)$) to perform the operations. A graph retirement epoch lasts at least $c_2 n \log^2 n$ operations, where $c_2 > 1$ is a constant. At the end of a graph retirement epoch, a total reconstruction is attempted. If it is successful, a new graph activation epoch is entered. Otherwise, the graph retirement epoch continues for another set of $c_2 n \log^2 n$ operations. The above procedure is repeated until a successful total reconstruction occurs.

During an activation epoch the algorithm maintains a forest of spanning trees, one tree per connected component of the graph. Moreover, the algorithm categorizes all edges of the graph into three classes: (1) pool edges, (2) tree edges, and (3) retired edges.

During an activation insertion operation, if the edge joins vertices of the same tree, the edge is marked as a retired edge and the appropriate data structures are updated; otherwise, the edge joins vertices of different trees, and the component name of the smaller tree is updated. During an activation deletion operation, if the edge is a pool or a retired edge, the edge is deleted by all data structures. Otherwise, if the edge is a tree edge of a small tree and the deletion of the edge disconnects the corresponding graph component, the tree is split into two small trees and the smaller of the two trees is relabeled. If on the opposite, a reconnection is possible, the tree is reconnected. If the edge is an edge of the tree of the giant component, a specific procedure, called NeighborhoodSearch, is executed.

Let the deleted edge be $e = \langle u, v \rangle$. Moreover, let $T(u), T(v)$ be the two pieces in which the tree is split by the removal of e, such that $u \in T(u)$ and $v \in T(v)$. NeighborhoodSearch executes a sequence of phases. A phase starts when a new node is visited. NeighborhoodSearch starts two BreadthFirstSearch procedures, the one out of u and the other out of v and executes them in an interleaved way; that is, the algorithm visits nodes by executing one step by each BFS. The visit of a node, independently of which of the two searches reaches the node, indicates the start of a new phase. In any phase an attempt for reconnection occurs. If this reconnection is achieved, the phase returns success; otherwise, it returns failure. The total number of nodes visited by each BFS procedure equals $c_3 \log n$, where $c_3 > 1$ is a constant. After this number of nodes have been visited, NeighborhoodSearch ends, independently of whether it achieves to reconnect the giant tree or not. If any phase returns success, NeighborhoodSearch also returns success. If more than one phases returns success, the edge which is "closer" to the root is selected for the reconnection. Thus, NeighborhoodSearch may: (1) finish with no success, in which case the algorithm undergoes total reconstruction and the current activation epoch ends. NeighborhoodSearch reports failure in this case; (2) the nodes connected to u may be exhausted before the search finishes, so that the one of the two components is a small component of $O(\log n)$ nodes; let this small component be $G(u)$. We search all edges emanating from nodes of $G(u)$ and if a reconnection is impossible, the giant component is just disconnected to a still-giant component and a midget component of $O(\log n)$ nodes. In this case the midget component is renamed, while NeighborhoodSearch reports success; (3) several phases may report a number of successes. In this case, the algorithm chooses for reconnection the edge which is "closer" to the root of the tree and reconnects the two pieces.

We now describe what happens during a phase initiated by the visit of some node w. Assume that w is reached due to the execution of the BFS initiated at node u, so that $w \in T(u)$. Roughly speaking, the algorithm checks if a randomly chosen (if any exists) pool edge out of node w reconnects the two pieces of the giant tree and if this reconnection is a "good" one. A good reconnection is one that does not increase the diameter of the tree. If a good reconnection is found, procedure Phase reports success; otherwise, it reports failure.

Graph activation epochs are partitioned into edge deletion intervals of $c_4 \log n$ deletions each, where $c_4 > 1$ is a constant. All edges marked as retired during an edge deletion interval are returned into the pool after a delay of one more subsequent edge deletion interval. Edges of small components are not reactivated by the algorithm.

Nikoletseas, Reif, Spirakis and Yung have proved that the amortized expected time of algorithm NRSY for update operations on the graph is $O(\log^3 n)$ with high probability; moreover, they have proved that the query time of their algorithm is $O(1)$.

3.2 Implementation

The algorithm has been implemented as a data structure (a class called nrsy_connectivity, in C++). For its implementation the Library of Efficient Data Types and Algorithms (LEDA), as well as the LEDA Extension Package-Dynamic Graph Algorithms (LEPDGA) have been used. It is worth pointing out that our implementation has been done using the new base class of LEDA for dynamic graphs [3]. The public methods of class nrsy_connectivity that are efficiently supported are: insert_new_edge, delete_edge and all queries ("global" or not). A few more public methods are provided for collecting statistics, as well as for demonstrating how the algorithm works (that is, a graphical user interface).

In the activaton epoch, each node maintains both a set of pool edges and a priority queue of retired edges incident to it. Each node is labeled by the identification of the component it belongs.

A forest of spanning trees is maintained during the execution of activaton epochs of the algorithm. Each node of the tree maintains a pointer to its parent, its left and right children (pointers lc, rc, respectively), as well as to its left and right siblings (pointers ls, rs, respectively). If a node does not have children pointers lc, rc are nil. If a node does not have right or left sibling, the corresponding pointer points to the node's parent. The trees are constructed in $O(n)$ time by using a slightly modified version of Karp and Tarjan's algorithm. For random dense graphs, these trees have diameter of expected length logarithmic on the number of nodes in the graph.

Clearly, the way that the nodes of each tree are connected allows fast traversal of parts of the tree (which is required e.g., for updating labels during insert/delete operations). It is worth pointing out that an inserted edge that connects nodes of two different trees may cause an *evert* operation to take place in one of the two trees, so that some particular node to become the new root of this tree. In this case, all nodes starting from the new root up to the old root of the tree should be updated. However, since the diameter of the tree is of logarithmic length, such operations can be performed in logarithmic time.

Each node maintains the size of its subtree (that is, the subtree emanated by the node). When insertions/deletions that change the forest of the spanning trees occur, the smallest of the two trees is updated (e.g., the size and the label of the nodes of the smallest tree are updated).

Retired edges are maintained in the retired priority queues of their incident nodes. A counter measuring the number of operations in each epoch is maintained and when an edge becomes retired, its priority takes the value of this counter. Reactivation of retired edges occurs only before the deletion of *tree* edges, that is, edges that disconnect one of the spanning trees in the forest. More specifically, reactivations occur only before the execution of procedure NeighborhoodSearch, since it is only for this procedure that it is important several pool edges to exist. Reactivation is performed if the difference of the priority of a retired edge from the current value of the counter expressing the number of operations performed in the current epoch is larger than $\log n$.

During the retirement epoch, the algorithm by Henziger and King (that is, the data structure dc_henzinger_king of LEPDGA [3, 17]) is used for performing each operation.

The correctness of the implementation has been checked by comparing the answers to the queries in sequences of random operations among NRSY and other already implemented graph connectivity algorithms in LEDA. Additionally, we have performed tests with random updates, and every time a specified number of updates had been done, we checked whether all possible queries were answered according to the component labels of the vertices. These labels were computed by the static algorithm COMPONENTS or GCOMPONENTS.

3.3   Performance Evaluation

In this section, we present the results of comparative experiments with NRSY and other algorithms currently implemented in LEDA, using different types and

Algorithm	Insert	Delete	Query
nrsy_connectivity	$O(\log^3 n)$	$O(\log^3 n)$	$O(1)$
dc_henzinger_king	$O(\log^3 n)$	$O(\log^3 n)$	$O(\log n)$
dc_simple	$O(n + m)$	$O(n + m)$	$O(1)$

Fig. 2. Theoretical bounds on the performance of each algorithm

sizes of inputs. More specifically, we measure the performance of the following dynamic connectivity data structures: nrsy_connectivity, dc_henzinger_king and dc_simple. The second is the algorithm by M. Henzinger and V. King [1, 15], while the third is a simple approach to dynamic connectivity [3, 17]. Algorithm dc_simple maintains and recomputes if necessary component labels for the vertices and edge labels for the edges in a current spanning forest. Inserting or deleting edges which do not belong to the current spanning forest takes constant time, while the deletion/insertion of forest edges takes time proportional to the size of the affected component(s) (thus, $O(|V| + |E|)$ in the worst case). Figure 2 summarizes information about the performance of the above three algorithms. The last two algorithms have been implemented in LEDA and their performance has been studied experimentally in [1].

Since NRSY algorithm guarantees good performance only on random inputs, we concentrate our study only on random sequences of updates (and queries) on random graphs. We are mainly interested in the average case performance of the algorithm on such inputs. The random inputs consisted of random graphs with different densities, and sequences of random operations. Each sequence of random operations consisted of random insertions, deletions and queries on the corresponding graph.

We conducted a series of tests for graphs on $n = 50, 100, 300, 500, 750$, and 1000 vertices and different sizes of edge sets. More specifically, for each such n we tested the performance of the above algorithms on graphs with $m = n/2$, n, $n \ln n$, $n^{3/2}$ and $n^2/4$ edges. For every pair of values for n and m we did five experiments and took averages. In every experiment, the same input was used for all algorithms. Since the NRSY algorithm may execute several total reconstructions during its execution, we didn't measure separately the time that the algorithm spent in preprocessing, that is, for building the data structure for the initial graph, and processing, that is for performing the sequence of operations. For the other two algorithms (dc_henzinger_king, dc_simple) we measure the time spent only for performing the operations (and not for the initialization of the data structures). We measure the performance of the algorithms for long update sequences e.g., of 100000 operations, for medium sequences e.g., of 10000 operations and for short sequences e.g., of 1000 operations. We did all experiments on an ULTRA ENTERPRISE 3000 SUN SPARC station with two processors working at 167MHz with 256 Mbytes of memory.

The most difficult inputs for the NRSY algorithm are inputs with m equal or slightly larger to n. In this case, the theory of random graphs implies that the graph is with high probability disconnected, but it may exist some component of size in the order of $n^{2/3}$. Thus, a total reconstruction with high probability does not succeed to find a giant component, or if it finds one it is highly expected that it will become disconnected in the next few operations. Thus, either the algorithm does not run in activation epoch, or it gets in activation epoch and after a few operations it has to run a total reconstruction again. Recall that a total reconstruction is an expensive operation. Even if the algorithm gets and remains in the activation epoch for some operations, almost all deletions/insertions are expected to be tree updates. Thus, in this case most of the operations to be performed are expensive.

Difficult inputs appear also to be those where $m < n$. In this case, the theory of random graphs implies that with high probability the graph consists of several components of size $O(\log n)$. Thus, with very high probability a total reconstruction will not succeed. At this point the algorithm uses the dc_henzinger_king data structure to execute the retirement epoch. If a total reconstruction succeeds after the retirement epoch the algorithm gets into an activaton epoch and so on. We expect that nrsy_connectivity and dc_henzinger_king perform in a similar way in this case. However, nrsy_connectivity periodically performs a total reconstruction and checks if a giant component exists in the graph. Since a total reconstruction is expensive, it would be also natural if nrsy_connectivity behaves in a slightly slower way than dc_henzinger_king in this case.

For inputs with $m > n$, that is, if $m \in O(n \ln n)$ or $m \in O(n^2)$, the theory of random graphs implies that a giant component exists in the graph, so that nrsy_connectivity should spend most of its time in activation epochs. Moreover, most of the updates are non-tree updates, so that nrsy_connectivity should perform them very fast (in $O(1)$ time). We further expect that the behavior of the algorithm should be better for long/medium sequences of operations than for small sequences, since the theoretical results hold asymptotically.

The experiments that we perform prove all the above observations. More specifically, for long sequences of operations, our implementation is proved to be very fast, faster than the Henzinger and King's algorithm in most cases. For medium sequences of operations our implementation has similar performance to the one of dc_henzinger_king for values of m close to n, while it is better for denser graphs. For short sequences of updates, the performance of our implementation becomes worse than the other two algorithm for sparse graphs. Recall that this is expectable, since the algorithm alternates between retirement epochs and total reconstructions, while all operations performed are with high probability expensive operations. Notice that this bad behavior of the algorithm in this case does not contradict the theoretical results, since all bounds provided in [20] have been derived assuming dense random input graphs. Moreover, even in the case of short sequences, our implementation has better performance than the other two algorithms for dense graphs (that is, if m is at least $O(n \ln n)$).

For the behavior of the other two algorithms we derive similar observations as those presented in [1]. Roughly speaking, dc_henzinger_king benefits for small components as they appear for $m < n$; moreover, the most difficult inputs are

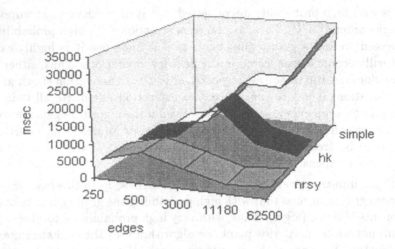

Fig. 3. Performance of NRSY, Henzinger-King and Simple for a random graph of 500 nodes and a sequence of 10000 random operations.

again those for which m is close to n, while the algorithm gets better and better if the number of edges increases (still remaining worse than nrsy_connectivity though). For dc_simple, if m increases the expected number of tree updates decreases, but each such tree update becomes very expensive. In theory, these two effects cancel out, so that it appears that the expected running of dc_simple is $O(n)$ per update.

The performance of dc_henzinger_king has been compared in [1] to the one of a sequence of other algorithms, among others sparsification [8], and it has been proved that dc_henzinger_king is the fastest algorithm among all for random inputs. Since NRSY performs better than Henzinger and King's algorithm for such inputs (especially if the graph is dense) we conclude that nrsy_connectivity is the fastest algorithm for random sequences of updates on dense random graphs, implemented under LEDA thus far. Moreover, the performance of NRSY is very good (comparable or better than the one of Henzinger and King's algorithm) even for non-dense random graphs, if the sequence of updates is long.

We didn't test the performance of nrsy_connectivity for non-random inputs, since we believe that the results would not be of theoretical interest. Recall that theory does not guarantee any bound on the update time of NRSY for non-random inputs on non-random graphs. Some of the above observations are illustrated in Figures 3 and 4[2].

---

[2] More diagrams demonstrating the above observations are provided in [10].

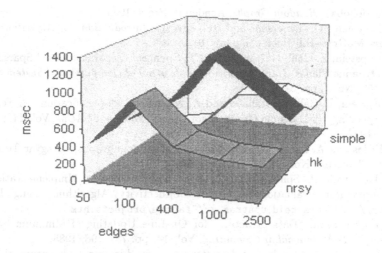

Fig. 4. Performance of NRSY, Henzinger-King and Simple for a random graph of 100 nodes and a sequence of 1000 random operations.

4   Discussion

We have described elegant, robust and efficient implementations of two graph connectivity algorithms, based on **LEDA C++** and **LEPDGA**. Moreover, we have shown with experimental data that these algorithms are fast and efficient in practice.

We are currently working on testing the performance of modified versions of the **NRSY** algorithm. One such direction is, for example, to let the algorithm remain in the activation epoch till the giant component is disconnected, instead of alternating between the two epochs after the execution of a specific number of operations. We believe that such modifications may yield better experimental results.

References

1. D. Alberts, G. Cattaneo and G. Italiano, "An Empirical Study of Dynamic Algorithms," *ACM Journal of Experimental Algorithmics*, Vol. 2, No. 5, pp. 192–201, 1997.
2. G. Amato, G. Cattaneo and G. Italiano, "Experimental Analysis of Dynamic Minimum Spanning Tree Algorithms," *Proceedings of the 8th Annual ACM-SIAM Symposium on Discrete Algorithms*, pp. 314–323, January 1997.
3. G. Amato, G. Cattaneo, G. F. Italiano, U. Nanni and C. Zaroliagis, "A Software Library of Dynamic Graph Algorithms," *Proceedings of the Workshop on Algorithms and Experiments (ALEX'98)*, pp. 129–136, February 1998.
4. D. Alberts and M. Henzinger, "Average Case Analysis of Dynamic Graph Algorithms," *Proceedings of the 6th Annual ACM-SIAM Symposium on Discrete Algorithms*, pp. 312–321, 1995.

5. Bela Bollobas, *Random Graphs*, Academic Press, 1985.

6. T. Cormen, C. Leiserson and R. Rivest, *Introduction to Algorithms*, MIT Press/McGraw-Hill Book company, 1990.

7. D. Eppstein, Z. Galil, G. Italiano and T. Spencer, "Separator Based Sparsification for Dynamic Planar Graph Algorithms," *Journal of Computer and System Science*, Vol. 52, No. 1, pp. 3–27, 1996.

8. D. Epstein, Z. Galil, G. Italiano and A. Nissenweig, "Sparsification - A Technique for Speeding Up Dynamic Graph Algorithms," *Journal of ACM*, Vol. 44, pp. 669–696, 1997.

9. P. Erdos and A. Renyi, "On the evolution of random graphs," Magyar Tud. Akad. Math. Kut. Int. Kozl., 5, pp. 17-61, 1960.

10. P. Fatourou, P. Spirakis, P. Zarafidis and A. Zoura, "Implementation and Experimental Evaluation of Graph Connectivity Algorithms using LEDA," http://students.ceid.upatras.gr/~faturu/projects.htm

11. G. Frederickson, "Data Structures for On-Line Updating of Minimum Spanning Trees," *SIAM Journal of Computing*, Vol. 14, pp. 781–798, 1985.

12. G. Frederickson, "Ambivalent data structures for dynamic 2-edge connectivity and k-smallest spanning trees," *Proceedings of the 32nd IEEE Symposium on Foundations of Computer Science (FOCS'91)*, pp. 632-641, 1991.

13. D. Frigioni, T. Miller, U. Nanni, G. Pasqualone, G. Schaefer and C. Zaroliagis, "An Experimental Study of Dynamic Algorithms for Directed Graphs," *Proceedings of the 6th Annual European Symposium on Algorithms (ESA'98)*, pp. 368–380, August 1998.

14. Z. Galil and G. Italiano, "Fully dynamic algorithms for edge-connectivity problems," *Proceedings of the 23rd Annual ACM Symposium on Theory of Computing (STOC'91)*, pp. 317–327, May 1991.

15. M. Henzinger and V. King, "Randomized Graph Algorithms with Polylogarithmic Time per Operation," *Proceedings of the 27th Annual ACM Symposium on Theory of Computing (STOC'95)*, pp. 519–527, 1995.

16. R. Karp and R. Tarjan, "Linear Expected Time for Connectivity Problems," *Journal of Algorithms*, Vol. 1, pp. 374–393, 1980.

17. "LEDA Extension Package Dynamic Graph Algorithms", User Manual, Version 3.8, 1998.

18. K. Mehlhorn and S. Naher, "LEDA, A platform for combinatorial and geometric computing " *Communications of the ACM*, 1995.

19. K. Mehlhorn and S.Naher, *The LEDA Platform of Combinatorial and Geometric Computing*, Cambridge University Press, 1999.

20. S. Nikoletseas, J. Reif, P. Spirakis and M. Yung, "Stochastic Graphs Have Short Memory: Fully Dynamic Connectivity in Poly-Log Expected Time," *Proceedings of the 22nd International Colloquium on Automata, Languages, and Programming (ICALP'95)*, pp. 159-170, 1995.

On-Line Zone Construction in Arrangements of Lines in the Plane*

Yuval Aharoni, Dan Halperin, Iddo Hanniel,
Sariel Har-Peled, and Chaim Linhart

Department of Computer Science
Tel-Aviv University, Tel-Aviv 69978, ISRAEL
{halperin, hanniel, sariel, chaim}@math.tau.ac.il

Abstract. Given a finite set $\mathcal{L}$ of lines in the plane we wish to compute the zone of an additional curve γ in the arrangement $\mathcal{A}(\mathcal{L})$, namely the set of faces of the planar subdivision induced by the lines in $\mathcal{L}$ that are crossed by γ, where γ is not given in advance but rather provided *on-line* portion by portion. This problem is motivated by the computation of the area bisectors of a polygonal set in the plane. We present four algorithms which solve this problem efficiently and exactly (giving precise results even on degenerate input). We implemented the four algorithms. We present implementation details, comparison of performance, and a discussion of the advantages and shortcomings of each of the proposed algorithms.

1 Introduction

Given a finite collection $\mathcal{L}$ of lines in the plane, the *arrangement* $\mathcal{A}(\mathcal{L})$ is the subdivision of the plane into vertices, edges and faces induced by $\mathcal{L}$. Arrangements of lines in the plane, as well as arrangements of other objects and in higher dimensional spaces, have been extensively studied in computational geometry [9, 15, 22], and they occur as the underlying structure of the algorithmic solution to geometric problems in a large variety of application domains. The *zone* of a curve γ in an arrangement $\mathcal{A}(\mathcal{L})$ is the collection of (open) faces of the arrangement crossed by γ (see Figure 1 for an illustration).

In this paper we study the following algorithmic problem: Given a set $\mathcal{L}$ of n lines in the plane, efficiently construct the zone of a curve γ in the arrangement $\mathcal{A}(\mathcal{L})$, where γ consists of a single connected component and is given on-line, namely γ is not given in its entirety as part of the initial input but rather given (contiguously) piece after piece and at any moment the algorithm has to report all the faces of the arrangement crossed by the part of γ given so far.

* This work has been supported in part by ESPRIT IV LTR Projects No. 21957 (CGAL) and No. 28155 (GALIA), by The Israel Science Foundation founded by the Israel Academy of Sciences and Humanities, by a Franco-Israeli research grant (monitored by AFIRST/France and The Israeli Ministry of Science), and by the Hermann Minkowski – Minerva Center for Geometry at Tel Aviv University. Dan Halperin has been also supported in part by an Alon fellowship and by the USA-Israel Binational Science Foundation.

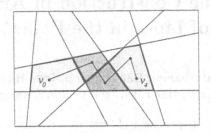

Fig. 1. An example of a simple arrangement induced by 7 lines and a bounding box, and the zone of a polygonal curve in it. The polyline $v_0, \ldots, v_4$ crosses 4 faces (*shaded faces*)

There is a straightforward solution to our problem. We can start by constructing $\mathcal{A}(\mathcal{L})$ in $\Theta(n^2)$ time, represent it in a graph-like structure $\mathcal{G}$ (say, the half-edge data structure [8, Chapter 2]), and then explore the zone of γ by walking through $\mathcal{G}$. However, in general we are not interested in the entire arrangement. We are only interested in the zone of γ, and this zone can be anything from a single triangular face to the entire arrangement.

The *combinatorial complexity* (complexity, for short) of a face in an arrangement is the overall number of vertices and edges on its boundary. The complexity of a collection of faces is the sum of the face complexity over all faces in the collection. The complexity of an arrangement of n lines is $\Theta(n^2)$ (and if we allow degeneracies, possibly less). The complexity of the zone of an arbitrary curve in the arrangement can range from $\Theta(1)$ to $\Theta(n^2)$. The simple algorithm sketched above will always require $\Omega(n^2)$ time, which may be too much when the complexity of the zone is significantly below quadratic. (Note that even an optimal solution to our problem may require more than $\Theta(n^2)$ running time for certain input curves, since we do not impose any restriction on the curve and it may cross a single face an arbitrary large number of times.)

If the whole curve γ were given as part of the initial input (together with the lines in $\mathcal{L}$) then we could have used one of several worst-case near-optimal algorithms to solve the problem. The idea is to transform the problem into that of computing a single face in an arrangement of segments, where the segments are pieces of our original lines that are cut out by γ [10]. Denote by k the overall number of intersections between γ and the lines in $\mathcal{L}$, the complexity of the zone of γ in $\mathcal{A}(\mathcal{L})$ (or the complexity of the single face containing γ in the modified arrangement) is bounded by[1] $O((n + k)\alpha(n))$ and several algorithms exist that compute a single face in time that is only a polylogarithmic factor above the worst-case combinatorial bound, for example [7].

However, since in our problem we do not know the curve γ in advance, we cannot use these algorithms, and we need alternative, on-line solutions. Our problem, the on-line version of the zone construction, is motivated by the study of area bisectors of polygonal sets in the plane [3] which is in turn motivated by algorithms for part orienting using Micro Electro Mechanical Systems [4].

---

[1] $\alpha(\cdot)$ denotes the extremely slowly growing inverse of the Ackermann function.

The curve γ that arises in connection with the area bisectors of polygons is determined by the faces of the arrangement through which it passes, it changes from face to face, and its shape within a face f is dependent on f.

In the work reported here we assume that γ is a *polyline* given as a set of $m+1$ points in the plane $v_0, v_1, \ldots, v_m$, namely the collection of m segments $\overline{v_0 v_1}, \overline{v_1 v_2}, \ldots, \overline{v_{m-1} v_m}$, given to the algorithm in this order. In the motivating problem γ is a piecewise algebraic curve where each piece can have an arbitrarily high degree. We simplify the problem by assuming that the degree of each piece is one since our focus here is on the on-line exploration of the faces of the arrangement. The additional problems that arise when each piece of the curve can be of high degree are (almost completely) independent of the problems that we consider in this paper and we discuss them elsewhere [1]. We further assume that we are given a *bounding box* B, i.e., the set $\mathcal{L}$ contains four lines that define a rectangle which contains $v_0, \ldots, v_m$ (see Figure 1).

An efficient solution to the on-line zone problem is given in [3], and it is based on the well-known algorithm of Overmars and van Leeuwen for the dynamic maintenance of halfplane intersection [20]. The data structure described in [20] is rather intricate and we anticipated that it would be difficult to implement. We resorted instead to simpler solutions which are nevertheless non-trivial.

We devised and implemented four algorithms for on-line zone construction. The first algorithm is based on halfplane intersection and maintains a balanced binary tree on a set of halfplanes induced by $\mathcal{L}$ (it is reminiscent of the Overmars-van Leeuwen structure, but it is simpler and does not have the good theoretical guarantee of running time as the latter). The second algorithm works in the dual plane and maintains the convex hull of the set of points dual to the lines in $\mathcal{L}$. Its efficiency stems from a heuristic to recompute the convex hull when the hull changes (the set of points does not change but their contribution to the lower or upper hull changes according to the face of the arrangement that γ visits). Algorithms 1 and 2 could be viewed as simple variants of the algorithm described in [3]. The third algorithm presents a novel approach to the problem. It combines randomized construction of the binary plane partition induced by the lines in $\mathcal{L}$ together with maintaining a doubly connected edge list for the faces of the zone that have already been built. We give two variants of this algorithm, where the second variant (called Algorithm 3b) handles conflict lists [19] more carefully than the first (Algorithm 3). Finally, the fourth algorithm is also a variant of Algorithm 3; however it differs from it in a slight but crucial manner: it refines the faces of the zone as they are constructed such that in the refinement no face has more than some small constant number of edges on its boundary.

The four algorithms are presented in the next section, together with implementation details and description of optimizations. In Section 3 we describe a test suite of five input sets on which the algorithms have been examined, followed by a chart of experiment results. Then in Section 4 we summarize the lessons we learned from the implementation, optimization and experiments. Concluding remarks and suggestions for future work are given in Section 5.

2   Algorithms and Implementation

2.1   Algorithm 1: Halfplane Intersection

Given a point p and a set $\mathcal{L}$ of n lines, the face that contains p in the arrangement $\mathcal{A}(\mathcal{L})$ can be computed in $O(n \log n)$ time by intersecting the set of halfplanes induced by the lines and containing p. The idea is to divide the set of halfplanes into two subsets, recursively intersect each subset, and then use a linear-time algorithm for the intersection of the two convex polygons (see, for example, [8, Chapter 4]). We shall extend this scheme to the on-line zone construction by maintaining the *recursion tree*.

The General Scheme Let us assume we have a procedure `ConvexIntersect` that intersects two convex polygons in linear time. Given the first point v_0 of γ, we construct the recursion tree in a bottom-up manner in $O(n \log n)$ time, as follows. The lowest level of the tree contains n leaves, which hold the intersection of the bounding box with each halfplane induced by a line of $\mathcal{L}$ and containing v_0. The next levels are constructed recursively by applying `ConvexIntersect` in postorder. The resulting data structure is a balanced binary tree, in which each internal node holds the intersection of the convex polygons of its two children. The first face the algorithm reports is the one held in the root of the tree.

When the polyline moves from one face of the arrangement to its neighbor it crosses a line in $\mathcal{L}$. All we need to do is update the leaf in the recursion tree that corresponds to this line (namely, take the other halfplane it determines), and then reconstruct the polygons of its ancestors, again in a bottom-up manner. Clearly, for each update the number of calls to `ConvexIntersect` is $O(\log n)$. The time it takes to move from one face to its neighbor therefore depends on the total complexity of the polygons in the path up the tree. In the worst case this can add up to $O(n)$ even if the returned face (i.e., the root of the tree) is of low complexity. However, if the complexity of each polygon along the path is constant, then the update time is only $O(\log n)$.

Convex-Polygon Intersection The main "building block" of the algorithm is the `ConvexIntersect` procedure. We first implemented a variant of the algorithm described in [8, Chapter 4]. The idea is to decompose each convex polygon into two chains — *left* and *right*, and use a sweep algorithm to sweep down these four chains, handling the various possible cases in each intersection event (this algorithm corresponds, with minor modifications, to the Shamos-Hoey method described in [21]). This algorithm turned out to be quite expensive — in each event, several cases of edge intersections have to be checked, although some of them cannot appear more than twice.

Therefore, we implemented a different algorithm that sweeps the two left chains and the two right chains separately, and then sweeps through the resulting left and right chains to find the top and bottom vertices of the intersection polygon. These sweeps are very simple, and can be implemented efficiently. For two polygons with 40 vertices, whose intersection contains 80 vertices, the new algorithm runs 4 times faster than the original one.

Optimization In addition to the improved intersection algorithm described above, we applied several other optimization techniques. The most important of these was the use of *floating-point filters* (see Section 4.2), which reduced the running time by a factor of 2. Another method we used to accelerate the intersection function was the use of a hashing scheme to avoid re-computation of intersections already computed (see Section 4.3). However, this did not yield a significant speedup — for an input of 1000 random lines and a hash table with 60000 entries, we got an improvement of only 10-15 percent.

2.2 Algorithm 2: Dual Approach

The algorithm is based on the duality between the problems of computing the intersection of halfplanes and calculating the convex hull of a set of points. The duality transform [9] maps lines (points) in the primal plane to points (lines) in the dual plane, preserving above/below relations — for example, a primal point above a primal line is mapped to a dual line which is above the line's dual point.

Given a set $\mathcal{L}$ of lines in the plane and a point p, we can find the intersection of halfplanes induced by the lines and containing p by first transforming $\mathcal{L}$ to a set of points P in the dual plane, and transforming p to a dual line ℓ. Next, we need to partition the set P into two subsets according to whether a point lies above or below ℓ. We then compute the convex hull of each of the two subsets. Now, by spending an additional linear amount of work (specifically, finding the tangents connecting the two convex hulls) we obtain a list of points whose original primal lines make up the boundary of the respective intersection of halfplanes and in the desired order. For more details on the duality between convex hulls and halfplane intersection see, e.g., [8, Section 11.4].

Our algorithm is based on a modification by Andrew [2] of Graham's scan algorithm [14]. The first step of this algorithm calls for sorting the set of points, and then the actual convex hull is computed in an $O(n)$ scan. Since our set of points is fixed, we can pre-sort them and, as the traversal progresses, shift points from one subset to the other without destroying the implied order – all we have to do is keep a tag with each point indicating the subset it currently belongs to. When moving from a face to one of its neighbors through an edge, a single dual point (whose primal line contains the crossed edge) must be moved from one subset to the other, and the convex hulls should be recomputed. Thus, after an initial $O(n \log n)$ work for sorting the points, the algorithm reports each face of the zone in $O(n)$ time. A few simple observations enable us to effectively reduce the needed amount of work in many cases. Due to lack of space, we will not give a detailed description of the algorithm in this paper.

2.3 Algorithm 3: Binary Plane Partition

As explained in Section 1, a straightforward solution to the zone problem would be to construct the arrangement $\mathcal{A}(\mathcal{L})$, and then explore the zone of γ, face by face. However, constructing $\mathcal{A}(\mathcal{L})$ takes $\Theta(n^2)$ time (and space), whereas the actual zone may be of considerably lower complexity. In this section we shall describe an algorithm that maintains a partial arrangement, namely a subset of the faces induced by $\mathcal{L}$. The idea would be to construct only the parts of $\mathcal{A}(\mathcal{L})$ that are intersected by γ, adding faces to the partial arrangement on-line, as we advance along the curve. The algorithm we have developed is a variant of the randomized *binary plane partition* (BPP) technique. The partial arrangement $\mathcal{A}^*(\mathcal{L})$ is represented in a data structure that is a combination of a *doubly connected edge list* (DCEL) and the BPP's binary search tree, both of which are described in [8]. We call this data structure a *face tree*.

The Face Tree Data Structure The face tree is a binary search tree whose nodes correspond to faces in the plane. Each face is split into two sub-faces, using one of the input lines that intersect it. We assign an arbitrary orientation to each line in $\mathcal{L}$. The sub-face that is to the left of the splitting line is set as the node's left child, and the right sub-face as its right child.

Formally, denote by $f(u)$ the face that corresponds to node u. The set of *inner segments* of u is the intersection of the lines $\mathcal{L}$ with the face $f(u)$: $I(u) = \{ \ell \cap f(u) \mid \ell \in \mathcal{L}, \ \ell \cap f(u) \neq \emptyset \}$; the set $I(u)$ is often referred to as the *conflict list* of $f(u)$. The face $f(u)$ is partitioned into two sub-faces using one of these inner segments, which we shall call the *splitter* of $f(u)$, and is denoted $s(u)$. We say that a face is *final* if $I(u) = \emptyset$, in which case it always corresponds to a leaf in the face tree. Locating a point p in a tree whose depth is d takes $O(d)$ time, using a series of calls to the SideOfLine predicate, i.e., simple queries of the form – "is p to the left of $s(u)$?".

As we have already mentioned, the faces are not only part of a binary search tree – they also form a DCEL. Each face is described by a doubly connected cyclic linked list of its edges (or half-edges, as they are called in [8]). For each half-edge in the DCEL, we maintain its *twin half-edge* and a pointer to its *incident face*, as in a standard DCEL. These pointers enable us to follow γ efficiently along adjacent faces, which is useful when γ revisits faces it has already traversed before.

Let u be a leaf corresponding to a non-final face in $\mathcal{A}^*(\mathcal{L})$, with n_u inner segments. In order to split the face $f(u)$, we first choose a splitter $s(u)$ from $I(u)$ at random; we then prepare the lists that describe the edges of the two sub-faces in constant time; and, finally, we prepare the inner segments list of each sub-face. Thus, splitting a face takes a total of $O(n_u)$ time.

The General Scheme The algorithm starts by initializing the root r of the face tree as the bounding box B, and its set of inner segments $I(r)$ as $\{ \ell \cap B \mid \ell \in \mathcal{L} \}$. Given the first point v_0 of the polyline, we locate it in the face tree. Obviously, we will find it in $f(r)$. Since r is not a final leaf, we split it, and continue recursively in the sub-face that contains v_0, as illustrated in Figure 2. The search ends when we reach a final leaf, that is, a face f_0 that is not intersected by any line in $\mathcal{L}$.

Suppose we are now given the next vertex v_1 along γ. If $\overline{v_0 v_1}$ exits f_0 at point v_e on edge e_e (see Figure 3), then we skip to the opposite face using e_e's twin-edge pointer – e_t, and locate v_e in it (actually, we locate a point infinitesimally

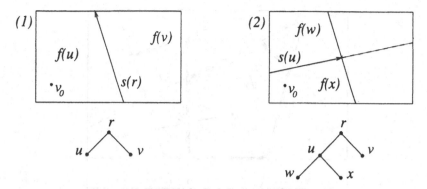

Fig. 2. An illustration of the construction of the face tree given the first point v_0 of the polyline: first, the root face $f(r)$ is set as the bounding-box; then, $f(r)$ is split into two sub-faces – $f(u)$, which contains v_0, and $f(v)$ (*1*); similarly, $f(u)$ is split into two sub-faces – $f(w)$ and $f(x)$ (*2*); and so on until no input line crosses the face f_0 which contains v_0

close to v_e and into the face) – as long as we are in an internal node of the face tree, we simply check on which side of the corresponding splitter lies v_e, and continue the search there; once we reach a leaf f', we repeat the same splitting algorithm as described above, until we find the face f_1 that contains v_e (we now update the twin-edge pointers of both e_e and $e_t \cap f_1$, so that next time γ moves from f_0 to f_1, or vice versa, we could follow it in $O(1)$ time).

Improved Maintenance of the Conflict Lists The main shortcoming of the algorithm we have just described is its worst-case behavior of $\Theta(n^2)$ time for preparing a face with $\Theta(n)$ edges, for example input 5 in Section 3. To overcome this shortcoming, we developed a data structure that enables us to split a face without having to check all its inner segments. Instead of maintaining the inner segments in a simple array, we distribute them among the vertices of the face in *conflict lists*. Denote by p the point that is being located (p is either the first vertex of the polyline, or its exit point from the previous face). An inner segment s in face f is said to be in *conflict* with the vertex v if v and p lie on different sides of s. According to our new approach, we maintain a list of inner segments in each vertex, so that all the segments that are held in the list of a vertex v are in conflict with v. Each inner segment is kept in only one list, at one of the vertices that it is in conflict with.

In the first stage of the algorithm, we wish to locate the first vertex v_0 of the polyline. To this end, we first prepare the conflict lists of the bounding box by simply checking for each input line with which vertices it is in conflict, and adding the line to one of the corresponding conflict lists at random. We then split the bounding box using a randomly chosen splitter, and continue recursively until we reach a final face. It remains to show how the conflict lists can be updated when the face is split. Let f be a face that contains v_0, and denote by s its splitter. s divides f into two sub-faces — f_g, or the *green face*, that contains v_0, and f_r, the *red face*. Let u_1 and u_2 be the endpoints of s. In order to construct

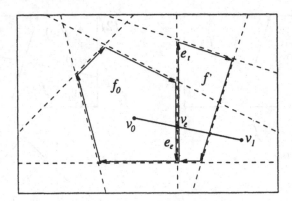

Fig. 3. Following γ along $\mathcal{A}^*(\mathcal{L})$ after the first face (f_0) has been built: first, we find the exit point v_e (on edge e_e) and skip to the adjacent face f' using the twin edge e_t; then, from f' we build the face f_1 through which γ passes using the same recursive procedure we used for building f_0 from $f(r)$ (as shown in Figure 2)

the final face that contains v_0, we need to work (i.e., recursively split) only on the green face. We therefore update its conflict lists — for each segment in one of f_r's conflict lists, we add it to the conflict list of u_1 (if it is in conflict only with u_1), u_2 (similarly), one of u_1 and u_2 at random (if it is in conflict with both vertices), or none (in case it is in conflict with neither u_1 nor u_2, which means that it does not intersect f_g). This algorithm guarantees that the face that contains v_0 is constructed in expected time $O(n \log n)$ (see [19]).

The rest of the algorithm is very similar to the original scheme – given the next vertex along the polyline, we first locate the exit point from the current face, and construct the next face γ intersects. However, the conflict lists in the red faces need to be updated, since the definition of a conflict depends on the point being located, and it has changed. Furthermore, some of the inner segments might be missing, and we need to collect them from the relevant green face, i.e., the sibling of the red face. The solution we implemented is to gather all the inner segments from the current face and from the sub-tree of the sibling node, and prepare new conflict lists, as we have done for the bounding box. Once the conflict lists are built, we continue as explained above.

This new algorithm, which we will denote 3b, gave a substantial performance improvement, especially for the cases of complex faces (see Section 3). Both algorithm 3 and 3b use floating-point filters to avoid performing exact computations when possible.

2.4   Algorithm 4: Randomized Incremental Construction

Recently, Har-Peled [17] gave an $O((n + k)\alpha(n) \log n)$ expected time algorithm for the on-line construction of the zone, where k is the number of intersections between γ and the input lines. This improves by almost a logarithmic factor the application of Overmars and van Leeuwen [20] for this restricted case [3].

The algorithm of [17] relies on a careful simulation of an off-line randomized incremental algorithm (which uses an oracle) that constructs the zone. We had implemented a somewhat simpler variant, which is similar to algorithm 3 (Section 2.3), with the difference that we split complex faces so that every internal node in the face tree T will hold a polygon of constant complexity. For a node v, let R_v be the polygonal region that corresponds to it (this may be a whole face in the arrangement, as in algorithm 3, or part of a face), and let $I(v)$ be its *conflict list*, i.e., the list of input lines that intersect R_v.

Computing a leaf of T which contains a given point p is performed by carrying out a point-location query, as in algorithm 3. We divert from this algorithm, in the following way: if a face R_v created by the algorithm has more than c vertices (where c is an arbitrary constant), we split it into two regions, by a segment s that connects two of its vertices (chosen arbitrarily), thus generating two new children in the tree T, such that the complexity of each of their polygons is at most $\lceil c/2 \rceil + 1$ (note that, in general, s does not lie on one of the lines in $\mathcal{L}$). Now, a node in T corresponds to a polygon having at most c vertices (a similar idea was investigated in [16]). This seems to reduce the expected depth of the tree, which we conjecture to be logarithmic.

To compute a face, we first compute a leaf v of T that contains our current point p, as described above. We next compute all the leaves that correspond to the face that contains p, by performing a point-location query in the middle of a splitter lying on the boundary of R_v. By performing a sequence of such queries, we can compute all the leaves in T that correspond to the face that contains p. Reconstructing the whole face from those leaves is straightforward.

The walk itself is carried out by computing for each face the point where the walk leaves the face, and performing a point-location query in T for this exit point, as in algorithm 3.

Computing the Conflict Lists Using Less Geometry When splitting a region R_v into two regions R_{v+}, R_{v-}, we have to compute the conflict lists of R_{v+}, R_{v-}. For a line $\ell \in I(R_v)$, this can be done by computing the intersection between ℓ and R_{v+}, R_{v-}, but this is rather expensive. Instead, we do the following: for each line in the conflict list of R_v, we keep the indices of the two edges of ∂R_v that ℓ intersects. As we split R_v into R_{v+} and R_{v-}, we compute for each edge of R_v the edges of R_v^+, R_v^- it is being mapped to. Thus, if the line ℓ does not intersect the two edges of R_v that are crossed by the splitting line, then we can decide, by merely inspecting this mapping between indices, whether ℓ intersects either R_{v+}, R_{v-}, or both, and what edges of R_v^+, R_v^- are being intersected by ℓ. However, there are situations where this mapping mechanism is insufficient. In such cases, we use the geometric predicate SideOfLine, which determines on which side of a line a given point lies, as illustrated in Figure 4. While there is a non-negligible overhead in computing the mapping, the above technique reduces the number of calls to the SideOfLine predicate by a factor of two (the predicate SideOfLine is an expensive operation when using exact arithmetic, even if floating-point filtering is applied).

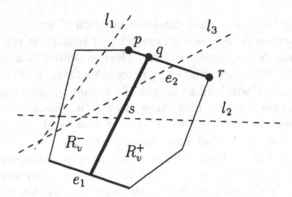

Fig. 4. Computing the conflict lists of R_v^+, R_v^-: The lines l_1, l_2 can be classified by inspecting their edge indices and the indices mapping. Classifying l_3 is done by inspecting its edge indices and computing SideOfLine of the points p, q (or q, r) relative to l_3.

Optimization In addition to caching results of computations (see Section 4.2), massive filtering was performed throughout the program. The lines are represented both in rational and floating-point representations. The points are represented as two pointers to the lines whose intersection forms the point. Whenever the result of a geometric predicate is unreliable (i.e., the floating-point result lies below a certain threshold) the exact representation of the geometric entities involved are calculated, and the predicate is recomputed using exact arithmetic. This filtering results in that almost all computations are carried out using floating-point arithmetic, which gives a speedup by a factor of two with respect to the standard floating-point filters.

3   Experiments

In order to test the programs that implement the four algorithms we described in the previous section, we have created 5 input files (see Figure 5). In inputs 1 and 2, the lines form a grid; inputs 3 and 4 consist of random lines; and in input 5 there is a very complex face to which each input line contributes one edge. The polyline in inputs 1, 3, and 5 is a segment; in inputs 2 and 4, it consists of 50 and 35 segments respectively. We ran the programs on the test suite and measured average running times, based on 25 executions, for $n = 2000$, namely each input file contained 2000 lines (see Table 1).

In general, the best results were achieved by algorithms 3b and 4, whose running times were considerably faster than those of algorithms 1 and 2. It is important to note that each algorithm was implemented by a different programmer, applying different optimization techniques, as will be discussed in Section 4. Comparing the performance of the programs might therefore be misleading with regard to the algorithms' potential performance. Nevertheless, some conclusions can be drawn from our experiments.

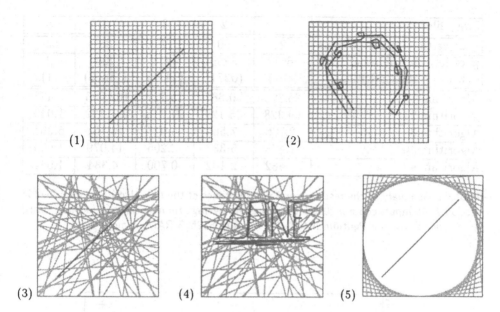

Fig. 5. The five inputs whose details, as well as the running times of the algorithms on them, are given in Table 1. The figures show the input lines (*bright lines*) and the polyline (*darker*). The arrangements are depicted here with only 50 lines, for reasons of clarity

Algorithm 2 is very slow overall, but performs quite well for the first face, as input 5 demonstrates. Perhaps the reason is that most of the work it performs in order to compute the first face is done by the function that sorts the dual points, and this function is relatively fast. As mentioned earlier, the improved maintenance of the conflict lists reduced the running time of algorithm 3b, compared to algorithm 3, on input 5. Interestingly, it also performs better on the other inputs. Algorithm 4 gave similar results to algorithm 3b, but after massive optimization (Section 2.4) it surpassed it by a factor of up to 3.

Figure 6 shows the performance of algorithm 3b on input 4, for various n's. Empirically, it seems that the algorithm computes the first face in an arrangement of random lines in linear time, and reports each of the following faces in $O(\log n)$ time on average. Similar results were obtained for input 2 (the grid).

4 Discussion

In this section, we summarize the major conclusions drawn from our experience, especially with regard to optimization techniques.

4.1 Software Design

Motivated by the structure of the CGAL basic library [6], in which geometric predicates are separated from the algorithms (using the traits mechanism [12]),

Input file	1	2	3	4	5
# of vertices in the polyline	2	50	2	35	2
# of faces in the zone (without multiplicities)	667 (667)	6289 (6271)	1411 (1411)	12044 (11803)	2 (1)
Algorithm 1	25.896	50.687	15.217	102.905	3.436
Algorithm 2	69.328	85.128	27.715	232.122	1.049
Algorithm 3	2.706	7.565	6.017	23.161	5.384
Algorithm 3b	1.478	5.338	2.205	14.079	1.036
Algorithm 4	0.482	2.102	0.700	4.334	1.047

Table 1. Summary of the test data and running times of the four algorithms described in Section 2. All inputs contain 2000 lines. Times are average running times in seconds, based on 25 executions on a Pentium-II 450MHz PC with 512MB RAM memory

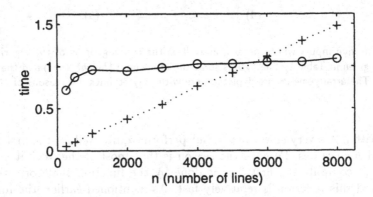

Fig. 6. Running times of algorithm 3b on input 4 — a random arrangement, for various n's. The figure shows average times, based on 25 executions, for computing the first face (*crosses*, in seconds) and the rest of the faces (*circles*, in milliseconds)

we restricted all geometric computations in the programs to a small set of geometric predicates. This enabled us to debug and profile the programs easily, and deploy caching, filtering, and exact arithmetic effortlessly. Writing such predicates is not always trivial, but fortunately LEDA [18] and CGAL [11] provide such implementations.

4.2 Number Type and Filtering

To get good performance, one would like to use the standard floating-point arithmetic. However, this is infeasible in geometric computing, where exact results are required, due to precision limitations. Since our input is composed of lines and vertices in rational coordinates, we can perform all computations exactly using LEDA rational type. However, LEDA rational suffers from several drawbacks: (i) computations are slow (up to 20-40 times slower than floating point), (ii) they consume a lot of memory, and (iii) the bit length of the representation of the numbers doubles with each operation, which in turn causes a noticeable slowdown in program execution time.

One possible approach to improve the efficiency of the representation of a LEDA rational number is to normalize the number explicitly. However, this operation is expensive, and the decision where to do such normalization is not straightforward. A different approach is to use filtering [13]. In general, filtering is a method of carrying out the computations using floating-point (i.e., fast and inexact) arithmetic, and performing the computations using exact arithmetic only when necessary. LEDA [5] provides a real number type, which facilitates such filtered computations (it is not restricted to rational operators or geometric computations). Additionally, LEDA provides a fine-tuned computational geometry kernel that performs filtering. Algorithms 1, 3, and 3b used LEDA's filtered predicates (e.g., `SideOfLine`), which resulted in a speedup by a factor of two or more, compared to the non-filtered rational computations. Furthermore, one can also implement the filtering directly on the geometric representation of the lines and points (for example, computing the exact coordinates of the points only when necessary), as was done in algorithm 4 (see Section 2.4). As mentioned earlier, this technique saves a considerable amount of exact computations, and gives an additional speedup of two. Overall, usage of filtering resulted in the most drastic improvement in the running times of the programs, and was easy to implement.

4.3   Caching

One possible approach to avoid repeated exact computations is to cache results of geometric computations (i.e., intersection of lines, `SideOfLine` predicate, etc.). Such a caching scheme is easy to implement, but does not result in a substantial improvement. Especially, as the filtered computations might require less time than the lookup time. Caching was implemented in algorithms 1 and 4.

4.4   Geometry

The average number of vertices of a face in an arrangement of lines is about 4 (this follows directly from Euler's formula). Classical algorithms rely on vertical decomposition, which have the drawback of splitting all the faces of the arrangement into vertical trapezoids. An alternative approach is to use constant complexity convex polygons instead of vertical trapezoids (as was done in algorithm 4). The results indicate that this approach is simple and efficient. This idea was suggested by Matoušek, and was also tested out in [16].

4.5   Miscellaneous

Additional improvement in running time can be achieved by tailoring predicates and functions to special cases, instead of using ready-made library ones.

For example, knowing that two segments intersect at a point, liberates us from the necessity to check whether they overlap. Such techniques were used in algorithms 1, 3, and 3b. Significant improvement can also be accomplished by improving "classical" algorithms for performing standard operations (see, for example, Section 2.1).

5   Conclusions

We have presented four algorithms for on-line zone construction in arrangements of lines in the plane. All algorithms were implemented, and we have also presented experimental results and comparisons between the algorithms.

A major question raised by our work is what could be said theoretically about the on-line zone construction problem. As mentioned in the Introduction, [3] proposes a near-optimal output-sensitive algorithm based on the Overmars-van Leeuwen data structure for dynamic maintenance of halfplane intersection. It would be interesting to implement this data structure and compare it with the algorithms that we have implemented.

If the number of faces in the zone of the curve γ is small (constant), then algorithm 3b described in Section 2.3 is guaranteed to run in expected time $O(n \log n)$, since in that case it is an almost verbatim adaptation of the randomized incremental algorithm for constructing the intersection of halfplanes, whose running time analysis is given in [19, Section 3.2]. Indeed it performs very well on input 5 (Section 3). An interesting open problem is to extend the analysis of [19] for the case of a single face to the case of the on-line zone construction.

Algorithm 4 is an attempt to implement a practical variant of a theoretical result giving a near-optimal output-sensitive algorithm for the on-line zone construction [17] (the result in [17] was motivated by the good results of algorithm 3 and differs from it only slightly but, as mentioned above, in a crucial factor — by subdividing large faces into constant size faces). Still, the practical variant which was implemented has no theoretical guarantee. Can the analysis of [17] be extended to explain the performance of algorithm 4 as well?

Finally, as mentioned earlier, the algorithms have been implemented independently, and hence there are many factors that influence their running time on the test suite beyond the fundamental algorithmic differences. We are currently considering alternative measures (such as the number of basic operations) that will allow for better comparison of algorithm performance.

References

1. Y. Aharoni. Computing the area bisectors of polygonal sets: An implementation. In preparation, 1999.
2. A. M. Andrew. Another efficient algorithm for convex hulls in two dimensions. *Information Processing Letters*, 9:216–219, 1979.
3. K.-F. Böhringer, B. Donald, and D. Halperin. The area bisectors of a polygon and force equilibria in programmable vector fields. In *Proc. 13th Annu. ACM Sympos. Comput. Geom.*, pages 457–459, 1997. To appear in Disc. and Comput. Geom.

4. K.-F. Böhringer, B. R. Donald, and N. C. MacDonald. Upper and lower bounds for programmable vector fields with applications to MEMS and vibratory plate parts feeders. In J.-P. Laumond and M. Overmars, editors, *Robotics Motion and Manipulation*, pages 255–276. A.K. Peters, 1996.

5. C. Burnikel, K. Mehlhorn, and S. Schirra. The LEDA class real number. Technical Report MPI-I-96-1-001, Max-Planck Institut Inform., Saarbrücken, Germany, Jan. 1996.

6. *The CGAL User Manual, Version 1.2*, 1998.

7. B. Chazelle, H. Edelsbrunner, L. J. Guibas, M. Sharir, and J. Snoeyink. Computing a face in an arrangement of line segments and related problems. *SIAM J. Comput.*, 22:1286–1302, 1993.

8. M. de Berg, M. van Kreveld, M. Overmars, and O. Schwarzkopf. *Computational Geometry: Algorithms and Applications*. Springer-Verlag, Berlin, 1997.

9. H. Edelsbrunner. *Algorithms in Combinatorial Geometry*, volume 10 of *EATCS Monographs on Theoretical Computer Science*. Springer Verlag, Heidelberg, Germany, 1987.

10. H. Edelsbrunner, L. J. Guibas, J. Pach, R. Pollack, R. Seidel, and M. Sharir. Arrangements of curves in the plane: Topology, combinatorics, and algorithms. *Theoret. Comput. Sci.*, 92:319–336, 1992.

11. A. Fabri, G. Giezeman, L. Kettner, S. Schirra, and S. Schönherr. The CGAL kernel: A basis for geometric computation. In M. C. Lin and D. Manocha, editors, *Proc. 1st ACM Workshop on Appl. Comput. Geom.*, volume 1148 of *Lecture Notes Comput. Sci.*, pages 191–202. Springer-Verlag, 1996.

12. A. Fabri, G. Giezeman, L. Kettner, S. Schirra, and S. Schönherr. On the design of CGAL, the Computational Geometry Algorithms Library. Technical Report MPI-I-98-1-007, Max-Planck-Institut Inform., 1998.

13. S. Fortune and C. J. van Wyk. Static analysis yields efficient exact integer arithmetic for computational geometry. *ACM Trans. Graph.*, 15(3):223–248, July 1996.

14. R. L. Graham. An efficient algorithm for determining the convex hull of a set of points in the plane. *Information Processing Letters*, 1:132–133, 1972.

15. D. Halperin. Arrangements. In J. E. Goodman and J. O'Rourke, editors, *Handbook of Discrete and Computational Geometry*, chapter 21, pages 389–412. CRC Press LLC, 1997.

16. S. Har-Peled. Constructing cuttings in theory and practice. In *Proc. 14th Annu. ACM Sympos. Comput. Geom.*, pages 327–336, 1998.

17. S. Har-Peled. Taking a walk in a planar arrangement. Manuscript, http://www.math.tau.ac.il/~sariel/papers/98/walk.html, 1999.

18. K. Mehlhorn and S. Näher. *LEDA: A Platform for Combinatorial and Geometric Computing*. Cambridge University Press, New York, 1999. To appear.

19. K. Mulmuley. *Computational Geometry: An Introduction Through Randomized Algorithms*. Prentice Hall, Englewood Cliffs, NJ, 1994.

20. M. H. Overmars and J. van Leeuwen. Maintenance of configurations in the plane. *J. Comput. Syst. Sci.*, 23:166–204, 1981.

21. F. P. Preparata and M. I. Shamos. *Computational Geometry: An Introduction*. Springer-Verlag, New York, NY, 1985.

22. M. Sharir and P. K. Agarwal. *Davenport-Schinzel Sequences and Their Geometric Applications*. Cambridge University Press, New York, 1995.

The Design and Implementation of Planar Maps in CGAL*

Eyal Flato, Dan Halperin, Iddo Hanniel, and Oren Nechushtan

Department of Computer Science
Tel Aviv University, Tel Aviv 69978, Israel
{flato, halperin, hanniel, theoren}@math.tau.ac.il

Abstract. Planar maps are fundamental structures in computational geometry. They are used to represent the subdivision of the plane into regions and have numerous applications. We describe the planar map package of CGAL[1] — the Computational Geometry Algorithms Library. We discuss problems that arose in the design and implementation of the package and report the solutions we have found for them. In particular we introduce the two main classes of the design—*planar maps* and *topological maps* that enable the convenient separation between geometry and topology. We also describe the *geometric traits* which make our package flexible by enabling to use it with any family of curves as long as the user supplies a small set of operations for the family. Finally, we present the algorithms we implemented for point location in the map, together with experimental results that compare their performance.

1 Design Overview

We describe the design and implementation of a data structure for representing planar maps. The data structure supports traversal over faces, edges and vertices of the map, traversal over a face and around a vertex, and efficient point location. Our design was guided by several goals, among them: (i) ease-of-use, for example, easy insertion of new curves into the map; (ii) flexibility, for example, the user can define his/her own special curves as long as they support a predefined interface; (iii) efficiency, for example, fast point location.

Our representation is based on the *Doubly Connected Edge List* (DCEL) structure [3, Chapter 2]. This representation belongs to a family of edge-based data structures in which each edge is represented as a pair of opposite *halfedges* [6, 14]. Our representation supports inner components (holes) inside the faces

* This work has been supported in part by ESPRIT IV LTR Projects No. 21957 (CGAL) and No. 28155 (GALIA), by The Israel Science Foundation founded by the Israel Academy of Sciences and Humanities, by a Franco-Israeli reserach grant (monitored by AFIRST/France and The Israeli Ministry of Science), and by the Hermann Minkowski – Minerva Center for Geometry at Tel Aviv University. Dan Halperin has been also supported in part by an Alon fellowship and by the USA-Israel Binational Science Foundation.
[1] http://www.cs.uu.nl/CGAL/

J.S. Vitter and C.D. Zaroliagis (Eds.): WAE'99, LNCS 1668, pp. 154–168, 1999.
© Springer-Verlag Berlin Heidelberg 1999

making it more general and suitable for a wide range of applications (e.g., geographical information systems). A planar map is not restricted to line segments, and our package can be used for any collection of bounded x-monotone curves (including vertical segments).

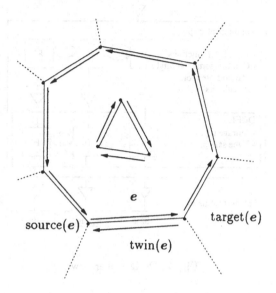

Fig. 1. Source and target vertices, and twin halfedges in a face with a hole

The package is composed of two main classes: `CGAL_Topological_map` which is described in Section 2.1 and `CGAL_Planar_map_2` which is described in Section 2.2.

The design, depicted in Figure 2, follows CGAL's polyhedron design introduced in [6]. The bottom layer holds base classes for vertices, halfedges and faces. Their responsibilities are the actual storage of the incidences, the geometry and other attributes. Addition of attributes (such as color) by the user is easily done by defining one's own bottom-layer classes (possibly by deriving from the given base classes and adding the attributes).

The next layer, the `Dcel` layer, is a *container*[2] that stores the classes of the bottom layer and adds functionality for manipulating between them and traversing over them.

The topological map layer adds high-level functions, high-level concepts[3] for accessing the items, i.e., handles, iterators and circulators (unlike the `Dcel` layer

---

[2] We use the term container as it is used in the C++ Standard Template Library, STL [11, 13], i.e., a class that stores a collection of other classes.

[3] By *concepts* we mean classes obeying a set of predefined requirements, which can therefore be used in generic algorithms. For example, our iterators obey the STL (Standard Template Library [11, 13]) requirements and can therefore be used in STL's generic algorithms.

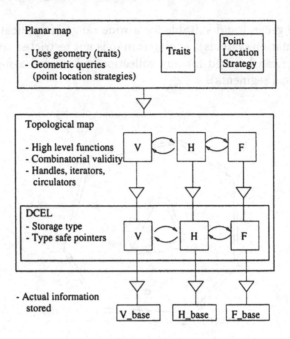

Fig. 2. Design overview

pointers are no longer visible at this interface), and protection of combinatorial validity.

The top layer, the planar map layer adds geometry to the topological map layer using a `Traits` template parameter (Section 3). We provide high level functionality based on the geometric properties of the objects. Geometric queries — point location and vertical ray shooting — are also introduced in this class. We provide three implementations, and a mechanism (the so-called *strategy pattern* [5]) for the users to implement their own point location algorithm. The abstract strategy class `CGAL_Pm_point_location_base <Planar_map>` is a pure virtual class declaring the interface between the algorithm and the planar map. The planar map keeps a reference to the strategy. The *concrete* strategy is derived from the abstract class and implements the algorithm interface.

We have derived three concrete strategies: `CGAL_Pm_default_point_location`, `CGAL_Pm_naive_point_location`, and `CGAL_Pm_walk_point_location` which is an improvement over the naive one. Each strategy is preferable in different situations. The *default* class implements the dynamic incremental randomized algorithm introduced by Mulmuley [9]. The *naive* algorithm goes over all the vertices and halfedges of the planar map. Namely, in order to find the answer to an upward vertical ray shooting query, we go over all the edges in the map and find the one that is above the query point and has the smallest vertical distance from it. Therefore, the time complexity of a query with the naive class is linear in the complexity of the planar map. The *walk* algorithm implements a walk over the zone in the planar map of a vertical ray emanating from the query point. This decreases the number of edges visited during the query, thus improving the time complexity. The main trade-off between the default strategy and the two other strategies, is between time and storage. The naive and walk algorithms need more time but almost no additional storage. Section 4 describes the point location classes.

2   Planar Maps and Topological Maps

A planar map has a combinatorial (or topological) component and a geometric one. The combinatorial component consists of what we refer to as combinatorial objects — vertices, halfedges and faces, and the functionality over them, for example — traversal of halfedges around a face, traversal of halfedges around a vertex or finding the neighboring face. The geometric component consists of geometric information like point coordinates and curve equations, and geometric functionality such as point location. We carry out this separation in our design with two classes — CGAL_Topological_map and CGAL_Planar_map_2 which is derived from it.

2.1   Topological Map

The topological map class is meant to be used as a base for geometric subdivisions (as we have done for 2D planar maps). It consists of vertices, edges and faces and an incidence relation on them, where each edge is represented by two halfedges with opposite orientations. A face of the topological map is defined by the circular sequences of halfedges along its inner and outer boundaries. The presence of a containment relationship between a face and its inner holes, is a topological characteristic which distinguished it from standard graph structures, and from other edge based combinatorial structures. This enables us to derive subdivisions with holes in them.

CGAL_Topological_map can be used as a base class for deriving different types of geometric subdivisions. We also regard it as provisory for implementing three-dimensional subdivisions induced by algebraic surfaces, for example, a two-dimensional map on a sphere or a polyhedral terrain. It can also be used almost as it is as a representation class for polygons with holes, by merely adding point coordinates as additional attributes to the vertices.

The following simple function combinatorial_triangle() demonstrates the use of CGAL_Topological_map. It creates an empty map (with one face corresponding to the unbounded face) and then inserts an edge e1 inside the unbounded face. It then inserts an edge e2 from the target vertex of e1 and finally inserts an edge between the target vertices of e2 and e1->twin(), closing a "combinatorial" triangle (i.e., a closed cycle of three vertices without coordinates).

```
typedef CGAL_Pm_dcel<CGAL_Tpm_vertex_base, CGAL_Tpm_halfedge_base,
                                      CGAL_Tpm_face_base> Dcel;
typedef CGAL_Topological_map<Dcel> Tpm;

void combinatorial_triangle() {
  Tpm t;
  Tpm::Face_handle uf=t.unbounded_face();
  Tpm::Halfedge_handle e1 = t.insert_in_face_interior(uf);
  Tpm::Halfedge_handle e2 = t.insert_from_vertex(e1);
  t.insert_at_vertices(e2,e1->twin());
}
```

Addition of attributes such as point coordinates is easy. The following example demonstrates how to add an attribute (in this case some Point type) to a vertex of a map. It creates a new vertex type My_vertex that derives from CGAL_Tpm_vertex_base and adds the attribute. The new vertex is then passed as a template parameter to the Dcel. After the insertion of the new edge the information in its incident vertices can be updated by the user. This can be used, for example, as a representation class for polygons with holes in them.

```
struct My_vertex : public CGAL_Tpm_vertex_base {
    Point pt;
}
typedef CGAL_Pm_dcel<My_vertex, CGAL_Tpm_halfedge_base,
                                     CGAL_Tpm_face_base> Dcel;
typedef CGAL_Topological_map<Dcel> Tpm;

void insert_with_info() {
  Tpm t;
  Tpm::Face_handle uf=t.unbounded_face();
  Tpm::Halfedge_handle e1 = t.insert_in_face_interior(uf);
  e1->source()->pt = Point(0,0);
  e1->target()->pt = Point(1,1);
}
```

2.2   Planar Map

The CGAL_Planar_map_2 class is derived from CGAL_Topological_map. It represents an embedding of a topological map T in the plane such that each edge of T is embedded as a bounded x-monotone curve and each vertex of T is embedded as a planar point. In this embedding no pair of edges intersect except at their endpoints[4].

CGAL_Planar_map_2 adds geometric information to the topological map it is derived from. The Traits template parameter (Section 3) defines the geometric

---

[4] We are currently implementing an *arrangement* layer on top of the planar map layer, in which the intersection of curves is supported, and in which curves need not be x-monotone.

types for the planar map and the basic geometric functions on them. The planar map implements its algorithms in terms of these basic functions.

The additional geometric information enables geometric queries, and an easier interface that uses the geometry, e.g., the users can insert a curve into the map without specifying where to insert it, and the map finds this geometrically. The modifying functions of the topological map (e.g., `remove_edge`) are overridden and are implemented using the geometric information. We have also overridden the combinatorial insertion functions (e.g., `insert_from_vertex`) so they use the geometric information. If the users have some combinatorial information (e.g., that one of the endpoints of the new curve is incident on one of the already existing vertices of the planar map) they can use the specialized insertion function instead of the more general one thus making the operation more efficient.

The following function `geometric_triangle()` creates an empty map and then inserts three segments into it, corresponding to a triangle. It resembles the example of the previous section and shows the difference between the interface of the topological map and the planar map.

```
typedef CGAL_Homogeneous<long> Rep;
typedef CGAL_Pm_segment_exact_traits<Rep> Traits;
typedef CGAL_Pm_default_dcel<Traits> Dcel;
typedef CGAL_Planar_map_2<Dcel,Traits> Pm;

void geometric_triangle() {
    Pm p;
    Traits::Point p1(0,0), p2(1,0), p3(0,1);
    Traits::X_curve cv1(p1,p2), cv2(p2,p3), cv3(p3,p1);
    p.insert(cv1);
    p.insert(cv2);
    p.insert(cv3);
}
```

2.3   Problems and Solutions

The separation of the design into a topological layer and a geometric layer creates some algorithmic problems. When designing a topological class such as the topological map, no geometric considerations can be used. This imposes constraints on the functions and function parameters of the topological map. For some functions (e.g., the `split_edge` function), the topological information suffices, while for others additional topological information is needed to avoid ambiguity. In the geometric layer this information can be deduced from the geometric information. In the remainder of this section we present examples of such algorithmic problems and their solutions in our design.

The Use of Previous Halfedges When inserting a new edge from a vertex v using the function `insert_from_vertex`, passing only v to the function will cause

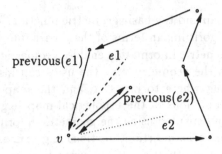

Fig. 3. Previous halfedges are necessary: Passing only v as a parameter to the insertion function cannot resolve whether $e1$ or $e2$ is the edge to be inserted into the map

ambiguity. Figure 3 shows an example of two possible edges that can be inserted if we had only passed the vertex. Topologically, what defines the insertion from a vertex uniquely is the *previous* halfedge to the edge inserted. Therefore, this is what is passed in our topological function. In the geometric layer, passing the vertex v is sufficient, since we can find the previous halfedge geometrically: we go over the halfedges around v and find geometrically the halfedge which is the first halfedge incident to v that is encountered when proceeding from the new inserted curve in counterclockwise order.

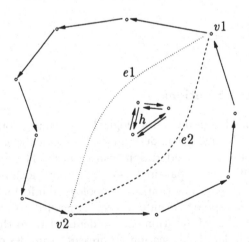

Fig. 4. The need for the *move_hole* function: After inserting a new edge between $v1$ and $v2$ the original face is split into two faces. Since we have no knowledge of the geometry of the curves in the map, we cannot define topologically which of the two faces contains the hole h

The Use of the *move_hole* **Function** When inserting a new edge between two vertices *v1* and *v2*, a new face might be created. Figure 4 shows an example of two possible edges that can be inserted into a topological map. If *e1* is inserted the hole *h* should be in the right face, if *e2* is inserted the hole *h* should be in the left face. The topological map has no geometric knowledge of the curves, so it cannot determine which of the two faces contains the hole. Therefore, this is done by the client (the derived class) of the topological map (e.g., the planar map), which calls the function `move_hole` if a hole needs to be moved to a new face created after insertion.

In the planar map class we go over the holes of the original face and check if they are inside the new face that was created, if so the hole is moved to the new face using the `move_hole` function. Since our planar map assumes that no curves intersect, we check whether a hole is inside a face by checking if one of its vertices, call it *v*, is inside the face. This is done using a standard algorithm. Conceptually, we shoot a ray from *v* vertically upwards and count the number of halfedges of the face that intersect it. If the number is odd the point is inside the face. In practice, we do not need to implement an intersection function (which can be expensive for some curves), we just need a function that defines if a curve is above or below a point. We then go over the halfedges of the face and use it to count how many of them are above the vertex.

3   Geometric Traits

The geometric traits class is an abstract interface of predicates and functions that wraps the access of an algorithm to the geometric (rather than combinatorial) inner representation.

In the planar map package, we tried to define the minimal geometric interface that will enable a construction and handling of a geometric map. Packing those predicates and functions under one traits class helped us achieve the following goals: flexibility in choosing the geometric representation of the objects (Homogeneous, Cartesian); flexibility in choosing the geometric kernel (LEDA[5], CGAL or a user-defined kernel); ability to have several strategies for robustness issues; extendibility to maps of objects other than line segments.

The documentation of the planar map class gives the precise requirements that every traits class should obey. We have formulated the requirements so they make as little assumptions on the curve as possible (for example, linearity of the curve is not assumed). This enables the users to define their own traits classes for different kinds of curves that they need for their applications. The only restriction is that they obey the predefined interface.

The first task of the planar map traits is to define the basic objects of the map: the point (`Point`) and the *x*-monotone curve (`X_curve`). In addition four types of predicates are needed: (i) access to the endpoints of the *x*-monotone curves; (ii) comparison predicates between points; (iii) comparisons between points and

---

[5] LEDA— Library of Efficient Data structures and Algorithms,
  http://www.mpi-sb.mpg.de/LEDA/leda.html

x-curves (e.g., whether the point is above the curve); (iv) predicates between curves (e.g., comparing the y-coordinate of two curves at a given x-coordinate). This interface of the four types of predicates satisfies the geometric needs of the planar map.

In the current version of the planar map package we implemented the following traits classes:

- CGAL_Pm_segment_exact_traits<R> — a class for planar maps of line segments that uses CGAL's kernel. The R template parameter enables the use of CGAL's homogeneous or Cartesian kernel. This class is robust when used with exact arithmetic.
- CGAL_Pm_leda_segment_exact_traits — also handles segment using exact arithmetic, but using LEDA [7] rational points and segments, the predicates become faster. One of the differences that makes this traits class more efficient is the use of LEDA's primitive predicates (e.g., orientation) that are implemented using floating point "filters" [4, 12] which speed up the use of exact computations.
- Arr_circles_real_traits — a class that introduces circular arcs as x-monotone curves. It uses LEDA *real* number type (to support robust square-root predicates) for calculation. This traits is mainly provided for arrangements of circles but can also be used with planar maps.

4   Point Location

The point location strategy enables the users to implement their own point location algorithm which will be used in the planar map. We have implemented three algorithms: (i) a naive algorithm — goes over all the edges in the map to find the location of the query point; (ii) an efficient algorithm (the default one) — Mulmuley's randomized incremental algorithm; and (iii) a "walk" algorithm that is an improvement over the naive one, and finds the point's location by walking along a line from "infinity" towards the query point. In the following sections we describe these algorithms and their implementation.

4.1   Fast Point Location

As mentioned above the default point location is based on Mulmuley's randomized, fully dynamic algorithm (see [1, 10]).

We remind the reader that our point location implementation handles general finite planar maps. The subdivision is not necessarily *monotone* (each face boundary is a union of x-monotone chains), nor connected and possibly contains holes. In addition the input may be x-degenerate.

Algorithm Our implementation supports insertions as well as deletions of map edges, while maintaining an efficient point location query time and linear storage space. This is achieved by a "lazy" approach that performs an occasional rebuilding step whenever the internal structure (the *history DAG — Directed Acyclic Graph*) passes predefined thresholds in size or in depth. The rebuilding step is an option that can be finetuned or disabled by the user.

Implementation Details Our implementation consists of two structures: An augmented *trapezoidal map* and a *search structure*. The trapezoidal map is a "uniform" collection of `X_trapezoids`,where each `X_trapezoid` corresponds to a subset of the plane bounded above and below by curves and from the sides by *vertical attachments*; see [3, Chapter 6] for more details.

Each `X_trapezoid` is one of four types: a non-degenerate `X_trapezoid`, a "curve like" `X_trapezoid`, a "point like" `X_trapezoid`, and a vertical one. The non-degenerate `X_trapezoid` corresponds to the area inside its geometric boundaries; the "curve like" `X_trapezoids` correspond to the interior of the input curves, with possibly many `X_trapezoids` corresponding to one curve; the "point like" `X_trapezoids` correspond to the endpoints of the input curves. The vertical degenerate `X_trapezoids` will be discussed later.

These four types are represented in the same way. Each `X_trapezoid` stores information regarding its geometric boundaries: *left* and *right* endpoints, *bottom* and *top* bounding curves, boundedness bit vector denoting, for example whether the left endpoint is infinite or not; its geometric neighborhood: *left bottom, left top, right bottom, right top* neighbors; and the node in the search structure that represents the `X_trapezoid`. In addition an `X_trapezoid` is either active or inactive. In the beginning we have only one active `X_trapezoid` representing the entire plane and no inactive `X_trapezoids`. As the structure thickens active `X_trapezoids` become inactive and new active ones are created. This is done while preserving the property that *the active* `X_trapezoid` *form a subdivision of the plane* (in the sense that their union covers the plane, and they do not overlap).

The active non-degenerate `X_trapezoids` at any stage are independent of the order in which the map was updated, that is, any update order generates the same decomposition. The invariant decomposition is called a *vertical decomposition* or a *trapezoidal decomposition*. The planar map induced by the decomposition is also known as a *trapezoidal map*.

In contrast, the search structure, also referred to as the *history DAG*, is dependent on the update order. The inner nodes are the "curve like", "point like" and non-active nodes while the leaves are the currently active non-degenerate and degenerate nodes.

We separate between geometric and combinatorial data, namely the user can supply his/her own search structure as a parameter when instantiating an object of the `CGAL_Trapezoidal_decomposition` class. The default is a specially designed class called `CGAL_Pm_DAG`.

x-Degeneracies There is an inherent difficulty working with vertical decomposition when the input is *x*-degenerate, for example, when two points have the same x coordinate. The difficulty arises from the definition, as the vertical attachments are assumed to be disjoint in their interiors. The algorithm solves this by using a symbolic shear transformation as shown in [3, Section 6.3]. This solution involves creating new nodes in the history DAG with corresponding vertical X_trapezoids of zero area.

Consider, for example, a vertical segment s inserted into the planar map. This segment corresponds to a degenerate vertical X_trapezoid in the search structure. It is treated as if it were non-degenerate by performing a symbolic shear transform on the input, $(x, y) \rightarrow (x + \epsilon y, y)$, for ϵ so small ($\forall r \in \mathbb{R}, \epsilon < r$) that the order on the x coordinates is linear. This scheme allows to distinguish between the vertical attachments bounding the segment s.

Bounding Box A *Bounding box* is a standard tool used to enable dealing with infinite objects. The idea is to keep the ordinary data structures used by the algorithm alongside with an additional data structure called a bounding box that contains in its interior all the "interesting" points of the map (e.g., segments' endpoints). This enables us to deal with infinite objects as if they were finite. In our framework the infinite objects are the active X_trapezoids representing infinite portions of the plane and the *interesting* points are the endpoints of the input curves as well as the intersections of the vertical attachments emanating from them with the curves.

Consider a vertical decomposition algorithm that uses a bounding box. At the beginning we have a unique active X_trapezoid representing a fixed bounding box. Whenever the property that *all the interesting points are in the current bounding box* becomes invalid the bounding box is enlarged to remedy this. This is done before the update or point location query takes place.

As a result all operations can take up to $\Theta(n)$ where n is the number of already inserted x-monotone curves. This is due to the possibly large (up to linear) number of X_trapezoids that are resized in a single bounding box update. The way we overcome this problem is by using a symbolic representation of infinity instead of a bounding box. For each X_trapezoid we keep the geometric information about its boundedness.

Analysis The algorithm offers on-line insert_edge, delete (remove_edge), split_edge and merge_edge operations with expected time of $O(log^2(n))$ where n is the number of update operations (insertions and deletions of curves) under the model suggested by Mulmuley [9, 10]. The expected storage requirement is $O(n)$.

It was also shown [10] that the update time is dominated by the point location part of the operation, and that other than the point location each update takes expected $O(1)$ time.

4.2   Walk-Along-a-Line

Unlike the naive strategy which traverses over all the edges of the map, the walk strategy starts with the unbounded face as the current face (unless the map is empty and we are done) and finds the hole that contains the query point q. If no such hole exists we know that the q belongs to the current face. Otherwise we find the halfedge on the current face boundary that is vertically closest to q. If the face incident to this halfedge does not contain q, we "walk" towards the query point q along the vertical ray emanating from q passing from a face to its adjacent neighbor along this ray, till the face containing the query point is located. This simple procedure is performed recursively until we cannot continue. It follows that the last face in the procedure is the required one.

Unlike other implementations of the walk algorithm (see for example [8]), the planar map does not necessarily deal with line segments. Our implementation assumes x-monotone curves, and does not assume an intersection predicate between lines and an x-monotone curve (this predicate is not part of the traits requirements). This is why we need to walk specifically along a vertical ray. For this reason we cannot store a starting point on the unbounded face, from which to start the walk, but need to find one which is vertically above the query point for every query. Since our package supports holes the query point can be in a hole inside a face. We would not like to walk over all the holes inside the faces we encounter, only through the ones that contain the point. This is achieved in our algorithm as described in the previous paragraph — we go over the holes of a face f only if we know that the query point is contained inside the outer boundary of f.

Whenever we use exact traits the computation is robust. Nevertheless, we pay special attention to degenerate cases such as walking over a vertical segment or having the query point on a vertical segment.

Remark: Due to the fact that the walk class has no internal data structures it suites perfectly for debugging purposes. Any users writing their own strategies can maintain the walk strategy alongside with their own class and use it for validity verifications.

4.3 Experimental Results

Figure 5 shows construction time[6] of a planar map defined by segments with different traits classes using the default strategy. The input was constructed using a sample of random segments with integer endpoints which were intersected using exact arithmetic. Then a conversion to the different number types (e.g., the built-in `double`) was done. It can be seen that the choice of traits strongly influences the running time. As expected, construction with floating point arithmetic is fastest, and the LEDA traits (see Section 3) perform very well compared with a Cartesian representation with a rational number type (in this case the *leda_rational* type[7]). [12] suggests that using other representations such as homogeneous coordinates with the `leda_integer` number type or Cartesian with

---

[6] All experiments were done on a Pentium-II 450MHz PC with 528MB RAM memory, under Linux.

[7] The reader should not confuse the *leda_rational* number type with LEDA's rational geometry kernel which uses a different representation, employs floating point filters, and is used by us in the LEDA traits.

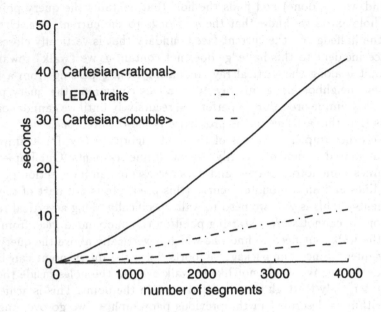

Fig. 5. Construction time with different traits

the `leda_real` number type (or other filtered number types) will give better results. It should be noted that on some of the larger inputs the program crashed when using floating point arithmetic (as noted in Section 3 we do not guarantee robustness or correctness when using floating point arithmetic), therefore we did not compare construction of a map with more than 4000 segments in this experiment.

Figure 6 demonstrates construction time with different point location strategies. We used the LEDA traits (see Section 3) for the experiments to maintain robustness with an acceptable running time. It can be seen that the point location query in the insertion function dominates the construction time. Therefore, the default strategy has much faster construction time than the naive and walk strategies. We do not show the results for the naive strategy beyond 4000 segments because the running time becomes very big. It can also be seen that there is no noticeable difference in the construction times when we disabled the rebuild option. This depends, of course, on the thresholds used for the rebuild option. We plan to experiment more with different thresholds for the default strategy.

Figure 7 shows average deletion time with different point location strategies, using the LEDA traits. We can see that the walk strategy performs this operation efficiently (as does the naive strategy which is not shown in the figure). Since they have no internal search structure to update, time is spent only on updating the topological map after the deletion. For small faces this takes constant time. The default strategy (with and without the rebuild option) needs to update its

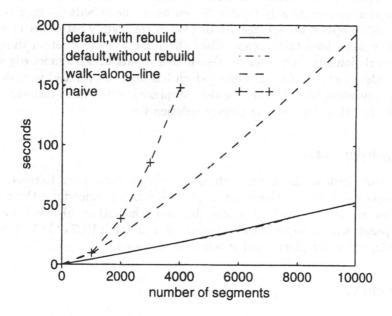

Fig. 6. Construction time with different strategies using the LEDA traits; the graphs for the default strategies with and without the rebuild option are almost overlapping

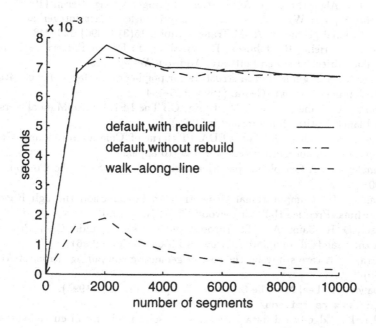

Fig. 7. Average deletion time with different strategies using the LEDA traits

internal structure in addition to updating the topological map. Therefore its deletion time is considerably greater. When using the default strategy without the rebuilding option the average deletion times are slightly better because some rebuilding steps have taken place which effect the average deletion time. The average deletion time is greater for the smaller inputs because more edges that were incident to the unbounded face (which has high complexity) were deleted. Since the deletion time depends on the complexity of the faces incident to the deleted edge, this increases the average deletion time.

Acknowledgment

The authors wish to thank Sariel Har-Peled, Sigal Raab, Lutz Kettner, Hervé Brönnimann and Michal Ozery for helpful discussions concerning the package described in this paper and other contributions. The authors also wish to thank all the people who attended the CGAL meeting in October 1997 at INRIA, Sophia Antipolis, where the planar map design was discussed.

References

1. Boissonnat, J.-D., Yvinec, M.: Algorithmic Geometry. Cambridge University Press (1998)
2. The CGAL User Manual, Version 1.2 (1998)
3. de Berg, M., van Krevald, M., Overmars, M., Schwarzkopf, O.: Computational Geometry: Algorithms and Applications. Springer-Verlag, Berlin (1997)
4. Fortune, S., van Wyk, C. J.: Static analysis yields exact integer arithmetic for computational geometry. ACM Trans. Graph., 15(3) (1996) 223–248
5. Gamma, E., Helm, R., Johnson, R., Vlissides, J.: Design Patterns – Elements of Reusable Object Oriented Software. Addison-Wesley (1995)
6. Kettner, L.: Designing a data structure for polyhedral surfaces. Proc. 14th Annu. ACM Sympos. Comput. Geom. (1998) 146–154
7. Melhorn, K., Näher, S., Seel, M., Uhrig, C.: The LEDA User Manual Version 3.7. Max-Planck Institut für Informatik (1998)
8. Melhorn, K., Näher, S.: The LEDA Platform of Combinatorial and Geometric Computing. Cambridge University Press (to appear)
9. Mulmuley, K.: A fast planar partition algorithm, I. J. Symbolic Comput., 10(34) (1990) 253–280
10. Mulmuley, K.: Computational Geometry: An Introduction through Randomized Algorithms. Prentice-Hall, Englewood Cliffs, NJ (1996)
11. Musser, D. R., Saini, A.: STL Tutorial and Reference Guide: C++ Programming with the Standard Template Library. Addison-Wesley (1996)
12. Schirra, S.: A case study on the cost of geometric computing. Proc. of ALENEX, (1999)
13. Stepanov, A., Lee, M.: The Standard Template Library (1995). http://www.cs.rpi.edu/~musser/doc.ps
14. Weiler, K.: Edge-based data structures for solid modeling in curved-surface environments. IEEE Computer Graphics and Application, 5(1) (1985) 21–40

An Easy to Use Implementation
of Linear Perturbations within CGAL[*]

Jochen Comes[1] and Mark Ziegelmann[2][**]

[1] Universität des Saarlandes, FB 14 Informatik
Im Stadtwald, 66123 Saarbrücken, Germany
comi@cs.uni-sb.de
[2] Max-Planck-Institut für Informatik
Im Stadtwald, 66123 Saarbrücken, Germany
mark@mpi-sb.mpg.de

Abstract. Most geometric algorithms are formulated under the non-degeneracy assumption which usually does not hold in practice. When implementing such an algorithm, a treatment of degenerate cases is necessary to prevent incorrect outputs or crashes. One way to overcome this nontrivial task is to use perturbations. In this paper we describe a generic implementation of efficient random linear perturbations within CGAL and discuss the practicality of using it examining the convex hull problem, line segment intersection and Delaunay triangulation.

1 Introduction

Implementing geometric algorithms is a difficult and errorprone task [14]. One reason for this is that most of the existing algorithms are described for non-degenerate input to simplify presentation. However, using input data from real world applications or random input, degenerate cases are very likely to occur. When implementing such algorithms, we are faced with the problem to identify and treat degenerate cases which leads to additional coding and often lets the structure of the program deteriorate. If one simply doesn't care about degenerate cases one is often faced with incorrect output or crashes. Another approach to deal with degeneracies which is often used in papers to state that the result also holds for general inputs, is the method of perturbation which suggests to move the input by an infinitesimal amount such that degeneracies are removed. More or less general perturbation methods that have been proposed are Edelsbrunner and Mücke's Simulation of Simplicity scheme (SOS)) [11, 17], Yap's symbolic scheme [23, 22], the efficient linear scheme of Canny and Emiris [6, 7], the randomized scheme of Michelucci [16] and the scheme for Delaunay triangulations of Alliez et al. [1]. For an excellent survey consult the paper of Seidel [21].

The question of usability of the perturbation approach divides the computational geometry community and the opinions changed over the years. In the late eighties Yap called perturbations "a theoretical paradise in which degeneracies

---

[*] This work was partially supported by ESPRIT IV LTR project 28155 (GALIA).
[**] Supported by a Graduate Fellowship of the German Research Foundation (DFG).

are abolished" and Edelsbrunner and Mücke believed that their SOS scheme will become standard in geometric computing. In the mid nineties, driven by increasing experience in the implementation of geometric algorithms more and more opinions questioned the practical use of perturbations [5, 20]. Burnikel et al. [5] argued that a direct thought treatment of degenerate cases is much more efficient in terms of runtime and only moderately more complicated in terms of additional coding. They argue that this results from the overhead of computing with perturbed objects and the need of a nontrivial postprocessing step to retain the original nonperturbed solution. Additional overhead may result from output sensitive algorithms like most algorithms for segment intersection. Considering input segments that all intersect in one point, the perturbed version would have to detect $n(n-1)/2$ intersections. We come back to these objections in our experimental section.

In this paper we describe a generic implementation of random linear perturbations (based on [21]) within the computational geometry software library CGAL [8, 12]. It enables the user to perturb the input objects and hence being able to only code the original algorithm without bothering about degeneracies. Contrary to previous implementations of perturbation schemes [17, 7], this is the first general and easy to use implementation requiring just to perturb the input rather than each test function.

After briefly reviewing the theoretical framework, we discuss our generic implementation of perturbations and finally illustrate and evaluate its practical use by examining three basic examples in computational geometry: convex hulls, line segment intersection and Delaunay triangulation.

2   An Efficient Linear Perturbation Scheme

In this section we will briefly review the efficient linear perturbation scheme as discussed in [21]. Proofs are omitted due to lack of space and can be found in [21] or [9].

The goal of the perturbation method is the following: Given a program for a certain problem that works correctly for all non-degenerate inputs, we want a purely syntactical transformation into a program that works correctly for all possible inputs.

More formally speaking: We are given a function F (describing an algorithm) from some input space $\mathcal{I}$ to some output space $\mathcal{O}$. We will think of $\mathcal{I}$ as $\mathbb{R}^N$ (n points in $\mathbb{R}^d$ if we take $N = dn$) and of $\mathcal{O}$ as $\mathbb{R}^M$ like this is the case in most geometric applications.

Definition 1. *For an input $q \in \mathbb{R}^N$ a linear perturbation of q is a linear curve π_q starting in q. It is a continuous mapping $\pi_q : [0, \infty) \to \mathbb{R}^M$ with $\pi_q(\epsilon) = q + \epsilon a_q$, where $0 \neq a_q \in \mathbb{R}^N$. A linear perturbation scheme Q assigns each input $q \in \mathbb{R}^N$ a linear perturbation π_q.*

Definition 2. *A linear perturbation scheme Q induces for every function F : $\mathbb{R}^N \mapsto \mathbb{R}^M$ a perturbed function $\overline{F}^Q : \mathbb{R}^N \mapsto \mathbb{R}^M$, defined by $\overline{F}^Q(q) = \lim_{\epsilon \to 0+} F(\pi_q(\epsilon))$.*

We will assume that this limit exists and sometimes just write $\overline{F}$ for the perturbed function. When do F and $\overline{F}$ agree ?

Lemma 1. *If F is continuous at q, then $F(q) = \overline{F}(q)$.*

The Lemma gives us a simple condition. However, if F is not continuous at q, we hope that there is some reasonable relationship between $F(q)$ and $\overline{F}(q)$, because the whole idea of the perturbation method is to compute $\overline{F}(q)$ instead of $F(q)$, thus being able to neglect degenerate cases.

Consider two examples: The problem of computing the convex hull area (CHA) and the convex hull sequence (CHS) of a set of points. CHA is continuous everywhere, so the result of the perturbed program $\overline{CHA}$ is always identical to the original result. Unfortunately this is not the case for CHS which is discontinuous for inputs with more than two collinear points on an edge of the convex hull (see Fig. 1). However, in this case $\overline{CHS}(q)$ is a subsequence of $CHS(q)$, thus the original output can be recovered quite easily. Not all discontinuous functions admit such an easy postprocessing step. We will come back to that problem in the experimental section.

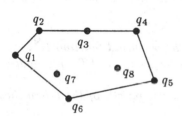

CHS with degenerate input

Output: $(q_1, q_2, q_4, q_5, q_6)$

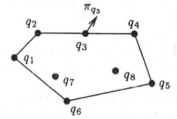

$\overline{CHS}$ with same input

Output: $(q_1, q_2, q_3, q_4, q_5, q_6)$

Fig. 1. CHS is discontinuous.

Now we want to turn to the computation of $\overline{F}$. Therefore we first choose a model of computation. A geometric algorithm A can be viewed as a decision tree[1] T where the decision nodes test the sign ($+,-$ or 0) of some test function (usually a low degree polynomial like geometric primitives as the orientation test or in-circle test) of the input variables. An input $q \in \mathbb{R}^N$ is called degenerate if the computation of A on input q contains a test with outcome zero.

Assume now, we have a decision tree T that computes some function F. How is it possible to compute the perturbed function $\overline{F}^Q$ for some perturbation scheme Q ? We simply perform a perturbed evaluation of T.

Definition 3. *Let $f : \mathbb{R}^N \mapsto \mathbb{R}$ be continuous, let $q \in \mathbb{R}^N$, and let π_q be a perturbation of q. We say π_q is valid for f iff $\lim_{\epsilon \to 0+} sign\ f(\pi_q(\epsilon)) \neq 0$.*

---

[1] A precise definition can be found in [21].

A perturbation scheme Q is valid for f iff π_q is valid for f for each $q \in \mathbb{R}^N$. If $\mathcal{F}$ is a family of test functions, then we call a perturbation (scheme) valid if it is valid for each $f \in \mathcal{F}$.

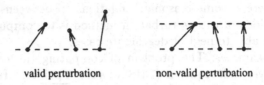

valid perturbation   non-valid perturbation

Fig. 2. A valid and a non-valid perturbation of three collinear points.

This means that we don't have to care about the 0-branches if we have a valid perturbation. The following theorem encompasses all benefits:

Theorem 1. *Let T be a correct decision tree computing some function $F :$ $\mathbb{R}^N \mapsto \mathcal{O}$, and let Q be a perturbation scheme that is valid for the set of test functions appearing in T.*

1. *A perturbed evaluation of T computes the perturbed function $\overline{F}^Q$.*
2. *If F is continous at q, then the perturbed evaluation of T with input q yields $F(q)$.*
3. *The above statements remain true, if some, or all, of the 0-branches of T are removed.*

Now it remains to show how we actually get valid perturbations for each possible input and how to evaluate the perturbed test functions: Linear perturbations are interesting since they allow relatively simple evaluation of perturbed test functions:

Theorem 2. *Let f be a multivariate polynomial of total degree at most D, and let B_f be a "black box algorithm" computing f. Let π_q be a linear perturbation of q that is valid for f. Then $\lim_{\epsilon \to 0^+} sign(f(\pi_q(\epsilon)))$ can be determined using at most $D + 1$ calls to B_f plus a small overhead.*

We are left with the problem how to actually come up with a valid perturbation for all inputs: It turns out that choosing the perturbation direction a_q randomly results in a valid perturbation for all inputs with very high probability as this result in [21] suggests:

Theorem 3. *Let T be a decision tree with a set $\mathcal{F}$ of S different test functions, each a multivariate polynomial of total degree at most D, and let $q \in \mathbb{R}^N$ be a fixed input to T. If direction a is chosen uniformly at random from $\{1, 2, \ldots m\}^N$, then the linear perturbation $\pi_q(\epsilon) = q + \epsilon a$ fails to be valid with probability at most DS/m.*

Now we know how to come up with a valid perturbation of the input. Of course, if a randomly chosen direction a turns out to be bad, i.e. during the evaluation of T a 0-branch is taken, we have to abort the computation and restart with a new randomly chosen a.

A deterministic construction of a valid linear perturbation seems to be rather difficult for the general case [7, 21].

3 An Implementation of Perturbations within CGAL

In this section we will describe how we implemented the perturbation approach of the last section within CGAL. The solution is surprisingly simple since CGAL uses the generic programming paradigm. All geometric objects of the CGAL kernel are parameterized by a number type NT. Hence, we developed a new number type CGAL_Epsilon_polynomial<NT> representing the linear perturbations while offering polynomial arithmetic[2] over NT. This frees us from identifying and perturbing every test function[3] since now *all* computations and comparisons are perturbed. The random perturbation direction (a random number between 1 and MAXINT) is assigned to the components of a geometric object p by calling the function CGAL_perturb(p).

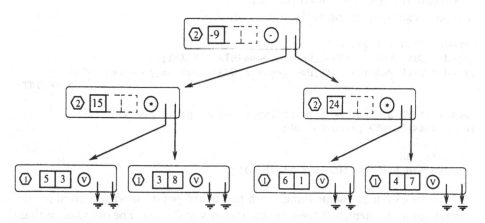

Fig. 3. Computation dag for $x \cdot y - v \cdot w$ if x is the ϵ-polynomial $5 + 3\epsilon$, y is $3 + 8\epsilon$, v is $6 + \epsilon$, and w is $4 + 7\epsilon$.

The ϵ-polynomials are represented with a vector of their coefficients. Initially we have linear polynomials with the original coordinate as constant coefficient and the random perturbation direction as linear coefficient. The sign of an ϵ-polynomial is the sign of the first nonzero coefficient.

---

[2] Polynomial addition, subtraction and multiplication as well as scalar operations are available (division is avoided in CGAL using CGAL_Quotient<NT>).

[3] A shortcome of previous methods and implementations [11, 17, 7].

Since the sign of an ϵ-polynomial is often determined by the constant coefficient or by low degree coefficients, we chose to pursue a lazy evaluation approach by recording its computation history in a directed acyclic graph and determining non-constant coefficients only once they are needed from the computation dag. An arithmetic operation simply constructs a new node in the graph representation, computes the new constant coefficient, determines the new degree, establishes pointers to subexpressions and labels the node with the type of the arithmetic operation (see Fig. 3). Comparisons are reduced to sign computations. The sign is determined by first looking at the (nonperturbed) constant coefficient. If this is zero, the linear coefficient is determined recursively from the subexpressions[4], and so on until we have reached a nonzero coefficient or the maximum degree of the polynomial. In that sense, the sign computation is adaptive, easy sign computations are fast, difficult ones are more expensive. If the leading coefficient is also zero, we have the case of a non-valid perturbation scheme, thus we throw an exception. This enables the user to catch it at runtime and restart the computation with new random perturbations.

The random perturbation coordinate is randomly chosen from the interval 1 to MAXINT. The bitlength of the perturbed coordinate may be chosen smaller than the bitlength of MAXINT[5], as is for example necessary when we use doubles with limited bitlength to assure exact computation. In such a case, the perturbation has to be limited similarly.

A typical example setting is the following:

```
typedef leda_integer NT;
typedef CGAL_Point_2<CGAL_Homogeneous<NT> > POINT;
typedef CGAL_Point_2<CGAL_Homogeneous<CGAL_Epsilon_polynomial<NT>
                                                > > ePOINT;

leda_list<POINT> pL; // list of input points pL is given
leda_list<ePOINT> perturbed_pL;
POINT p;
forall(p, pL)
  perturbed_pL.append(CGAL_perturb(p));
```

Now we can do the same things with perturbed points as with normal points, we may compare them, test them for collinearity and so on. The outcome of these test functions with perturbed points in general won't be zero. If we encounter a zero outcome then our random perturbation was bad and we have to start over. However, there is another possibility for a zero outcome, e.g. if we make a local copy of a perturbed input point and test both for equality[6]. Thus the reason for an exception or warning that is automatically produced encountering a zero test has to be carefully studied.

---

[4] Each recursively computed coefficient is stored in the coefficient vector thus does not have to be recomputed.

[5] The bitlength of MAXINT is 32 in our experiments.

[6] In CGAL and LEDA points and segments are handle types, i.e there is a pointer to their representation. Testing for identity of the pointers eliminates this undesired zero outcome.

4 Experiments

To evaluate the practical use of our implementation of perturbations we will discuss three examples: computing the convex hull of a planar point set, computing the intersections of line segments and computation of the Delaunay triangulation of a planar point set. We take efficient, well designed algorithms of LEDA [15, 13][7] and CGAL for those problems which work on all input instances by treating degenerate cases. We will compare these algorithms with the corresponding perturbed ones without the treatment of degeneracies in terms of running time[8] and simplification of the code. Since perturbations assume the use of exact arithmetic, we compare the use of doubles with suitably restricted bitlength and the type **integer** of LEDA to represent the homogeneous coordinates of points and segments.

4.1 Convex Hulls

Given a set S of points in the plane, its *convex hull* is the smallest convex set containing S. A natural representation is a minimal cyclic list of the vertices on the hull. The possible degenerate cases are the identity of two points or the collinearity of three points. We take two algorithms from LEDA for this problem (see [15, 13]), one based on sweep (a modification of Graham's scan) and one based on randomized incremental construction[9]. The highlevel LEDA code for the sweep is 28 lines long where 7 lines deal with degeneracies. The RIC is slightly more involved and consists of 106 lines of code, 19 of them dealing with degeneracies (see [15] and [10]). In our perturbed versions we can omit those lines since we use a list of perturbed points. Planar convex hulls are an easy example where it is not very difficult to handle degenerate cases. Hence there is only a modest gain in terms of lines of code and structure of the code. Figure 4 shows the price we have to pay for using perturbed points[10] instead of normal points:

The coordinates of the points had been randomly chosen from $[0, \ldots, 10000]$ which results in few degeneracies. To enforce degeneracies we also performed experiments where the random coordinates had been rounded to a grid with gridwidth 100 and 500 respectively (the two dashed lines).

In the first two diagrams we used doubles to represent the coordinates. For normal random input (solid lines) this results in a large overhead factor of around

---

[7] CGAL provides a large number of geometric algorithms but not yet an algorithm for line segment intersection, so we CGALized the existing LEDA code.

[8] All experiments measuring CPU time in seconds on a Sun Enterprise 333MHz, 6Gb RAM running Solaris 2.6 using LEDA 3.7.1 and CGAL 1.2 compiled with g++ -O2

[9] CGAL also offers a number of implementations of different convex hull algorithms [18] where degeneracies are handled implicitly in most cases.

[10] The input points are now polynomials of degree one, so operations on polynomials replace the previous operations on doubles and **integers** respectively. The maximum degree in intermediate computations is 2 (in the left/right-turn and orientation primitives).

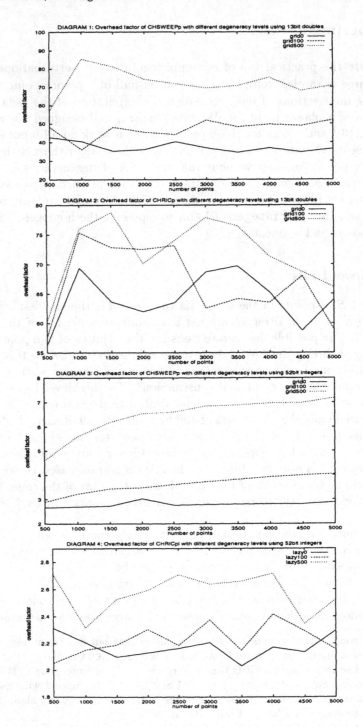

Fig. 4. Overhead factor between the original and perturbed convex hull algorithms for random points in $[0, \ldots, 10000]$.

40 for the perturbed sweep (Diagram 1) and an even larger overhead factor of around 65 for the perturbed RIC (Diagram 2). This huge performance loss arises since the arithmetic part of the convex hull algorithms is very large such that the computation with polynomials over doubles incurs a significant slowdown compared to the extremely fast computation with doubles.

The overhead rises further if we take more degenerate inputs (dashed lines). This stems from the fact that we now have many identical and collinear points that may let the intermediate hull of the perturbed algorithms grow but not that of the unperturbed ones. Moreover the sign computations are more "difficult" now for the perturbed case.

Diagrams 3 and 4 show the same experiments where the point coordinates had been "pumped up" to 52 bits. We now have to use exact **integers** for the coordinates to obtain the correct result.

For normal random inputs (solid lines), the overhead of the perturbed sweep (Diagram 3) is reduced to a constant factor of 2.8, whereas the perturbed RIC (Diagram 4) has an constant overhead factor of around 2.2. This can be explained as follows: The use of exact integers increases the running time of the arithmetic part in both the perturbed and the unperturbed versions. Since the random perturbation direction has bitlength at most 32, not all operations are between 52-bit numbers in the perturbed versions, so the overhead of polynomial arithmetic is reduced.

If we look at highly degenerate inputs (dashed lines), we again observe higher overheads of up to 7 for the sweep and around 2.6 for the RIC. In the RIC the difference between random and highly degenerate input is not as pronounced as in the sweep since the points are taken into consideration in random order[11].

Postprocessing to retain the original nonperturbed solution takes neglectable time since the number of points on the hull is expected to be logarithmic in the total number of points in this case (only for points on a circle it takes about as long as preprocessing, i.e. perturbing the points) but the savings in terms of lines of code are lost.

4.2 Line Segment Intersection

Now we turn to another basic problem in computational geometry, the *line segment intersection* problem. We investigate the performance of the LEDA implementation of the Bentley-Ottmann sweep that can cope with degeneracies and takes $O((n + s) \log n)$ time (where s is the number of intersections). In contrast to convex hulls, there are a large number of possible degenerate cases: Multiple intersections, overlapping segments, segments sharing endpoints, segments degenerated to a single point, vertical segments and endpoints or intersection points with same x-coordinate[12].

---

[11] Before adopting the lazy evaluation approach we simply performed the "complete" operation when computing with ε-polynomials. The lazy evaluation approach reduced the overhead by around 80% for random inputs but only slightly for highly degenerate inputs.

[12] The latter two are often called algorithm induced degeneracies.

A perturbed version of the algorithm is easily obtained using a list of perturbed segments as input. Since the algorithm is more complicated it is not straightforward to identify the treatment of degenerate cases that we may remove. Therefore we implemented an own perturbed version of the sweep which saves 71 lines of 201 and is easier to understand since we followed the original formulation (see [15] and [10]). Hence, the savings in terms of lines of code and development time are quite significant.

What penalty do we have to pay for this? The maximum degree of the ϵ-polynomials is 5 in the sweep (intersection computation, orientation test of input segment with endpoint or intersection point and using homogeneous coordinates). Figure 5 shows the results of our experiments for segments with random coordinates in $[0, \ldots, 1000]$ (a grid rounding (for gridwidth with 20 and 50 respectively) has again been used to obtain varying levels of degeneracy).

In the first diagram we used doubles to represent the coordinates. For normal random input (solid line) we observe an overhead factor of about 9.5. The overhead using doubles is much smaller as in the convex hull algorithms since the arithmetic part of the sweep is not so dominant.

To assure exact computation we had to limit the perturbation direction to 10 bits. This sometimes resulted in zero signs despite perturbation, thus we had to restart the computation. From a certain input size on it was not possible anymore to come up with a valid perturbation scheme for highly degenerate inputs .

In Diagram 2 we pumped up the coordinates to 52-bits and used **integers** to represent them. The overhead factor is smaller again:
For normal random input (solid line) the overhead factor is about 4.1. Using highly degenerate inputs (dashed lines) the overhead grows to about 4.5. This comes from the fact that in this case we have many overlapping segments and many multiple intersection points which are "better" for the output sensitive original sweep; moreover we again get more difficult sign computations in the perturbed case.

Experiments with n segments all intersecting in one point showed us that the perturbed sweep performs badly since it has to detect $n(n-1)/2$ intersection points, whereas the original output-sensitive version takes $O(n \log n)$ and is faster than usual [10]. This drawback of the perturbation approach was already pointed out in [5].

Postprocessing for the perturbed segment sweep is nontrivial since the perturbation may cause an intersection to vanish. We implemented a "postprocessing" step that returns the (unperturbed) list of intersection points. This requires to test whether endpoints of segments lie within another segment during the sweep and real postprocessing (unperturbing the intersection points and eliminating duplicates) after the sweep . Apart from additional coding the running time overhead factor increases slightly for random input (around 15%) and more drastically for highly degenerate inputs (up to 70%) where the difference of the number of intersections is the perturbed and the original version is very high.

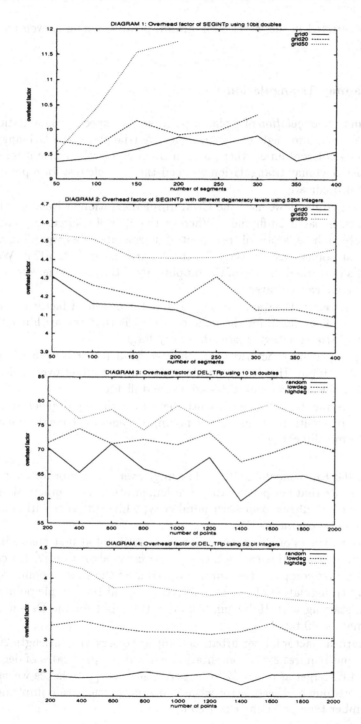

Fig. 5. Overhead factor between the original and perturbed sweep algorithm for random segments in $[0, \ldots, 1000]$ and between the original and perturbed Delaunay Triangulation algorithm for random points in $[0, \ldots, 1000]$.

If we are interested in the intersection graph, postprocessing is even more challenging [3].

4.3   Delaunay Triangulation

The *Delaunay triangulation* of a planar point set is a special triangulation maximizing the minimum angle of the triangles. A triangle of the Delaunay triangulation does not contain another point in its circumcircle, hence if we want to compute the Delaunay triangulation we need the in-circle test for 4 points which is a degree 4 predicate.

In our experiments we used the CGAL-implementation of Delaunay triangulations which is based on flipping. Whereas the case of cocircular points is not really a problem here, a careful treatment of degenerate cases regarding collinear and identical points was necessary for the "march-locate" step [24]. We didn't implement a perturbed version of the flipping algorithm but believe that at least 10% of the code can be saved.

It can be seen in Diagram 3 of Fig. 5 that the overhead factor is around 65 for doubles[13]. The reason for the huge overhead is that the arithmetic part of the Delaunay triangulation algorithm is very large.

Using pumped up exact **integers** the overhead factor is reduced to about 2.5 for random input. Highly degenerate inputs have a larger overhead factor of up to 4 since the sign computations are more difficult.

Postprocessing can be done efficiently here by shrinking degenerate triangles resulting from identical points and removing degenerate triangles on the hull resulting from collinear points.

A Note on the Experimental Setting. The huge overhead factor for **doubles** may seem to indicate that our perturbation implementation is not practical, however, going over from **doubles** to an exact number type like **integers** without filtering results in similar overheads [19].

Input coordinates of 52 bits may also seem artificial at first glance but there are algorithms where this can occur, e.g. the crust algorithm [2] for curve reconstruction first computes the Voronoi diagram of the sample points and then a Delaunay triangulation of the Voronoi vertices and the sample points. Hence even when starting with 10 bit input points, the input for the Delaunay triangulation may be 40 bits long.

The overhead factor for perturbation using **integers** with bitlength 20 (which is realistic and requires exact computation when using predicates of degree ≥ 3) is only 20-40% larger as in our 52 bit experiments (see [10]), hence we can really speak of a medium overhead factor when using our perturbation implementation with a number type providing exact computation[14].

---

[13] The bitlength of the perturbation direction has to be limited to 13 to assure exact computation. This resulted in a few zero signs for highly degenerate inputs.

[14] Experiments with **reals** [4] show similar results [10].

5   Summary and Conclusion

We presented a generic and easy to use implementation of the linear perturbation approach of [21] within CGAL. Using our implementation, it is possible to perturb the input which abolishes the degenerate cases[15].

As we have seen in the three applications, this introduces a medium overhead factor for the running time which depends on the runtime fraction of the arithmetic part of an algorithm as well as on the used number type. The performance on highly degenerate inputs increases even more.

We face additional problems if we want to have the original result and not the perturbed one, as it is the case when our (perturbed) implementation is part of a larger system and involves interaction between other components. The necessary postprocessing step often amounts to comparable work as treating the degenerate cases in the first place.

We conclude that the perturbation approach using our ϵ-polynomial implementation is an important tool for rapid prototyping of geometric algorithms. It enables us to implement difficult algorithms in quite reasonable time if we don't care about a medium runtime penalty. We hope that this might be an aid that more of the theoretical work in computational geometry will find its way into practice. However, if we have the original non-perturbed output in mind or if we cannot afford overhead and don't care about the time for the implementation process, it will be necessary to code a stable nonperturbed version.

More details of the implementation and the experiments can be found in [9, 10]. The ϵ-polynomial code including documentation and demo programs can be downloaded from http://www.mpi-sb.mpg.de/~mark/perturbation - it will be included in a future release of CGAL.

Acknowledgements

We would like to thank Stefan Schirra, Michael Seel and Raimund Seidel for many helpful discussions.

References

[1] P. Alliez, O. Devillers, and J. Snoeyink. Removing degeneracies by perturbing the problem or perturbing the world. In *Proc. 10th Canadian Conference on Computational Geometry (CCCG98)*, 1998.

[2] N. Amenta, M. Bern, and D. Eppstein. The crust and the β-skeleton: Combinatorial curve reconstruction. *Graphical Models and Image Processing*, pages 125–135, 1998.

[3] U. Bartuschka, M. Seel, and M. Ziegelmann. Sweep segments easily - a perturbed approach. Manuscript.

---

[15] Theorem 3 guarantees this with very high probability. We never encountered occurrence of degenerate cases in our experiments with the perturbed algorithms when choosing the random perturbation directions between 1 and MAXINT.

[4] C. Burnikel, R. Fleischer, K. Mehlhorn, and S. Schirra. Efficient exact geometric computation made easy. In *Proc. 15th ACM Symposium on Computational Geometry (SCG99)*, 1999. to appear.

[5] C. Burnikel, K. Mehlhorn, and S. Schirra. On degeneracy in geometric computations. In *Proc. 5th Annual ACM-SIAM Symp. on Discrete Algorithms*, pages 16–23, 1994.

[6] J. Canny and I. Emiris. A general approach to removing degeneracies. *SIAM J. Comput.*, 24:650–664, 1995.

[7] J. Canny, I. Emiris, and R. Seidel. Efficient perturbations for handling geometric degeneracies. *Algorithmica*, 19(1–2):219–242, 1997.

[8] CGAL Web page: http://www.cs.uu.nl/CGAL.

[9] J. Comes. Implementierung von Perturbationen für geometrische Algorithmen. Master's thesis, Universität des Saarlandes, FB 14 Informatik, 1998. (german).

[10] J. Comes and M. Ziegelmann. An easy to use implementation of linear perturbations within CGAL. Technical report, Max-Planck-Institut für Informatik, Saarbrücken, 1999. to appear.

[11] H. Edelsbrunner and E. Mücke. Simulation of simplicity: A technique to cope with degenerate cases in geometric algorithms. *ACM Trans. Graphics*, 9(1):67–104, 1990.

[12] A. Fabri, G.-J. Giezeman, L. Kettner, S. Schirra, and S. Schönherr. On the design of CGAL, the computational geometry algorithms library. Technical Report MPI-I-98-1-007, Max-Planck-Institut für Informatik, Saarbrücken, 1998.

[13] LEDA Web page: http://www.mpi-sb.mpg.de/LEDA.

[14] K. Mehlhorn and S. Näher. The implementation of geometric algorithms. In *13th World Computer Congress IFIP94*, volume 1, pages 223–231, 1994.

[15] K. Mehlhorn and S. Näher. *The LEDA platform for combinatorial and geometric computing*. Cambridge University Press, 1999. in press.

[16] D. Michelucci. An ε-arithmetic for removing degeneracies. In *Proc. 12th IEEE Symposium on Computer Arithmetic*, pages 230–237, 1995.

[17] E. Mücke. SoS - A first implementation. Master's thesis, Department of Computer Science Univ. of Illinois at Urbana-Champaign, Urbana Ill, 1988.

[18] S. Schirra. Parameterized implementations of classical planar convex hull algorithms and extreme point computations. Technical Report MPI-I-98-1-003, Max-Planck-Institut für Informatik, Saarbrücken, 1998.

[19] S. Schirra. A case study on the cost of geometric computing. In *Proc. Workshop on Algorithm Engineering and Experimentation (ALENEX99)*, 1999.

[20] P. Schorn. Limits of the perturbation approach in geometric computing. *The Computer Journal*, 37(1):35–42, 1994.

[21] R. Seidel. The nature and meaning of perturbations in geometric computing. *Discrete and Computational Geometry*, 19(1):1–19, 1998.

[22] C.-K. Yap. A geometric consistency theorem for a symbolic perturbation scheme. *J. Comput. Syst. Sci.*, 40:2–18, 1990.

[23] C.-K. Yap. Symbolic treatment of geometric degeneracies. *Journal of Symbolic Computation*, 10:349–370, 1990.

[24] M. Yvinec, 1999. Personal communication.

Analysing Cache Effects in Distribution Sorting

Naila Rahman and Rajeev Raman*

Department of Computer Science
King's College London
Strand, London WC2R 2LS, U. K.
{naila, raman}@dcs.kcl.ac.uk

Abstract. We study cache effects in *distribution* sorting algorithms. We note that the performance of a recently-published distribution sorting algorithm, Flashsort1 which sorts n uniformly-distributed floating-point values in $O(n)$ expected time, does not scale well with the input size n due to poor cache utilisation. We present a two-pass variant of this algorithm which outperforms the one-pass variant and comparison-based algorithms for moderate to large values of n. We present a cache analysis of these algorithms which predicts the cache miss rate of these algorithms quite well. We have also shown that the integer sorting algorithm MSB radix sort can be used very effectively on floating point data. The algorithm is very fast due to fast integer operations and relatively good cache utilisation.

1 Introduction

Most algorithms are analysed on the random-access machine (RAM) model of computation [1], using some variety of unit-cost criterion. In particular, the RAM model postulates that accessing a location in memory costs the same as a built-in arithmetic operation, such as adding two word-sized operands. However, over the last 20 years or so CPU clock rates have grown explosively, with an average annual rate of increase of $35 - 55\%$ [3]. As a result, nowadays even entry-level machines come with CPUs with clock frequencies of 400 Mhz or above. Unfortunately, the speeds of main memory have not increased as rapidly, and today's main memory typically has a latency of about 70ns. Hence, a conservative estimate is that a memory access can take 30+ CPU clock cycles.

In order to overcome this difference in speeds, modern computers have a *memory hierarchy* which inserts multiple levels of *cache* between CPU and main memory. A cache is a fast associative memory which holds the values of some main memory locations. If the CPU requests the contents of a memory location, and the value of that location is held in some level of cache, the CPU's request is answered by the cache itself (a cache *hit*); otherwise it is answered by consulting the main memory (a cache *miss*). A cache hit has small or no penalty (1-3 cycles is fairly typical) but a cache miss is very expensive.

---

* Supported in part by EPSRC grant GR/L92150

J.S. Vitter and C.D. Zaroliagis (Eds.): WAE'99, LNCS 1668, pp. 183–197, 1999.
© Springer-Verlag Berlin Heidelberg 1999

Nowadays a typical memory hierarchy has CPU registers (the highest level of the hierarchy), L1 cache, L2 cache and main memory (the lowest level). The number of registers and the size of caches are limited by several factors including cost and speed [2]. Normally, the L1 cache holds more data than CPU registers, and L2 cache much more than L1 cache. Even so, L2 cache capacities are typically 512KB to 2MB [1], which is considerably smaller than the size of main memory.

It should be added that the memory hierarchy continues beyond main memory to disk storage [3]. As the cost of servicing 'cache misses' in this context is not included in the running times, we are not primarily concerned with these levels of the hierarchy. However, most CPUs provide hardware support for managing these levels of the hierarchy in the form of a *translation look-aside buffer (TLB)* [3]. The TLB can affect running times, as a cache miss may require *two* memory accesses to be made[2]. Hence a cache miss can easily cost 60+ cycles.

Programs have much faster running times if they have a high cache *hit rate*—i.e. if most of their memory references result in a cache hit. By tuning an algorithm's cache performance, one can obtain substantial improvements in the running time of its implementations, as we demonstrate in the context of *distribution* sorting algorithms.

Distribution sorting is a popular technique for sorting data which is assumed to be randomly distributed, and involves distributing the n input keys into m *classes* based on their value. The classes are chosen so that all the keys in the ith class are smaller than all the keys in the $(i+1)$st class, for $i = 0, \ldots, m-2$, and furthermore, the class to which a key belongs can be computed in $O(1)$ time. In $O(n)$ time, the problem is reduced to sorting the m classes. Distribution sorting can be used to sort randomly distributed keys in $O(n)$ time on average [4, Ch 5.2, 5.2.1], which is asymptotically faster than the $\Theta(n \log n)$ running time of comparison-based approaches. Neubert [7] presented an implementation of a distribution sorting algorithm *Flashsort1*, which used a combination of a well-known counting method and an in-place permutation method similar to one described in [4, Soln 5.2-13]. With Neubert's choice of parameters Flashsort1 uses only $n/10$ words of memory in addition to the memory required by the data, and sorts n uniformly (and independently) distributed floating-point keys in $O(n)$ time. Neubert's experiments showed that his implementation of Flashsort1 was twice as fast as Quicksort when sorting about 10,000 keys. In the RAM model, the lower asymptotic growth rate of Flashsort1 and the fact that Flashsort1 outperforms Quicksort for $n = 10,000$ would indicate that Flashsort1 would continue to outperform Quicksort for larger values of n. Unfortunately, this is not the case. We translated Neubert's FORTRAN code for Flashsort1 into C and performed extensive experiments which clearly indicated that although Flashsort1 was significantly faster than Quicksort for n in the range 4K to 128K, Quicksort caught up with and surpassed Flashsort1 at 1M keys. Note that Quicksort has $O(n \log n)$ running time even on average, whereas Flashsort1 is a linear-time

---

[1] K = 1024 and M = 1024K in this paper, except that M in MHz equals 10^6.

[2] This happens in case of a *TLB miss*, i.e. the TLB does not hold the page table entry for the page which was accessed [3]

algorithm. In fact, the ratio of the running times of Flashsort1 to Quicksort continued to grow with n, up to $n = 64M$.

Our analyses and simulations verify that this is due to the poor cache performance of Flashsort1. By adapting Flashsort1 to run in *two* passes we get an algorithm *Flashsort2P*: because Flashsort2P makes better use of the cache, it out-performs both Flashsort1 for $\geq$ 1M items, and Quicksort for $\geq$ 4M items. We present a cache analysis of both Flashsort variants, and validate the formula by simulations. We also show that by using the MSB (most-significant-bit first) radix sort [4] on the integer representation of floating point numbers we can clearly out-perform Quicksort, Flashsort1 and Flashsort2P for all values of n that we tested. We note that understanding the cache behaviour for the Flashsorts does not directly say anything about MSB radix sort. However, our simulations show that it maintains a low miss rate, and hence is able to benefit from the extra speed of integer operations.

2 Preliminaries

This section introduces some terminology and notation regarding caches. The size of the cache is normally expressed in terms of two parameters, the *block size* (B) and the number of *cache blocks* (C). We consider main memory as being divided into equal-sized *blocks* consisting of B consecutively-numbered memory locations, with blocks starting at locations which are multiples of B. The cache is also divided into blocks of size B; one cache block can hold the value of exactly one memory block. Data is moved to and from main memory only as blocks.

In a *direct-mapped* cache, the value of memory location x can only be stored in cache block $c = (x \text{ div } B) \bmod C$. If the CPU accesses location x and cache block c holds the values from x's block the access is a cache hit; otherwise it is a cache miss and the contents of the block containing x are copied into cache block c, *evicting* the current contents of cache block c. For our purposes, cache misses can be classified into *compulsory misses*, which occur when a memory block is accessed for the first time, and *conflict misses*, which happen when a block is evicted from cache because another memory block that mapped to the same cache block was accessed.

An important consideration is what happens when a value is written to a location stored in cache. If the cache is *write-through* the value is simultaneously updated in the cache and in the next lower level of the memory hierarchy; if the cache is *write-back*, the change is recorded only in cache. Of course, when a block is evicted from a write-back cache it must be copied to the next lower level of the hierarchy.

We performed our experiments on a Sun UltraSparc-II, which has a blocksize of 64 bytes, a L1 cache size of 16KB and a L2 cache size of 512KB. Both L1 and L2 caches are direct-mapped, the L1 cache is write-through and the L2 cache is write-back. As our programs hardly ever read a memory location without immediately modifying it, the L1 cache is ineffective for our programs and we focus on the L2 cache. Hence, for our machine's L2 cache we have $C = 8192$. This paper deals mainly with single-precision floating-point numbers and integers, both of which are 4 bytes long on this system. It is useful to express B in terms of the number of 'items' (integers or floats) which fit in a block; hence we use $B = 16$ in what follows.

3   Overview of Algorithms

We now describe the main components of the two Flashsort variants, and explain why Flashsort1 may have poor cache performance. We then introduce Flashsort2P and MSB radix sort. While describing all these algorithms, the term *data* array refers to the array holding the input keys, and the term *count* array refers to an auxiliary array used by these algorithms.

3.1   Flashsort1

Flashsort1 has three main phases, the first two of which are a *count* phase and a *permute* phase. After the count and permute phases, the data array should have been permuted so that all elements of class k lie consecutively before all elements of class $k + 1$, for $k = 0, \ldots, m - 2$. The data array is then sorted using insertion sort. Fig 1 gives pseudo-code for the count and permute phases; it is assumed that value of m has been set appropriately before the count phase, and that the function classify maps a key to a class numbered $\{0, \ldots, m - 1\}$ in $O(1)$ time. For example, if the values are uniformly distributed over $[0, 1)$ then classify(x) can return $\lfloor m \cdot x \rfloor$.

(a) A count phase

```
1 for i := 0 to n - 1 do
     COUNT[classify(DATA[i])]++;
2 COUNT[m - 1] := n - COUNT[m -
1];
  for i := m - 2 downto 0 do
     COUNT[i] :=
        COUNT[i + 1] - COUNT[i];
```

(b) A permute phase

```
1 start := 0; nmoved := 0;
2 idx := start; x := DATA[idx];
3 c := classify(x); idx := COUNT[c];
   swap x and DATA[idx];
   nmoved := nmoved + 1; COUNT[c]++;
   if idx ≠ start go to 3;
4 Find start of next cycle and set
   start to this value. Go to 2.
```

Fig. 1. The two phases in both Flashsort1 and Flashsort2P. DATA holds the input keys and COUNT is an auxiliary array, initialised to all zeros. The permute phase terminates whenever in the above: (i) $nmoved \geq n - 1$ or (ii) *start* moves beyond the end of the array.

Fix some class k and let t be the value of COUNT$[k]$ at the start of the permute phase. During the course of the permute phase, an invariant is that locations

$t, t + 1, \ldots, \text{COUNT}[k] - 1$ contain elements of class k, i.e. $\text{COUNT}[k]$ points to the 'next available' location for an element of class k. The permute phase moves elements to their final locations along a cycle in the permutation; a little thought is needed to find cycle leaders correctly.

If the number of classes is $m \leq n$ the total expected cost of the insertion sort is easily seen to be $O(n^2/m)$, while the rest of the algorithm evidently runs in $O(n)$ time. The algorithm uses m extra memory locations. After experimentation Neubert chose $m = n/10$ to minimise extra memory while maintaining a near-minimum expected running time [7]. However, for large values of n, it is apparent that the cache performance of Flashsort1 will be poor: e.g., for $n = 64M$ the count array will be approximately 50 times the size of cache, which means that all accesses to the count array will be cache misses.

3.2   Flashsort2P

It appears from the preceding discussion that we should reduce m in order to get an algorithm with better cache performance. In *Flashsort2P*, the distribution is done in two phases. In the first, we use $m = \sqrt{n}/2$, giving classes with about $2\sqrt{n}$ keys on average. We apply distribution sort again to each sub-problem, this time with $m = \sqrt{n}$ classes. At the end, we have classes with an expected size of 2 keys, and we sort these using insertion sort. Note that the expected running time of Flashsort2P is $O(n)$. The potential benefits of this approach are:

• A smaller number of classes in the first phase may lead to a lower number of misses;

• The problems in the second phase will be of size $2\sqrt{n}$; for $n < 2^{32}$ the expected size of a problem in the second phase will be smaller than 512KB, which is our L2 cache size. Hence, each of the sub-problems in the second phase should fit entirely into cache, giving few misses in the second phase. We can also now perform the insertion sort for a sub-problem as soon as we finish with the count/permute step for this sub-problem, avoiding the compulsory misses for the global insertion sort of Flashsort1.

• Flashsort2P has much lower auxiliary space requirements than Flashsort1, since the count array is of size $O(\sqrt{n})$, rather than $\Theta(n)$ for Flashsort1.

• The insertion sort problems are smaller in Flashsort2P. However, the insertion sort is only a small fraction of the total running time so this improvement should not yield great benefits.

3.3   MSB Radix Sort

Most significant bit (MSB) radix sort treats the keys as integers by looking at the bit-string representation of the floating point numbers. It is well-known that if the floating point numbers are represented according to the IEEE 754 standard, then the bit strings that represent two floating point numbers have the same ordering as the numbers themselves, at least if both the numbers are non-negative [3]. In our implementation of MSB radix sort we first distribute all numbers according to their most significant r bits, where $r = \min\{\lceil \log n \rceil - 2, 16\}$, where n is the number of keys to be sorted. The permute phase is similar to the Flashsort variants. A sub-problem consisting of $n' \geq 16$ keys is then attacked in the analogous manner, i.e. we distribute the n' keys based on their next most significant r' bits, where $r' = \min\{\lceil \log n' \rceil - 2, 16\}$. Problems of size ≤ 16 are solved with insertion sort.

For lack of space we do not describe in detail the major differences between MSB radix sort and the Flashsort variants. The main point to be noted is that a uniform distribution on floating point numbers in the range $[0, 1)$ does *not* induce a uniform distribution on the representing integers, and we cannot analyse it the same way as the Flashsort variants. For example, half the numbers will have value in $[0.5, 1)$; after normalisation their exponents will all be 0. This means that in the IEEE 754 standard representation of single-precision floating-point numbers, half the keys will have the pattern 01111111 in their most significant 9 bits. However, the algorithm can be shown to be linear-time for random w-bit floating point numbers, if some assumptions are made about the size of the exponent.

The above argument also means that after the first pass, there will be several large sub-problems (e.g. with $r = 16$ there may be problems of expected size about $n/256$) which may not fit easily into L2 cache for very large n. Hence MSB radix sort does not have as good cache utilisation as Flashsort2P for large n. However, we find that MSB radix sort can out-perform Flashsort1, Flashsort2P and Quicksort over the range of values that we considered.

4   Experimental Results

We have implemented Flashsort1, Flashsort2P and MSB radix sort, and have used them to sort n uniformly distributed floating-point numbers, for $n = 2^i$, $i = 10, 11, \ldots, 26$. We have also tested a highly tuned implementation of recursive Quicksort from [6]. Our algorithms were coded in C (as was the Quicksort implementation) and all code was compiled using gcc 2.8.1. For each algorithm, we have measured actual running times as well as simulated numbers of cache misses. The running times were measured on a Sun Ultra II with 2×300 Mhz processors and 512MB main memory. As mentioned above, this machine has a 16KB L1 data cache, 512KB L2 cache, both of which are direct-mapped. Our simulator simulates only the L2 cache on this machine, and only reports cache hit/miss statistics. Each run time and simulation value reported in this section is the average of 100 runs.

Figure 2 summarises the running times for Flashsort1, Flashsort2P, Quicksort and MSB radix sort. For the smaller input sizes ($n \leq 512K$), the timing was obtained by repeatedly copying and sorting a given (unsorted) sequence of numbers and taking the average time. The running times reported include the time for copying. We observe that:

- For small values of n Flashsort1 gets steadily faster than Quicksort until it uses about 70-75% of the time for Quicksort for about 128K. After that the performance advantage narrows until at $n = 1M$ Quicksort overtakes Flashsort1. The gap between Quicksort and Flashsort1 grows steadily until the largest input value we considered. This is interesting given that Quicksort has a higher asymptotic running time than Flashsort1.

- Flashsort2P is slow for small values of n but starts to out-perform Flashsort1 at $n = 1M$ and Quicksort at $n = 4M$. At large values of n Flashsort2P is almost twice as fast as Flashsort1.

- MSB radix sort out-performs the other algorithms for all values of n shown.

Timings (s) on UltraSparc-2, single precision keys			
n	Flashsort1	Flashsort2P	Quicksort	MSB radix
1K	0.0004	0.0006	0.0004	0.0004
2K	0.0008	0.0011	0.0009	0.0008
4K	0.0016	0.0023	0.0020	0.0016
8K	0.0035	0.0049	0.0044	0.0033
16K	0.0073	0.0099	0.0095	0.0069
32K	0.0150	0.0199	0.0203	0.0137
64K	0.0313	0.0402	0.0429	0.0276
128K	0.0687	0.0818	0.0916	0.0611
256K	0.1626	0.1896	0.1925	0.1381
512K	0.3840	0.4077	0.3930	0.3022
1M	1.0121	0.9419	0.8516	0.6477
2M	2.4634	1.8245	1.8048	1.3262
4M	5.5477	3.7342	3.8523	2.7178
8M	12.630	7.6996	8.1271	5.5562
16M	27.335	15.641	17.123	11.490
32M	57.912	32.714	36.503	25.166
64M	131.01	66.322	77.206	53.493

Fig. 2. Experimental evaluation of Flashsort1, Flashsort2P, Quicksort and MSB radix sort on a Sun Ultra II using single precision floating point keys.

4.1 Cache Simulations

We also ran these algorithms on an L2 cache simulator for the Sun Ultra II. Figure 3 compares Quicksort and Flashsort1, while Fig 4 compares the three faster algorithms. These figures show the number of L2 cache misses per key for the four algorithms on single precision floating point keys. We observe that:

- when the problem is small and fits in L2 cache, for $n \leq 64K$, the number of misses per key are almost constant for each algorithm, these are the compulsory misses.

- in Flashsort1 for $n \geq 128K$ we see a rapid increase in the number of misses per key as n grows, appearing to level off at over 3 misses per key (as we will see, virtually every access that could be a miss is a miss). Clearly the misses per key are much lower for Flashsort2P, Quicksort and MSB radix sort than for Flashsort1.

- in Quicksort the number of cache misses per key increases very gradually as the problem size increases, reaching about 0.75 misses per key at 64M.[3]

- in Flashsort2P we see a very gradual increase in the misses per key, reaching 0.75 misses per key at 64M. However, the increase for Flashsort2P is not smooth.

- in MSB radix sort the number of cache misses per key increases very gradually reaching a maximum of 1.4. Other than for small n, the cache utilisation for MSB radix sort is much better than Flashsort1.

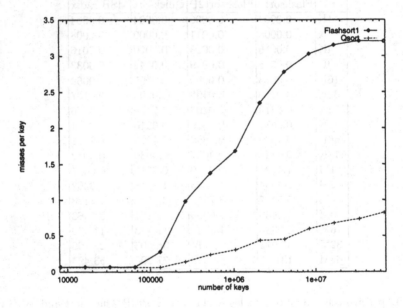

Fig. 3. L2 cache misses per key on single precision floating point keys: Flashsort1 vs Quicksort. Note that the x-axis is on a log scale.

5   Cache Analysis of Flashsort

In this section we analyse the cache misses made by the Flashsort variants. We will assume that the input size is an integral multiple of BC (in particular this means that we only worry about $n \geq 128K$) and that important arrays begin

---

[3] For Quicksort the number of misses per key will continue to grow very gradually as n grows as it performs $O(\log n)$ misses per key, but with a very small constant.

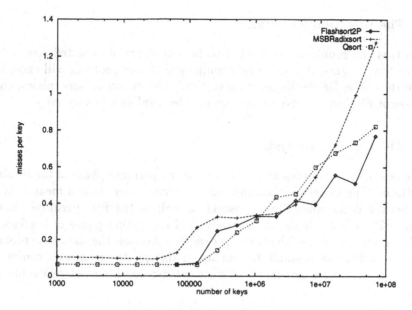

Fig. 4. L2 cache misses per key on single precision floating point keys: Flashsort 2P vs Quicksort vs MSB radix sort. Note that the x-axis is on a log scale.

at locations which are multiples of B. We also ignore rounding as the effect is insubstantial. For the sake of simplicity we will assume that the various phases are independent, i.e. the cache is emptied after each phase. This assumption causes inaccuracies for small input sizes, where a significant part of the input may stay in cache between say the count and permute phases, so we report only the predicted values for $n \geq 256K$. The analyses make extensive use of the idea that if there are k memory blocks $b_1, \ldots, b_k$ mapped to a cache block, and in the 'steady state' memory block b_i is accessed with probability p_i, then for $i = 1, \ldots, k$ the probability that memory block b_j is currently in the cache is $p_j / \sum_{i=1}^{k} p_i$ [6].

5.1   The Count Phase

Step 1 of the count phase consists of n sequential accesses to the data array and n accesses to random count array locations. [4] Step 2 is a sequential traversal through the count array. In addition to the compulsory misses, there may be conflict misses in the first step; these can be analysed by a direct application of the results of [5], and we do not go into greater detail here.

---

[4] Here and later we only count memory accesses which may result in L2 cache misses. For instance the statement COUNT[i] := COUNT[i] + 1 involves two memory accesses, one read and one write. However, the write cannot be a cache miss, so we ignore it.

5.2   The Insertion Sort Phase

Given that the problems which need to be solved are of expected size ≤ 10 in each of the variants, it is extremely unlikely that any problem will exceed the size of the cache. Hence the insertion sort only incurs compulsory misses, and in the case of Flashsort2P, even these are avoided (unless n is very large).

5.3   The Overall Analysis

In the next section we present an analysis of the permute phase of the Flashsort algorithms. The underlying assumptions and parameter choices mean that this explains the cache misses in Flashsort1 as well as the first permute phase of Flashsort2P. What remains is the analysis of the second phase of Flashsort2P, which is presented in the full paper. As expected, because the size of the problems in the second phase is small, the misses in this phase are mostly compulsory misses. However, as n gets large, we start getting small but noticeable numbers of conflict misses.

5.4   The Permute Phase

The permute phase can be viewed as alternating cycle-following with finding cycle leaders. During the cycle following phase we make $2n$ memory accesses which may lead to cache misses. This consists in alternating random access to the count array (to determine the class of x) with one of m 'active' locations in the data array—the pointers to the 'next available' locations for each class—in order to place x. In addition to the compulsory misses, we have the following potential conflicts which may occur in the cycle-following phase.

There are m/B active count blocks, and at most m active data blocks (the number of active data blocks may be much smaller than m if the class sizes are small). Whenever more than one active block is mapped to the same cache block, we potentially have conflict misses. In order to make precise the mapping of active blocks to cache blocks, we consider three cases; these cases cover the the possible parameter choices in our two Flashsort variants.

Case 1: $m < BC$ and $n/m < B$ In this case we assume that each data block contains $p = m/(n/B)$ data pointers (this may be a fraction). There are two regions in the cache: region R_1 which has only data blocks mapped to it and region R_2 which has both data and cache blocks mapped to it. Each cache block in R_1 has $\tau = n/(BC)$ data blocks mapped to it, whereas each block in R_2 has one count array block and τ data array blocks mapped to it. Since each count array block is accessed with probability B/m, and each data array block is accessed with probability $p/m = B/n$, we have that in region R_2:

$$\Pr[\text{Count access in } R_2 \text{ is a hit}] = \frac{B/m}{B/m + B/n \cdot \tau} = \frac{B}{B + m/C} \qquad (1)$$

$$\Pr[\text{Data access in } R_2 \text{ is a hit}] = \frac{B/n}{B/m + B/n \cdot \tau} = \frac{1}{n/m + \tau} \quad (2)$$

Again, each access to the first element in a data block must be a miss, and we add $1/B$ compulsory misses. Of the remaining $(B-1)/B$ fraction of data array accesses, $1 - m/(BC)$ go to region R_1, where the hit rate for data accesses is simply τ^{-1}, whereas m/BC go to R_2 where the hit rate is given by Eqn 2. Therefore, on average:

$$\#misses/n = \frac{1}{B} + \left(1 - \frac{B}{B + m/C}\right) +$$
$$\left(\frac{B-1}{B}\right)\left[\left(1 - \frac{m}{BC}\right)(1 - \tau^{-1}) + \frac{m}{BC}\left(1 - \frac{1}{n/m + \tau}\right)\right], (3)$$

where $\tau = n/(BC)$. For Flashsort1, this formula explains data points for 256K to 1M as shown in Fig 5.

Case 2: $m \geq BC$ and $n/m < B$ We omit the details of this case, which are similar to Case 1. The final formula is:

$$\#misses/n = 1/B + (1 - BC/(2m)) + ((B-1)/B) \cdot (1 - BC/(2n)) \quad (4)$$

For Flashsort1, this formula explains data points for 2M to 64M as shown in Fig 5.

Input size	256K	512K	1M	2M	4M	8M	16M	32M	64M
Predicted	0.8073	1.1495	1.4106	1.6895	1.8604	1.9458	1.9885	2.0099	2.0206
Simulated	0.7927	1.1695	1.4249	1.7330	1.8931	1.9826	2.0080	2.0191	2.0587

Fig. 5. Predicted and simulated miss rates for permute phase of Flashsort1. The predicted values include a term $1/(2B)$ which accounts for misses incurred while searching for cycle leaders. The justification for this term is deferred to the full paper.

Case 3: $m < BC$ and $n/m \gg B$ In this case we assume that there is at most one pointer per data block. Again, we divide the cache into region R_1 which has only data blocks mapped to it and region R_2 which has both data and cache blocks mapped to it. We will assume that each block in R_2 has $\rho = m/C$ active data blocks mapped to it (note that ρ can be fractional) in addition to the one count block. This gives $\Pr[\text{Count access in } R_2 \text{ a hit}] = B/(B + \rho)$, and $\Pr[\text{Data access in } R_2 \text{ a hit}] = 1/(B + \rho)$. This analysis is a bit coarse, but it will prove accurate enough for our purposes.

It is more interesting to study the hit rate in region R_1, which contains only data pointers. It will be convenient to assume that R_1 covers the entire cache, and has m data pointers mapped to it; as we will see, we can scale down the values without changing the hit rate. Let the number of pointers mapped to cache block i be m_i. If cache block i has $m_i \neq 0$, then the probability that a data array access reads cache block i is m_i/m, but the probability of this access being a hit is $1/m_i$. Hence, the probability of a hit given that cache block i was accessed is $1/m$. Summing over all i such that $m_i \neq 0$ gives the overall hit rate as simply ν/m, where ν is the number of cache blocks such that $m_i \neq 0$. Assuming that the pointers are independently and uniformly located in cache blocks, we get the expected value of ν as $C \cdot (1 - 1/C)^m \approx C \cdot (1 - e^{-\rho})$. Hence we have:

$$\Pr[\text{Data access in } R_1 \text{ a hit}] = \nu/m = \rho^{-1}(1 - e^{-\rho}), \tag{5}$$

and note that this is invariant to scaling C and m by the same amount. Hence, on average:

$$\#misses/n = \frac{1}{B} + \left(1 - \frac{B}{B+\rho}\right) +$$
$$\frac{B-1}{B}\left[\frac{m}{BC}\left(1 - \frac{1}{B+\rho}\right) + (1 - \rho^{-1}(1 - e^{-\rho}))\left(1 - \frac{m}{BC}\right)\right] \tag{6}$$

where $\rho = m/C$. Using this we predict the misses during the first permute phase of Flashsort2P below:

Input Size	256K	512K	1M	2M	4M	8M	16M	32M	64M
Predicted	0.112	0.119	0.130	0.145	0.165	0.192	0.230	0.279	0.345
Actual	0.076	0.092	0.134	0.128	0.210	0.181	0.332	0.272	0.511

(The predicted values again include a term of $1/(2B)$ to account for misses incurred while searching for cycle leaders.) As can be seen, the predictions are rather inaccurate for $n = 1M, 4M, 16M$ and $64M$. This is because the assumption of random pointer placement is a particularly poor one for these values. Note that the expected starting location of the k-th pointer in the data array is precisely the expected number of elements in classes $0, \ldots, k-1$. It is therefore a binomial random variable with expected value $k \cdot n/m$ and standard deviation $\sqrt{n \cdot (k/m) \cdot (1 - k/m)}$, from which the expected starting location of this pointer is easily deduced. If n is a multiple of BC then one can reason about this particularly simply. We view the cache as being 'continuous' and place m *marks* on the cache numbered $0, 1, \ldots, m-1$, with the i-th mark being $i \cdot (BC)/m$ words from the beginning of the cache. The expected location in the cache of the k-th pointer is easily seen to be mark $(k\tau) \bmod m$ where $\tau = n/(BC)$. This shows that the choice of $m = 4096$ for $n = 64M = 2^{26}$ is quite bad, since $\tau = 512$, and the expected starting locations of all pointers are marks numbered $512j$, for integer j. In other words, there are only eight different marks where the pointers expected locations lie, and one would expect ν to be very small at the outset.

To simplify the analysis, we make the assumption that the value of ν does not change much over the entire course of the permute phase. This is somewhat plausible: as the algorithm progresses, one would expect all pointers to move at roughly the same rate, hence maintaining the original pattern of pointers. To validate this, we performed extensive simulations with the following parameters: $n = 1M$, $C = 1K$, $B = 16$, and all values of m between 256 and 768 (including the value $m = 512$ which would have been chosen by Flashsort2P). The results are shown in Figure 6, where for each value of the parameters, we plot:

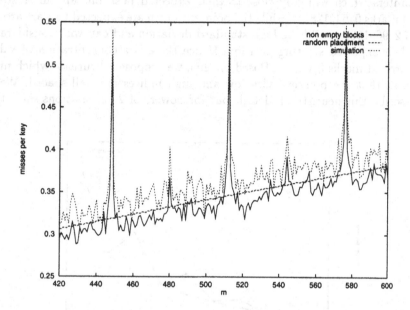

Fig. 6. Exhaustive simulation of various values of m.

- The average number of misses per key over 5 simulations of the permute algorithm ('simulation' in Figure 6);
- The number of misses per key predicted by Eq 6; ('random placement' in Figure 6);
- The number of misses per key predicted by Eq 6, but explicitly calculating ν at the start of the permute phase. ('non-empty blocks' in Figure 6);

This suggests Eq. 6 is a reasonable predictor for most values of m, and also that the initial value of ν does seem to be a fairly good predictor of the value of ν over the course of the permute phase. However, it appears that it is difficult to obtain a simple closed-form expression for the initial value of ν for arbitrary m. One way out might be to choose m so that the analysis is made convenient. For instance, if we choose m to be relatively prime to τ, then we know that all the pointers are expected to start at different marks. Unfortunately, the example of $m = 511$ in Fig 6 shows that there is yet another factor to consider. Notice that $\tau = 64$ in our experiment, and $\gcd(511, 64) = 1$. Although all the pointers are expected to start at different marks, the miss rate for $m = 511$ is still quite high. This is because the expected pointer starting locations are as follows:

mark	0	1	2	3	...	62	63	64	65	66	...
pointer	0	8	16	24	...	496	504	1	9	17	... ..

Observe that pointers with low or high indices have low standard deviation, so the pointers which will stay close to their expected positions are all clustered around marks $0, 64, 128 \ldots$, while the pointers which are expected to start around mark $32, 96, 160, \ldots$ all have high standard deviation and can vary considerably from their expected starting position. Hence there is a concentration of values again around marks $0, 64, \ldots$. Based on this, we propose a heuristic which modifies m so that the pointers with low and high indices are well-spaced. We do not describe this heuristic in detail but for powers of 2, simulations show that

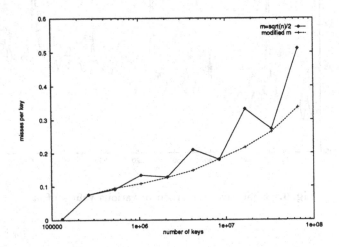

Fig. 7. Smoothing effect of heuristic when n is a power of 2.

this heuristic smooths out the variation seen previously (Fig 7) and indeed for arbitrary values of n the heuristic chooses a value of m close to $0.5\sqrt{n}$ such that the misses for this value of m are close to what would be predicted by random placement (Fig 8).

The improvements produced in actual running time by this heuristic are, however, not noticeable on our machine. This is mainly because we have *physical* caches [3] and virtually every data array access causes a TLB miss (see footnote 1 for a definition of a TLB miss) and hence a memory access, so the improvements apply to only half the memory accesses. In addition, the cache miss rate is already quite low, so waits for memory do not heavily dominate the running time. Our rough calculations suggest that the improvement should be at most 5% on our machine, but this heuristic should be more useful on a machine with a higher cache miss penalty.

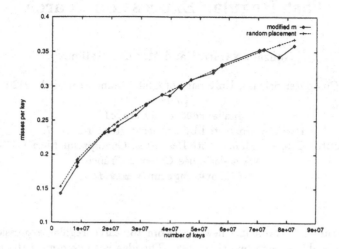

Fig. 8. Smoothing effect of heuristic for arbitrary values of n.

6   Conclusions

We have shown that Flashsort1 is fast when sorting a small number of keys, but due to poor cache utilisation it starts to perform poorly when the data is larger than the cache size. We have shown that a 2-pass variant of Flashsort1, called Flashsort2P, outperforms Flashsort1 and Quicksort for moderate to large values of n, as it make much fewer cache misses. We have analyzed the cache miss rates of the Flashsort variants and can accurately predict the miss rates in the permute phases of these algorithms. We have also shown that the integer sorting algorithm MSB radix sort can be used very effectively on floating point data. The algorithm is very fast due to fast integer operations and relatively good cache utilisation.

References

1. A. V. Aho, J. E. Hopcroft, and J. D. Ullman. *The Design and Analysis of Computer Algorithms*. Addison-Wesley, Reading, Massachusetts, 1974.
2. J. Handy. *The Cache Memory Handbook*. Academic Press, 1998.
3. J. L Hennessy and D. A. Patterson, *Computer Architecture: A Quantitative Approach, 2nd ed.* Morgan Kaufmann, 1996.
4. D. E. Knuth. *The Art of Computer Programming. Volume 3: Sorting and Searching, 3rd ed.* Addison-Wesley, 1997.
5. R. E. Ladner, J. D. Fix and A. LaMarca, The Cache performance of traversals and random accesses In *Proc. 10th ACM-SIAM Symposium on Discrete Algorithms*, pp. 613–622, 1999.
6. A. LaMarca and R. E. Ladner, The influence of caches on the performance of sorting. In *Proc.8th ACM-SIAM Symposium on Discrete Algorithms*, pp. 370–379, 1997
7. K. D. Neubert. The Flashsort1 algorithm. *Dr Dobb's Journal*, pp. 123–125, February 1998. FORTRAN code listing, p. 131, *ibid*.

Fast Regular Expression Search

Gonzalo Navarro[1] and Mathieu Raffinot[2]

[1] Dept. of Computer Science, University of Chile. Blanco Encalada 2120, Santiago,
Chile
gnavarro@dcc.uchile.cl
Partially supported by Fondecyt grant 1-990627.
[2] Institut Gaspard Monge, Cité Descartes, Champs-sur-Marne, 77454
Marne-la-Vallée Cedex 2, France
raffinot@monge.univ-mlv.fr

Abstract. We present a new algorithm to search regular expressions,
which is able to skip text characters. The idea is to determine the minimum length ℓ of a string matching the regular expression, manipulate
the original automaton so that it recognizes all the reverse prefixes of
length up to ℓ of the strings accepted, and use it to skip text characters
as done for exact string matching in previous work. As we show experimentally, the resulting algorithm is fast, the fastest one in many cases
of interest.

1 Introduction

The need to search for regular expressions arises in many text-based applications,
such as text retrieval, text editing and computational biology, to name a few.
A *regular expression* is a generalized pattern composed of (i) basic strings, (ii)
union, concatenation and Kleene closure of other regular expressions. Readers
unfamiliar with the concept and terminology related to regular expressions are
referred to a classical book such as [1].

The traditional technique [16] to search a regular expression of length m in
a text of length n is to convert the expression into a nondeterministic finite
automaton (NFA) with $O(m)$ nodes. Then, it is possible to search the text using
the automaton at $O(mn)$ worst case time. The cost comes from the fact that
more than one state of the NFA may be active at each step, and therefore all
may need to be updated. A more efficient choice [1] is to convert the NFA into a
deterministic finite automaton (DFA), which has only one active state at a time
and therefore allows to search the text at $O(n)$ cost, which is worst-case optimal.
The problem with this approach is that the DFA may have $O(2^m)$ states, which
implies a preprocessing cost and extra space exponential in m.

Some techniques have been proposed to obtain a good tradeoff between both
extremes. In 1992, Myers [13] presented a four-russians approach which obtains
$O(mn/\log n)$ worst-case time and extra space. The idea is to divide the syntax
tree of the regular expression into "modules", which are subtrees of a reasonable
size. These subtrees are implemented as DFAs and are thereafter considered as

leaf nodes in the syntax tree. The process continues with this reduced tree until a single final module is obtained.

The DFA simulation of modules is done using *bit-parallelism*, which is a technique to code many elements in the bits of a single computer word and manage to update all them in a single operation. In this case, the vector of active and inactive states is stored as bits of a computer word. Instead of (ala Thompson [16]) examining the active states one by one, the whole computer word is used to index a table which, together with the current text character, provides the new set of active states (another computer word). This can be considered either as a bit-parallel simulation of an NFA, or as an implementation of a DFA (where the identifier of each deterministic state is the bit mask as a whole).

Pushing even more on this direction, we may resort to pure bit-parallelism and forget about the modules. This was done in [19] by Wu and Manber, and included in their software *Agrep* [18]. A computer word is used to represent the active (1) and inactive (0) states of the NFA. If the states are properly arranged and the Thompson construction [16] is used, all the arrows carry 1's from bit positions i to $i + 1$, except for the ϵ-transitions. Then, a generalization of Shift-Or [3] (the canonical bit-parallel algorithm for exact string matching) is presented, where for each text character two steps are performed. First, a forward step moves all the 1's that can move from a state to the next one, and second, the ϵ-transitions are carried out. As ϵ-transitions follow arbitrary paths, an $E : 2^m \to 2^m$ function is precomputed and stored, where $E(w)$ is the ϵ-closure of w. Possible space problems are solved by splitting this table "horizontally" (i.e. less bits per entry) in as many subtables as needed, using the fact that $E(w_1 w_2) = E(w_1)$ *or* $E(w_2)$. This can be thought of as an alternative decomposition scheme, instead of Myers' modules.

The ideas presented up to now aim at a good implementation of the automaton, but they must inspect all the text characters. In many cases, however, the regular expression involves sets of relatively long substrings that must appear for the regular expression to match. In [17, chapter 5], a multipattern search algorithm is generalized to regular expression searching, in order to take advantage of this fact. The resulting algorithm finds all suffixes (of a predetermined length) of words in the language denoted by the regular expression and uses the Commentz-Walter algorithm [7] to search them. Another technique of this kind is used in *Gnu Grep v2.0*, which extracts a single string (the longest) which must appear in any match. This string is searched for and the neighborhoods of its occurrences are checked for complete matches using a lazy deterministic automaton. Note that it is possible that there is no such single string, in which case the scheme cannot be applied.

In this paper, we present a new regular expression search algorithm able to skip text characters. It is based on extending BDM and BNDM [9, 14]. These are simple pattern search algorithms whose main idea is to build an automaton able to recognize the reverse prefixes of the pattern, and examine backwards a window of length m on the text. This automaton helps to determine (i) when it is possible to shift the window because no pattern substring has been seen, and (ii) the next position where the window can be placed, i.e. the last time that a pattern prefix was seen. BNDM is a bit-parallel implementation of this automaton, faster and much simpler than the traditional version BDM which makes the automaton deterministic.

Our algorithm for regular expression searching is an extension where, by manipulating the original automaton, we search for any reverse prefix of a possible match of the regular expression. Hence, this transformed automaton is a compact device to achieve the same multipattern searching, at much less space. The automata are simulated using bit-parallelism. Our experimental results show that, when the regular expression does not match too short or too frequent strings, our algorithm is among the fastest, faster than all those unable to skip characters and in most cases faster than those based on multipattern matching. An extra contribution is our bit-parallel simulation, which differs from Agrep's in that less bits are used and no ϵ-transitions exist, although the transitions on letters are arbitrary and therefore a separate table per letter is needed (the tables can be horizontally split in case of space problems). Our simulation turns out to be faster than Agrep and the fastest in most cases.

Some definitions that are used in this paper follow. A *word* is a string or sequence of characters over a finite alphabet Σ. A word $x \in \Sigma^*$ is a *factor* (or substring) of p if p can be written $p = uxv$, $u, v \in \Sigma^*$. A factor x of p is called a *suffix* (resp. *prefix*) of p is $p = ux$ (resp. $p = xu$), $u \in \Sigma^*$. We call R our pattern (a regular expression), which is of length m. We note $L(R)$ the set of words generated by R. Our text is of size n.

We define also the language to denote regular expressions. Union is denoted with the infix sign "|", Kleene closure with the postfix sign "*", and concatenation simply by putting the sub-expressions one after the other. Parentheses are used to change the precedence, which is normally "*", ".", "|". We adapt some widely used extensions: $[c_1...c_k]$ (where c_i are characters) is a shorthand for $(c_1|...|c_k)$. Instead of a character c, a range c^1-c^2 can be specified to avoid enumerating all the letters between (and including) c^1 and c^2. Finally, the period (.) represents any character.

2   The Reverse Factor Search Approach

In this section we describe the general reverse factor search approach currently used for a single pattern [12, 9, 14] or multiple patterns [8, 15].

The search is done using a window which has the length of the minimum word that we search (if we search a single word, we just take its length). We note this minimum length ℓ.

We shift the window along the text, and for each position of the window, we search backwards (i.e from right to left, see Figure 1) for any factor of any length-ℓ prefix of our set of patterns (if we search a single word, this means any factor of the word). Also, each time we recognize a factor which is indeed a prefix of some of the patterns, we store the window position in a variable *last* (which is overwritten, so we know the last time that this happened). Now, two possibilities appear:

(i) We do not reach the beginning of the window. This case is shown in Figure 1. The search for a factor fail on a letter σ, i.e σu is not a factor of a length-ℓ prefix of any pattern. We can directly shift the window to start at position *last*, since no pattern can start before, and begin the search again.

(ii) We reach the beginning of the window. If we search just one pattern, then we have recognized it and we report the occurrence. Otherwise, we just recognized a length-ℓ prefix of one or more patterns. We verify directly in the text if there is a match of a pattern, with a forward (i.e left to right) scan. This can be done with a trie of the patterns. Next, in both cases, we shift the window according to position *last*.

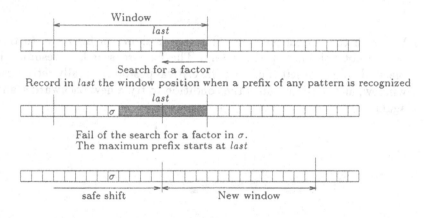

Fig. 1. The reverse factor search approach.

This simple approach leads to very fast algorithms in practice, such as BDM [9] and BNDM [14]. For a single pattern, this is optimal on average, matching Yao's bound [20] of $O(n \log(\ell)/\ell)$ (where n is the text size and ℓ the pattern length). In the worst case, this scheme is quadratic ($O(n\ell)$ complexity). There exists however a general technique to keep the algorithms sub-linear on average and linear in the worst case.

2.1   A Linear Worst Case Algorithm

The main idea used in [9, 14, 8, 15] is to avoid retraversing the same characters in the backward window verification. We divide the work done on the text in two parts: forward and backward scanning. To be linear in the worst case, none of these two parts must retraverse characters. In the forward scan, it is enough to keep track of the longest pattern prefix v that matches the current text suffix.

This is easily achieved with a KMP automaton [11] (for one pattern) or an Aho-Corasick automaton [2] (for multiple patterns). All the matches are found using the forward scan.

However, we need to use also backward searching in order to skip characters. The idea is that the window is placed so that the current longest prefix matched v is aligned with the beginning of the window. The position of the current text character inside the window (i.e. $|v|$) is called the *critical position*. At any point in the forward scan we can place the window (shifted $|v|$ characters from the current text position) and try a backward search. Clearly, this is only promising when v is not very long compared to ℓ. Usually, a backward scan is attempted when the prefix is less than $\lfloor \ell/\alpha \rfloor$, where $0 < \alpha < \ell$ is fixed arbitrary (usually $\alpha = 2$).

The backward search proceeds almost as before, but it finishes as soon as the critical position is reached. The two possibilities are:

(i) We reach the critical position. This case is shown in Figure 2. In this case we are not able to skip characters. The forward search is resumed in the place where it was left (i.e. from the critical position), totally retraverses the window, and continues until the condition to try a new backward scan holds again.

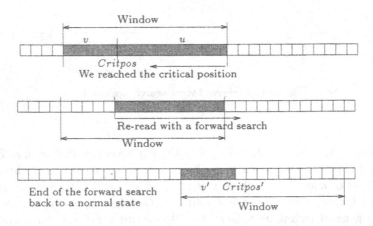

Fig. 2. The critical position is reached, in the linear-time algorithm.

(ii) We do not reach the critical position. This case is shown in Figure 3. This means that there cannot be a match in the current window. We start a forward scan from scratch at position *last*, totally retraverse the window, and continue until a new backward scan seems promising.

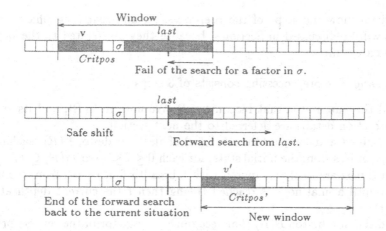

Fig. 3. The critical position is not reached, in the linear-time algorithm.

3 Extending the Approach to Regular Expression Searching

In this section we explain how to adapt the general approach of Section 2 to regular expression searching. We first explain a simple extension of the basic approach and later show how to keep the worst case linear. Recall that we search for a regular expression called R which is of size m and generates the language $L(R)$.

3.1 Basic Approach

The search in the general approach needs a window of length ℓ (shortest pattern we search). In regular expression searching this corresponds to the length of the shortest word of $L(R)$. Of course, if this word is ϵ, the problem of searching is trivial since every text position matches. We consider in the rest of the paper that $\ell > 0$.

We use the general approach of Section 2, consisting of a backward and, if necessary (i.e we reached the beginning of the window), a forward scan. To adapt this scheme to regular expression search, we need two modifications:

(i) The backward search step of the general approach imposes here that we are able recognize any factor of the reverse prefixes of length ℓ of $L(R)$. Moreover, we mark in a variable *last* the longest prefix of $L(R)$ recognized (of course this prefix will be of length less than ℓ).

(ii) The forward search (if we reached the beginning of the window) verifies that there is a match of the regular expression starting at the beginning of the window (however, the match can be much longer than ℓ).

We detail now the steps of the preprocessing and searching phases. Complexities will be discussed in Section 4 because they are related to the way the automata are built.

Preprocessing The preprocessing consists of 3 steps:

1. Build the automaton that recognizes R. We note it $F(R)$, and its specific construction details are deferred to the next section.
2. Determine ℓ and compute the set $P_i(R)$ of all the nodes of $F(R)$ reachable in i steps or less from the initial state, for each $0 \leq i \leq \ell$ (so $P_i(R) \subseteq P_{i+1}(R)$). Both things are easily computed with a breadth-first search from the initial state until a final node is reached (being then ℓ the current depth at that point).
3. Build the automaton $B(R)$ that recognizes any factor of the reverse prefixes of length ℓ of $L(R)$. This is achieved by restricting the original automaton $F(R)$ to the nodes of $P_\ell(R)$, reversing the arrows, taking as (the only) terminal state the initial state of $F(R)$ and all the states as initial states.

The most interesting part of the above procedure is $B(R)$, which is a device to recognize the reverse factors of prefixes of length ℓ of $L(R)$. It is not hard to see that any such factor corresponds to a path in $F(R)$ that touches only nodes in $P_\ell(R)$. In $B(R)$ there exists the same path with the arrows reversed, and since all the states of $B(R)$ are initial, there exists a path from an initial state that spells out the reversed factor. Moreover, if the factor is a prefix, then the corresponding path in $B(R)$ leads to its final state.

Note, however, that $B(R)$ can recognize more words than desired. For instance, if there are loops in $B(R)$ it can recognize words longer than ℓ. However, we can restrict more the set of words recognized by $B(R)$. The idea is that, if a state of $B(R)$ is active but it is farther than i positions to the final state of $B(R)$, and only i window characters remain to be read, then this state cannot lead to a match. Hence, if we have to read i more characters of the window, we intersect the current active states of $B(R)$ with the set $P_i(R)$.

It is easy to see that, with this modification, the automaton recognizes exactly the desired prefixes, since if a state has not been "killed" with the intersection with $P_i(R)$ it is because it is still possible to obtain a useful prefix from it. Hence, only the desired (reverse) factors can survive all the process until they arrive to the final state and become (reverse) prefixes.

In fact, an alternative method in this sense would be a classical multi-pattern algorithm to recognize the reverse factors of the set of prefixes of length ℓ of $L(R)$. However, this set may be large and the resulting scheme may need much more memory. The automaton $B(R)$ is a more compact device to obtain the same result.

Searching The search follows the general approach of Section 2. For each window position, we activate all the states of $B(R)$ and traverse the window backwards updating *last* each time the final state of $B(R)$ is reached (recall that after each step, we "kill" some states pf $B(R)$ using $P_i(R)$). If $B(R)$ runs out of active states we shift the window to position *last*. Otherwise, if we reached the beginning of the window, we start a forward scan using $F(R)$ from the beginning of the window until either a match is found[1], we reached the end of the text, or $F(R)$ runs out of active states. After the forward scan, we shift the window to position *last*.

3.2   Linear Worst Case Extension

We also extended the general linear worst case approach (see Section 2.1) to the case of regular expression searching.

The main difficulty to extend the general linear approach is where to place the window in order to not lose a match. The general approach considers the longest prefix of the pattern already recognized. However, this information cannot be inferred only from the active states of the automaton (for instance, it is not known how many times we have traversed a loop). We use an alternative concept: instead of considering the longest prefix already matched, we consider the shortest path to reach a final state. This value can be determined from the current set of states. We devise two different alternatives that differ on the use of this information.

We transform the forward scan automaton $F(R)$ of the previous algorithm by adding a self-loop at its initial state, for each letter of Σ (so now it recognizes $\Sigma^* L(R)$). This is the normal automaton used for classical searching, and the one we use for the forward scanning.

Prior to explaining both alternatives, we introduce some notation. In general, the window is placed so that it finishes ℓ' characters ahead of the current text position (for $0 \leq \ell' \leq \ell$). To simplify our explanation, we call this ℓ' the "forward-length" of the window.

In the first alternative the forward-length of the window is the shortest path from an active state of $F(R)$ to a final state (this same idea has been used for multipattern matching in [8]). In this case, we need to recognize any reverse factor of $L(R)$ in the backward scan (not only the factors of prefixes of length ℓ)[2]. Each time ℓ' is large enough to be promising ($\ell' \geq \alpha\ell$, for some heuristically fixed α), we stop the forward scan and start a backward scan on a window of forward-length ℓ' (the *critical position* being $\ell - \ell'$). If the backward automaton runs out of active states before reaching the critical position, we shift the window as in the general scheme (using the *last* prefix found) and restart a fresh forward scan. Otherwise, we continue the previous forward scan from the critical position, totally traversing the window and continuing until the condition to start a backward scan holds again.

The previous approach is linear in the worst case (since each text position is scanned at most once forward and at most once backwards), and it is able to

---

[1] Since we report the beginning of matches, we stop the forward search as soon as we find a match.

[2] A more strict choice is to recognize any reverse factor of any word of length ℓ' that starts at an active state in $F(R)$, but this needs much more space and preprocessing time.

skip characters. However, a problem is that *all* the reverse factors of $L(R)$ have to be recognized, which makes the backward scans longer and the shifts shorter. Also, the window forward-length ℓ' is never larger than our previous ℓ, since the initial state of $F(R)$ is always active.

The second alternative solves some of these problems. The idea now is that we continue the forward scan until all the active states belong to $P_i(R)$, for some fixed $i < \ell$ (say, $i = \ell/2$). In this case, the forward-length of the window is $\ell' = \ell - i$, since it is not possible to have a match after reading that number of characters. Again, we select heuristically a minimum $\ell' = \alpha\ell$ value. In this case, we do not need to recognize all the factors. Instead, we can use the already known $B(R)$ automaton. Note that the previous approach applied to this case (with all active states belonging to $P_i(R)$) yields different results. In this case we limit the set of factors to recognize, which allows to shift the window sooner. On the other hand, its forward-length is shorter.

4   Building an NFA from a Regular Expression

There exist currently many different techniques to build an NFA from a regular expression R of size m. The most classical one is the Thomson construction [16]. It builds an NFA with at most $2m$ states that present some particular properties. Some algorithms like that of Myers [13] and of Wu and Manber in Agrep [19, 18] make use of these properties.

A second one is the Glushkov's construction, popularized by Berry and Sethi in [4]. The NFA resulting of this construction has the advantage of having just $m + 1$ states (one per position in the regular expression). A lot of research on Glushov's construction has been pursued, like [5], where it is shown that the resulting NFA is quadratic in the number of edges in the worst case. In [10], a long time open question about the minimal number of edges of an NFA with linear number of states was answered, showing a construction with $O(m)$ states and $O(m(\log m)^2)$ edges, as well as a lower bound of $O(m \log m)$ edges. Hence, Glushkov construction is not space-optimal.

Some research has been done also to try to construct directly a DFA from a regular expression, without constructing an NFA, such as [6].

For our purpose, when we consider bit-parallelism, the most interesting is to have a minimal number of states, because we manage computer words of a fixed length w to represent the set of possible states. Hence, we choose the original Gluskov's construction, which leads to an NFA with $m+1$ states and a quadratic (in the worst case) number of edges. The number of edges is unimportant for our case.

In Gluskov's construction, the edges have no simple structure, and we need a table which, for each current set of states and each current text character, gives the new set of states. On the other hand, the construction of Wu and Manber uses the regularities of Thompson's construction so that they need only a table for the ϵ-transitions, not for every character. In exchange, we have $m + 1$ states instead of nearly $2m$ states, and hence their table sizes square ours. As we show later experimentally, our NFA simulation is faster than those based on the Thompson construction, so the tradeoff pays off.

5   Experimental Results

We compare in this section our approach against previous work. We divide this comparison in three parts. First, we compare different existing algorithms to implement an automaton. These algorithms process all the text characters, one by one, and they only differ in the way they keep track of the state of the search. The goal of this comparison is just to show that our simulation is competitive. Second, we compare, using our automaton simulation, a simple forward-scan algorithm against the different variants of backward search proposed, to show that backward searching is faster in general. Finally, we compare our backward search algorithm against other algorithms that are also able to skip characters.

We use an English text (writings of B. Franklin), filtered to lower-case and replicated until obtaining 10 Mb. A major problem when presenting experiments on regular expressions is that there is not a concept of "random" regular expression, so it is not possible to search, say, 1,000 random patterns. Lacking such good choice, we fixed a set of 10 patterns, which were selected to illustrate different interesting cases rather than more or less "probable" cases. In fact we believe that common patterns have long exact strings and our algorithm would behave even better than in these experiments. Therefore, the goal is not to show what are the typical cases in practice but to show how the scheme behaves under different characteristics of the pattern.

The patterns are given in Table 1. We also show their number of letters, which is closely related to the size of the automata recognizing them, the minimum length ℓ of a match for each pattern, and a their empirical matching probability (number of matches divided by n). The period (.) in the patterns matches any character except the end of line (lines have approximately 70 characters).

No.	Pattern	Size (# letters)	Minimum length ℓ	Prob. match (empirical)
1	benjamin|franklin	16	8	.00003586
2	benjamin|franklin|writing	23	7	.0001014
3	[a-z][a-z0-9]*[a-z]	3	2	.6092
4	benj.*min	8	7	.000007915
5	[a-z][a-z][a-z][a-z][a-z]	5	5	.2024
6	(benj.*min)|(fra.*lin)	15	6	.00003586
7	ben(a|(j|a)*)min	9	6	.009491
8	be.*ja.*in	8	6	.00001211
9	ben[jl]amin	8	8	.000007915
10	(be|fr)(nj|an)(am|kl)in	14	8	.00003586

Table 1. The patterns used on English text.

Our machine is a Sun UltraSparc-1 of 167 MHz, with 64 Mb of RAM, running Solaris 2.5.1. We measured CPU times in seconds, averaging 10 runs over the 10 Mb (the variance was very low). We include the time for preprocessing in the figures.

5.1 Forward Scan Algorithms

In principle, any forward scan algorithm can be enriched with backward searching to skip characters. Some are easier to adapt than others, however. In this experiment we only consider the performance of the forward scan method we adopted. The purpose of this test is to show that our approach is competitive against the rest. We have tested the following algorithms for the forward scanning (the implementations are ours except otherwise stated). See the Introduction for detailed descriptions of previous work.

DFA: builds the classical deterministic automaton and runs it over the text. We have not minimized the automaton.

Thompson: simulates the nondeterministic automaton by keeping a list of active states which is updated for each character read (this does not mean that we build the automaton using Thompson's method).

BP-Thompson: same as before, but the set of active states is kept as a bit vector. Set manipulation is faster when many states are active.

Agrep: uses a bit mask to handle the active states [19]. The software [18] is from S. Wu and U. Manber, and has an advantage on frequently occurring patterns because it abandons a line as soon as it finds the pattern on it.

Myers: is the algorithm based on modules implemented as DFAs [13]. The code is from G. Myers.

Ours: our forward algorithm, similar to that of Agrep except because we eliminate the ϵ-transitions and have a separate transition table for each character (Section 4).

Except for Agrep and Myers, which have their own code, we use the NFA construction of Section 4. Table 2 shows the results on the different patterns. As it can be seen, the schemes that rely on nondeterministic simulation (Thompson variants) worsen when the combination of pattern size and matching probability increases. The rest is basically insensitive to the pattern, except because all worsen a little when the pattern matches very frequently. If the pattern gets significantly larger, however, the deterministic simulations worsen as well, as some of them are even exponentially depending on the automaton size. Agrep, Myers and Ours can adapt at higher but reasonable costs, proportional to the pattern length. This comes not only from the possible need to use many machine words but also because it may be necessary to cut the tables horizontally.

With respect to the comparison, we point out that our scheme is competitive, being the fastest in many cases, and always at most 5% over the performance of the fastest. DFA is the best in the other cases. Our algorithm can in fact be seen as a DFA implementation, where our state identifier is the bit mask and the

Pattern	DFA	Thompson	BP-Thompson	Agrep	Myers	Ours	Ours/best
1	0.70	4.47	4.19	1.88	5.00	**0.68**	1.00
2	**0.73**	4.13	5.10	1.89	8.57	0.76	1.04
3	1.01	18.2	3.75	**0.98**	2.19	0.99	1.01
4	0.71	4.16	3.17	0.97	2.17	**0.68**	1.00
5	0.87	18.7	4.32	1.05	2.18	**0.82**	1.00
6	0.76	4.25	4.06	1.87	4.94	**0.72**	1.00
7	0.73	4.67	2.82	0.99	2.17	**0.72**	1.00
8	**0.72**	4.93	3.40	0.96	2.18	0.73	1.01
9	**0.66**	4.75	3.11	1.00	2.16	0.68	1.03
10	**0.71**	4.36	3.97	1.86	5.01	0.73	1.03

Table 2. Forward search times on English, in seconds for 10 Mb.

transition table is the one we use. However, the DFA has less states, since most of the bit combinations we store are in fact unreachable[3]. On the other hand, the bit-parallel implementation is much more flexible when it comes to adapt it for backward searching or to extend it to handle extended patterns or to allow errors.

We have left aside lazy deterministic automata implementations. However, as we show in Section 5.3, these also tend to be slower than ours.

5.2 Forward Versus Backward Scanning

We compare now our new forward scan algorithm (called **Fwd** in this section and **Ours** in Section 5.1) against backward scanning. There are three backward scanning algorithms. The simplest one, presented in Section 3.1, is called **Bwd**. The two linear variations presented in Section 3.2 are called **LBwd-All** (that recognizes all the reverse factors) and **LBwd-Pref** (that recognizes reverse factors of length-ℓ prefixes). The linear variations depend on an α parameter, which is between 0 and 1. We have tested the values $0.25, 0.50$ and 0.75 for α, although the results change little.

Table 3 shows the results. We obtained improvements in 7 of the 10 patterns (and very impressive in four cases). In general, the linear versions are quite bad in comparison with the simple one, although in some cases they are faster than forward searching. It is difficult to determine which of the two versions is better in which cases, and which is the best value for α.

5.3 Character Skipping Algorithms

Finally, we consider other algorithms able to skip characters.   Basically, the other algorithms are based in extracting one or more strings from the regular

---

[3] We do not build the transition table for unreachable states, but we do not compact reachable states in consecutive table positions as the DFA implementation does. This is the essence of the bit-parallel implementation.

Pattern	Fwd	Bwd	LBwd-All			LBwd-Pref		
			$\alpha = 0.25$	$\alpha = 0.50$	$\alpha = 0.75$	$\alpha = 0.25$	$\alpha = 0.50$	$\alpha = 0.75$
1	0.68	**0.28**	0.44	0.43	0.46	0.47	0.49	0.50
2	0.76	**0.65**	1.17	1.00	1.09	0.93	0.98	0.95
3	**0.99**	2.37	3.30	2.59	3.01	2.56	2.56	2.56
4	0.68	**0.56**	1.70	1.68	1.71	0.94	0.93	0.92
5	**0.82**	2.02	2.05	2.40	2.09	2.13	2.15	2.18
6	0.72	**0.70**	1.82	1.85	1.84	1.10	1.12	1.09
7	0.72	**0.30**	0.46	0.45	0.47	0.51	0.52	0.51
8	**0.73**	0.91	1.75	1.85	1.87	1.33	1.45	1.47
9	0.68	**0.24**	0.37	0.37	0.39	0.41	0.39	0.41
10	0.73	**0.29**	0.42	0.45	0.44	0.47	0.46	0.48

Table 3. Backward search times on English, in seconds for 10 Mb.

expression, so that some of those strings must appear in any match. A single- or multi-pattern exact search algorithm is then used as a filter, and only where some string in the set is found its neighborhood is checked for an occurrence of the whole regular expression. Two approaches exist:

Single pattern: one string is extracted from the regular expression, so that the string must appear inside every match. If this is not possible the scheme cannot be applied. We use *Gnu Grep v2.3*, which implements this idea. Where the filter cannot be applied, *Grep* uses a forward scanning algorithm which is 30% slower than our *Fwd*[4]. Hence, we plot this value only where the idea can be applied. We point out that *Grep* also abandons a line when it finds a first match in it.

Multiple pattern: this idea was presented in [17]. A length $\ell' < \ell$ is selected, and all the possible suffixes of length ℓ' of $L(R)$ are generated and searched for. The choice of ℓ' is not obvious, since longer strings make the search faster, but there are more of them. Unfortunately, the code of [17] is not public, so we have used the following procedure: first, we extract by hand the suffixes of length ℓ' for each regular expression; then we use the multipattern search of *Agrep* [18], which is very fast, to search those suffixes; and finally the matching lines are sent to *Grep*, which checks the occurrence of the regular expression in the matching lines. We find by hand the best ℓ' value for each regular expression. The resulting algorithm is quite similar to the idea of [17].

Our algorithms are called **Fwd** and **Bwd** and correspond to those of the previous sections. Table 4 shows the results. The single pattern filter is a very effective trick, but it can be applied only in a restricted set of cases. In some cases its improvement over our backward search is modest. The multipattern

---

[4] Which shows that our implementation is faster than a good lazy deterministic automaton implementation.

filter, on the other hand, is more general, but its times are higher than ours in general, especially where backward searching is better than forward searching (an exception is the 2nd pattern, where we have a costly preprocessing).

Pattern	Fwd	Bwd	Single pattern filter	Multipattern filter
1	0.68	**0.28**	—	0.31
2	0.76	0.65	—	**0.37**
3	**0.99**	2.37	—	1.65
4	0.68	0.56	**0.17**	0.87
5	**0.82**	2.02	—	2.02
6	0.72	**0.70**	—	1.00
7	0.72	0.30	**0.26**	0.44
8	0.73	0.91	**0.63**	0.66
9	0.68	0.24	**0.19**	0.31
10	0.73	**0.28**	0.98	0.35

Table 4. Algorithm comparison on English, in seconds for 10 Mb.

6 Conclusions

We have presented a new algorithm for regular expression searching able to skip characters. It is based on an extension of the backward DAWG matching approach, where the automaton is manipulated to recognize reverse prefixes of strings of the language. We also presented two more complex variants which are of linear time in the worst case. The automaton is simulated using bit-parallelism.

We first show that the bit-parallel implementation is competitive (at most 5% over the fastest one in all cases). The advantage of bit-parallelism is that the algorithm can easily handle extended patterns, such as classes of characters, wild cards and even approximate searching. We then compare the backward matching against the classical forward one, finding out that the former is superior when the minimum length of a match is not too short and the matching probability is not too high.

Finally, we compare our approach against others able to skip characters. These are based on filtering the search using multipattern matching. The experiments show that our approach is faster in many cases, although there exist some faster hybrid algorithms which can be applied in some restricted cases. Our approach is more general and performs reasonably well in all cases.

The preprocessing time is a subject of future work. In our experiments the patterns were reasonably short and the simple technique of using one transition table was the best choice. However, longer patterns would need the use of the table splitting technique, which worsens the search times. More work on minimizing the NFA could improve the average case.

Being able to skip characters and based on an easily generalizable technique such as bit-parallelism permits to extend our scheme to deal with other cases, such as searching a regular expression allowing errors, and being still able to skip characters. This is also a subject of future work.

References

1. A. Aho, R. Sethi, and J. Ullman. *Compilers: Principles, Techniques and Tools.* Addison-Wesley, 1985.
2. A. V. Aho and M. J. Corasick. Efficient string matching: an aid to bibliographic search. *Communications of the ACM*, 18(6):333–340, 1975.
3. R. Baeza-Yates and G. Gonnet. A new approach to text searching. *CACM*, 35(10):74–82, October 1992.
4. G. Berry and R. Sethi. From regular expression to deterministic automata. *Theor. Comput. Sci.*, 48(1):117–126, 1986.
5. A. Brüggemann-Klein. Regular expressions into finite automata. *Theoretical Computer Science*, 120(2):197–213, November 1993.
6. C.-H. Chang and R. Paige. From regular expression to DFA's using NFA's. In *Proc. of the CPM'92*, number 664 in LNCS, pages 90–110. Springer-Verlag, 1992.
7. B. Commentz-Walter. A string matching algorithm fast on the average. In *Proc. ICALP'79*, number 6 in LNCS, pages 118–132. Springer-Verlag, 1979.
8. M. Crochemore, A. Czumaj, L. Gasieniec, S. Jarominek, T. Lecroq, W. Plandowski, and W. Rytter. Fast practical multi-pattern matching. Rapport 93-3, Institut Gaspard Monge, Université de Marne la Vallée, 1993.
9. A. Czumaj, Maxime Crochemore, L. Gasieniec, S. Jarominek, Thierry Lecroq, W. Plandowski, and W. Rytter. Speeding up two string-matching algorithms. *Algorithmica*, 12:247–267, 1994.
10. Juraj Hromkovič, Sebastian Seibert, and Thomas Wilke. Translating regular expression into small ε-free nondeterministic automata. In *STACS 97*, LNCS, pages 55–66. Springer-Verlag, 1997.
11. D. E. Knuth, J. H. Morris, and V. R. Pratt. Fast pattern matching in strings. *SIAM Journal on Computing*, 6(1):323–350, 1977.
12. T. Lecroq. Experimental results on string matching algorithms. *Softw. Pract. Exp.*, 25(7):727–765, 1995.
13. E. Myers. A four-russian algorithm for regular expression pattern matching. *J. of the ACM*, 39(2):430–448, 1992.
14. G. Navarro and M. Raffinot. A bit-parallel approach to suffix automata: Fast extended string matching. In *Proc. CPM'98*, LNCS v. 1448, pages 14–33, 1998.
15. Mathieu Raffinot. On the multi backward dawg matching algorithm (MultiBDM). In *Proc. WSP'97*, pages 149–165. Carleton University Press, 1997.
16. K. Thompson. Regular expression search algorithm. *CACM*, 11(6):419–422, 1968.
17. B. Watson. *Taxonomies and Toolkits of Regular Language Algorithms.* Phd. dissertation, Eindhoven University of Technology, The Netherlands, 1995.
18. S. Wu and U. Manber. Agrep – a fast approximate pattern-matching tool. In *Proc. of USENIX Technical Conference*, pages 153–162, 1992.
19. S. Wu and U. Manber. Fast text searching allowing errors. *CACM*, 35(10):83–91, October 1992.
20. A. Yao. The complexity of pattern matching for a random string. *SIAM J. on Computing*, 8:368–387, 1979.

An Experimental Evaluation of Hybrid Data Structures for Searching*

Maureen Korda and Rajeev Raman

Department of Computer Science
King's College London
Strand, London WC2R 2LS, U. K.
{mo, raman}@dcs.kcl.ac.uk

Abstract. A common paradigm in data structures is to combine two different kinds of data structures into one, yielding a *hybrid* data structure with improved resource bounds. We perform an experimental evaluation of hybrid data structures in the context of maintaining a dynamic ordered set whose items have integer or floating-point keys. Among other things we demonstrate clear speedups over library implementations of search trees, both for predecessor queries and updates. Our implementations use very little extra memory compared to search trees, and are also quite generic.

1 Introduction

Solutions to data structuring problems often involve trade-offs: for example, some problems can be solved faster if the data structure is allowed more space, while in other problems faster updates may come at the expense of slower queries. In some cases, two different data structures for the same problem may lie on opposite extremes of the trade-off, and it may be possible to combine these data structures into a *hybrid* data structure which outperfoms each individual data structure. This principle has been used in the theoretical data structure literature [19, 23, 14, 7, 22, 24], in external-memory data structures [15] and in practical implementations of indices for text searching [16].

We study hybrid data structures in the context of the *dynamic predecessor* problem, which is to maintain a set S of keys drawn from an ordered universe U under the operations of insertion, deletion and the *predecessor query* operation, which, given some $x \in U$ as input, returns $pred(x, S) = \max\{y \in S \mid y \leq x\}$. The dynamic predecessor problem is a key step in a number of applications, and most libraries of data structures provide an ADT for this problem, e.g. LEDA [17] provides the *sorted sequence* ADT and STL [21] provides the *set* ADT which has similar functionality. These library implementations are based on balanced search trees [5] or skip lists [20] and obtain information about the relative order of two elements of U only through pairwise comparisons. However, additional

---

* Supported in part by EPSRC grant GR/L92150

J.S. Vitter and C.D. Zaroliagis (Eds.): WAE'99, LNCS 1668, pp. 213–227, 1999.
© Springer-Verlag Berlin Heidelberg 1999

knowledge about the universe U can lead to asymptotically more efficient data structures than would be possible in the comparison-based framework.

For example, a number of data structures for the RAM model have been proposed for this problem when the keys are integers. Some of these achieve $O(\sqrt{\log n})$ query and update times (see [10] for a survey of results). Although these algorithms appear to be too complex to be competitive, there is a simple alternative, the *digital search tree* or *trie* [13, p492ff]. There are many papers— both theoretical and experimental—that deal with tries (see e.g. [2, 18, 6]), but these mostly focus on the data type of variable-length strings. Also, the emphasis is mostly on *dictionary* queries (given a key x, does it belong to S?) which are easier than predecessor queries in the context of tries.[1] A common conclusion from these papers is that tries are slow to update and are memory-hungry. On the other hand, balanced search trees are (relatively) slow to search but are quick to update once the item has been located. For example, red-black trees and skiplists take $O(1)$ amortized or expected time to perform an update if the location of the update is known. Since the strengths of tries and search trees are complementary, one could try to combine the two into a hybrid data structure.

In a hybrid data structure for the dynamic predecessor problem, the keys are partitioned into a collection of *buckets*, such that for any two buckets B, B', $B \neq B'$, either it holds that $\max B < \min B'$ or vice versa. The keys belonging to each bucket are stored in a dynamic predecessor data structure for that bucket. Each bucket is associated with a *representative* key, and the representative keys of all buckets are stored in a *top-level* data structure, which itself is a dynamic predecessor data structure. The representative key k for a bucket B is such that $k \leq \min B$, and for any $B' \neq B$ either $k < \min B'$ or $k > \max B'$ (as a concrete example, the representative of a bucket could just be the smallest key stored in it, but in general k need not belong to B). From this it follows that if a key x is in the hybrid data structure, it must belong to the bucket whose representative is x's predecessor in the top-level data structure. Hence, searching in a hybrid is simply a predecessor search in the top-level followed by a search in the appropriate bucket.

The size of the buckets is determined by a parameter b which can be viewed as the 'ideal' bucket size. If a bucket gets too large or too small, it is brought back to a near-ideal size by operations such as: splitting a bucket into two, joining two buckets or transferring elements from one bucket to another. These operations may cause the set of representatives to change, but this is done so that a sequence of n inserts or deletes, starting from an empty data structure, result in $O(n/b)$ changes to the top level.

As can be seen, even relatively small bucket sizes should greatly reduce the number of operations on, and hence the number of keys in, the top level substantially. This suggests that a trie-search tree hybrid should have good all-round performance, as the update time and memory use of the trie should be greatly reduced, while the search would be slightly slower due to the use of search trees

---

[1] A few recent papers also deal with a kind of prefix-matching problem which appears in IP routers (see eg [3]).

in the (small) buckets. Hybrids of tries and trees have been shown to have reasonable asymptotic performance, e.g. [24] shows $O(\sqrt{w})$ time and $O(n)$ space, where w is the word size of the machine.

Since tries and search trees have each been studied extensively, it may seem that there is little need for an investigation of a data structure which combines the two. However, the use of these data structures in a hybrid introduces certain differences, so existing studies may not suffice:

(i) Buckets need to support relatively complex operations such as *split* or *join*, which are not as well studied in the literature.

(ii) Bucket sizes are much smaller than usual test input sizes, so we are interested in performance for 'small' n rather than 'large' n.

(iii) In most tests of search trees, the data structure being tested is the main entity in memory and it is likely to have use of much of the cache. However, in a hybrid data structure, the top-level data structure is more likely to occupy cache, and the data structure for each bucket will have to be brought into cache each time it is needed. Hence, data structures with poor cache performance when used as isolated data structures (e.g. splay trees) may prove more competitive in a hybrid data structure.

Also, a trie in a hybrid data structure may store keys drawn from somewhat unusual distributions. This is because the distribution of representatives in the top-level data structure is the result of applying the representative-selection process to a sequence of inputs drawn from the input distribution. This difference in distributions can lead to tangible effects in performance, as we note in Section 4. Since the process of choosing representatives can be arbitrarily complex, the average-case properties of the trie may even be intractable to exact analysis. Similarly, statistical properties of buckets can also be hard to obtain.

In addition, there is also some interest from the software interface design aspect, as it is important for data structures to be *generic*, i.e. to work unchanged with different data types [21]. Some thought needs to go into making the trie generic, as it works with the representation of a key rather than its value. Our current implementation supports the keys of type int, **unsigned int**, **float** or **double**. The hybrid data structure also provides another valuable kind of genericity, the ability to 'plug-and-play' with different top-level data structures, as well as to use various search tree implementations in the buckets.

2 Implementations

All implementations have been done in C++ (specifically **g++** version 2.8.1). There are four different main classes in the hybrid data structure at present:

The bucket base class The class **Bucket** is an empty class used for communication between the top-level data structure and the collection of buckets

The hybrid class The external interface to the user is provided by a class template Hybrid, which takes four arguments: the type of key to be stored, the type of the information associated with each key, the class which is to be used for the top level data structure and the class which is to be used at the bottom level. The hybrid data structure currently supports three main operations:

void insert(*key_type* k, *item_type* i);

Inserts key k and associated information i, if k is not already present. If k is present, then we change the information associated with k to i.

void locate_pred(*key_type* k, *key_type* &pred, *info_type* &pred_info);

Returns the predecessor of k in pred, and information associated with pred in pred_info. If k does not have a predecessor these values are undefined.

bool del(*key_type* k);

Deletes k together with its associated information i if k is present. Otherwise, the operation is null.

An instance of Hybrid contains an instance each of the top-level class and the bottom-level class. The top-level class and the bottom-level data structure are expected to implement certain interfaces. In particular, the top-level class should be generic, implement dynamic predecessor operations, should hold pairs of the type ⟨*key_type*, Bucket *⟩ and be searchable by the *key_type* component. The bottom-level class should be a *collection of buckets* class (described in greater detail below) and will hold pairs of the form ⟨*key_type*, *info_type*⟩, namely the pairs which are input by the user.

The hybrid data structure implements its externally-available functions by calling the appropriate functions in the top and bottom-level data structures, as the following example shows:

```
void locate_pred(key_type k, key_type &pred, info_type &pred_info) {
    Bucket* B;
    key_type Bkey;

    Top.locate_pred(k, Bkey, B);
    Bottom.locate_pred(B, k, pred, pred_info);
}
```

As all Hybrid functions are inline, an optimising compiler should eliminate the calls to Hybrid and hence avoid the overhead of the indirection.

The trie class The class template Trie takes two arguments, the type of key to be stored and the type of information associated with it. Internally, the trie operates on a fixed-length string of 'symbols'. A data type such as an integer or a floating-point number is viewed as such a string by chopping up its (bit-string) representation. into chunks of bits, each chunk being viewed as a symbol.

The problem with the string representation is that the representation $rep(x)$ of a value x need not be ordered the same way as x even for common data types x. For unsigned integers the representation and the value of a key obey the same ordering, that is $x < y$ iff $rep(x) < rep(y)$ (the bit-strings are compared lexi-cographically). Also, signed integers and floating point numbers which obey the IEEE 754 standard also satisfy the relation $x < y$ iff $rep(x) < rep(y)$ provided both x and y are non-negative.

However, signed negative integers are represented by their 2s complement, so that $x < y$ iff $rep(x) < rep(y)$ holds even when x and y are both negative, but if $x < 0$ and $y \geq 0$, then $rep(x) > rep(y)$. Finally, negative floating-point numbers use the signed-magnitude representation, so that $x < y$ iff $rep(x) > rep(y)$ holds whenever $x < 0$. Hence before processing each key in the trie, we must transform it as follows:

signed integers: complement MSB.
floating point numbers: complement MSB of non-negative numbers, complement all bits for negative numbers.

The trie class implements (overloaded) functions which convert a key of the types **int**, **unsigned int**, **float** and **double** into a string representation. When the template is instantiated, the compiler chooses the appropriate function from these, with a compile-time error resulting if the key type is not among the above data types.

We now summarise the main design choices in implementing a trie. We have implemented a *compressed* trie, i.e. where there are no internal nodes of out-degree 1. Although compression does not offer benefits for the input distributions and input sizes that we have used for testing [13], we still use it because of the better worst-case memory usage of a compressed trie. Consequently, we suffer from slightly more complex logic while searching.

We have also chosen to divide the key representation into equal-sized chunks[2], i.e., each trie internal node has the same branching factor. For specific distribu-tions, this may not be the best thing to do: for example, it may be better to have a larger branching factor near the root and a smaller branching factor near the leaves [1]. By choosing a fixed branching factor we aim to ensure that the implementation is not tailored to a particular distribution (we repeat here that the distribution of the representatives may be a non-standard one).

Finally, since our trie is used for predecessor queries, we need to maintain a dynamic predecessor data structure at each internal node in the trie. This predecessor data structure would typically store the first symbol of the edge label of each edge leaving that internal node. We currently use a bit-vector based data structure, combined with looking up small tables, for this purpose. For 'reasonable' values of the branching factor like 256 or 2048 this data structure seems quite fast. Finally, we maintain the leaf nodes in a doubly linked list (useful for predecessor queries).

The collection of buckets class The collection of buckets class template takes three arguments: the key type, the information type and the class that

---

[2] The sizes may vary slightly due to rounding.

implements each bucket. It takes care of searching in buckets and ensures that the bucket sizes satisfy some invariants. The class template which embodies the notion of 'bucket' must be derived from the empty `Bucket` class. This absolves the top-level data structure from knowing the precise implementation of the bottom level; the top-level can refer to pointers to the 'bucket' simply using the type `Bucket *`.

The collection of buckets implements dynamic predecessor operations, but with an additional parameter, for example:

```
void locate_pred(Bucket* b, key_type k, key_type &pred,
                 info_type &pred_info);
```

returns the predecessor of `k` in `pred`, and the information associated with `pred` in `pred_info`. If the predecessor is not defined these values are undefined as well. The search should be performed in the bucket `b`. Since an insertion may cause a bucket to split, the signature of an insert into a collection of buckets is as follows:

```
void insert(Bucket* b, key_type k,
            info_type i, key_type &splitkey, Bucket* &newbucket);
```

The new representative and the new bucket pointer are returned through `splitkey` and `newbucket` respectively (if an insertion caused a bucket to split.) Since a deletion may cause a bucket to join with a neighbour, the signature of a delete from a collection of buckets has extra parameters as follows:

```
bool del(Bucket* b, key_type k, key_type & top_delete);
```

If a delete from the bucket `b` causes it to 'join' with an adjacent bucket then one representative key must be deleted from the top level: it is returned in the parameter `top_delete`.

We have identified two ways of implementing a collection of buckets. The first is randomised, where each key is chosen to be a representative independently with some probability p. A bucket consists of the keys between a representative and the next larger representative. Clearly, the average bucket size is $b = 1/p$. One way of implementing this is by storing all the keys in a skip list, and choosing a level l so that the probability that a given key is at level l is approximately $1/b$. A representative then is simply an element at level l or greater[3]. A newly-inserted key is a representative with probability about $1/b$ and a deletion causes a change in the top-level data structure only if the key deleted is a representative, which happens only with probability about $1/b$. Hence a sequence of n updates leads to an expected $O(n/b)$ changes to the top level, as desired. We have implemented this idea and call it SkipBC below. Due to the parameter choices in the underlying skip list implementation, we can only support the average bucket sizes which are powers of 4.

The second is more traditional, where the buckets are dynamic predecessor data structures (usually search trees) that allow efficient splits and joins. We

---

[3] Actually, we can limit the skip list to level l and dispense with levels $l + 1$ onwards.

have created a generic implementation of such a bucket collection into which implementations of various search trees can be incorporated. The bucket sizes are maintained in one of two ways (b is the maximum bucket size):

- an *eager* method, where buckets are always required to have sizes in the range $\lfloor b/2 \rfloor$ to $b - 1$. An insertion can cause a bucket to have more than $b - 1$ elements: if so, this bucket is split into two buckets of equal size and a new representative is added. A deletion can cause a bucket to have fewer than $\lfloor b/2 \rfloor$ elements: if so, it is joined with an adjacent bucket and a representative is deleted. Furthermore, if the resulting bucket has b or more elements, it is split into two of equal size and a new representative is added.
- a *lazy* method, where the invariants are that each bucket should have between 1 and $b - 1$ elements, and that for every pair of adjacent buckets B and B', $|B| + |B'| \geq \lfloor b/2 \rfloor$. An insertion can cause a bucket to have more than $b - 1$ elements: if so, it is split into two equal buckets and a new representative is added. A deletion can empty a bucket: if so, the empty bucket is removed and its representative is deleted. A deletion can also cause two adjacent buckets to have have fewer than $\lfloor b/2 \rfloor$ elements in total: if so, they are joined and a representative is deleted. It is easy to verify that the invariants are restored.

It should be noted that to maintain good worst-case performance, the eager method should have a ratio of more than 2 between the minimum and maximum bucket sizes. Also note that the lazy method has a cleaner interface, as a deletion can only cause one change in the top level. There are currently three specific instantiations of this generic bucket collection:

- Using a traditional implementation of splay trees (with bottom-up splaying). We refer to this as SplayBC below. Searches, insertions and deletions which do not cause a split are handled in $O(\log b)$ amortized time. Splits and joins take $O(\log b)$ amortised time as well.
- Using a traditional implementation of red-black trees. We refer to this as RedBlackBC below. Searches, insertions and deletions which do not cause a split or join are handled in $O(\log b)$ time. Joins take $O(\log b)$ time but splits take $O((\log b)^2)$ time[4].
- Using sorted arrays and binary search. Searches take $O(\log b)$ time but insertions take $O(b)$ time. Joins and splits also take $O(b)$ time. All constants are quite small. We refer to this as ArrayBC below.

A few points are worth noting: as the number of splits and joins in a sequence of n operations on the hybrid should be $O(n/b)$, the cost of splits and joins are essentially negligible in all the above implementations. Also, the ArrayBC may not be able to support some operations that the other implementations can, such as deleting a key in $O(1)$ time given a pointer to it.

---

[4] This is because we recompute black heights $O(\log b)$ times during a split.

3   Experiments

We have performed a series of experiments evaluating the effect of the different parameters on performance, which are summarised below. All experiments were performed on a Sun Ultra-II machine with 2×300 MHz processors and 512MB of memory. It has a 16KB L1 data cache and 512KB L2 data cache; both are direct-mapped. The L1 cache is write-through and the L2 cache is write-back. As mentioned above the compiler used is g++ version 2.8.1. The LEDA speeds are based on LEDA 3.7 and for the SkipBC we also make use of the LEDA 3.7 memory manager.

Data types We considered 32-bit integer (int) and 64-bit floating-point keys (double). The floating-point keys were generated uniformly using the function drand48(). The integer keys were chosen from one of two distributions *uniform* or *biased-bit*. In the uniform distribution, we choose an integer from the full range of integers (-2^{31} to $2^{31} - 1$ in our case) using the function mrand48(). The biased-bit distribution generates integers where each bit—including the sign bit—is chosen to be 0 independently with probability p (we use $p = 0.25$ here). The searches were overwhelmingly for keys which were not in the data structure (unsuccessful searches). This is important because some data structures such as the trie perform successful searches very quickly.

The searches were were generated in two ways. In the first (*uniform access*) we generated a series of random keys from the same distribution as the keys already in the data structure, and used these keys as the search keys. In the second (*pseudo-zipf* access), we generate unsuccessful searches that induce an approximation to the zipf distribution[5] on the keys in the data structure. This is done as follows: we generate n keys $y_1, \ldots, y_n$ from the same distribution as the n keys currently in the data structure, and and generate a sequence of searches, where each search takes an argument drawn from $y_1, \ldots, y_n$ according to the zipf distribution. In both cases, as the size of the data structure is small compared to the range of possible inputs, most searches will be unsuccessful.

Data structures We tested the following data structures: *Hybrid (ARRAY)*, *Hybrid(RB)*, *Hybrid(SKIP)* and *Hybrid (SPLAY)*, which are hybrids of the trie with ArrayBC, RedBlackBC, SkipBC and SplayBC respectively; *Leda(RB)*, which is the LEDA sorted sequence with red-black tree implementation parameter and *Trie*, which is the raw trie. After preliminary experiments (not reported here) the branching factor of the trie was fixed at 256. The bucket size was fixed at 48—this choice ensures that the size of the top-level trie is no more than twice the size of a search tree with n nodes, if a bucket on average is

---

[5] Recall that the zipf distribution states that if keys $x_1, \ldots, x_n$, are present in the data structure, then x_i should be accessed with probability $\frac{1}{i \cdot H_n}$, where $H_n = \sum_{i=1}^{n} 1/i$.[13]

half-full. For SkipBC, the average bucket size is 64, which is the closest approximation to 48. Unless mentioned otherwise, Hybrid(ARRAY), Hybrid(RB) and Hybrid(SPLAY) should be considered as implementing lazy bucket rebalancing.

Test setup The main series of tests aimed to determine the performance of the various hybrid data structures under the standard metrics of run-time and memory usage. Starting from an empty data structure, the following operations were performed in turn:

(A) n inserts,
(B) a mixed sequence of $2n$ inserts/deletes, where each operation is an insert or delete with probability 0.5,
(C) n predecessor queries, generated as described above.

The time for each of these was measured, as was the memory required by the top-level data structure, plus any 'overheads' incurred by the bucket collection, such as space to store the sizes of buckets etc. The amount of memory required was measured both after step (A) and step (B). Each test was performed twice using the same random number seed, taking the minimum as the running time for this seed; the running times reported are the average over five seeds. We tested `ints` under three scenarios: uniform input and uniform access, biased input and uniform access and uniform input and pseudo-zipf access. We tested `doubles` only with uniform input and uniform access.

We have performed preliminary experiments which evaluate the effectiveness of various bucket sizes and branching factors; the results of these are not reported here but are reflected in the choices of parameters. We have also compared the effectiveness of the lazy versus the eager strategies for bucket rebalancing. Since the lazy and eager strategies are the same if we are doing only insertions, we perform the test primarily on a mixed sequence of inserts and deletes, i.e., in (A) to (C) above, we replace the $2n$ in (B) with $8n$, the idea being to reach a 'steady state' which was more representative of the mixed sequence.

4   Experimental Results and Conclusions

Main experiments Figs 1, 2 summarise the performance of the various data structures considered on uniform data and uniform access. The running times are per operation, and are given in microseconds. The update time reported in Figs 1 and 2 is the mean of the update times from (A) and (B). Some conclusions may be drawn:

1. Hybrid(ARRAY) is a clear winner, as it nearly matches the performance of the raw trie for searches, but is far superior for updates.
2. The hybrid running times seem to grow roughly logarithmically, but not very smoothly. The trie search times grow logarithmically and smoothly for `ints` but not as smoothly for `doubles`. The trie update times seem to grow irregularly, but again there is a difference between `ints` and `doubles`.

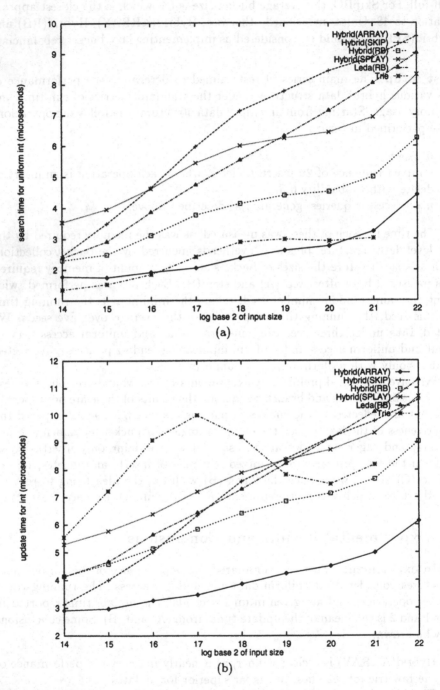

Fig. 1. Uniform ints: (a) search and (b) update times. The trie exceeded available memory at $n = 2^{22}$.

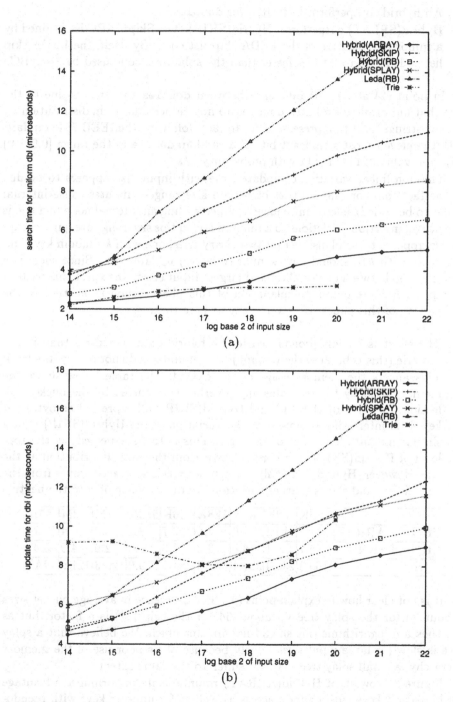

Fig. 2. Uniform doubles: (a) search and (b) update times. The trie exceeded available memory at $n = 2^{21}$.

3. All hybrids outperform Leda(RB) for `doubles`.
4. Hybrid(SPLAY) outperforms Hybrid(SKIP). The SkipListBC is obtained by a minor modification of the LEDA skip list code. By itself, the LEDA skip list code is 1.5 to 2 times faster than the splay tree code used by SplayBC.

In point (2) above, the difference between `doubles` and `ints` is due to the fact that uniformly distributed `doubles` do not induce a uniform distribution on the bit-strings used to represent them. In fact, following the IEEE 754 standard [11], the eleven most significant bits of a random `double` in the range $[0.0, 1.0)$ have the value 01111111111 with probability $1/2$.

The non-linear variation of update time with input size appears to be due to the fact that for random tries of certain size ranges, updates cause internal nodes to be added/deleted infrequently. Since adding an internal node in a trie is expensive, insertions/deletions to a random trie in this size range are quite cheap. It also appears to be plausible that two B-ary tries, one with k random keys and the other with kB, should have similar (relative) update costs. Since there is a factor of 256 between our smallest and largest input sizes, we expect this pattern to repeat. A more detailed explanation of this phenomenon is deferred to the full version of this paper. However, we note here that:

- The effect is far less pronounced in the hybrid data structures than in the raw trie (this is because the vast majority of updates do not change the trie);
- This affects the memory usage of the hybrid. The table below shows the memory used by the trie, plus any 'overheads' incurred by the bucket collection, for Hybrid(ARRAY) and Hybrid(SKIP), all expressed as bytes per key. It is interesting to note that the variation in the Hybrid(SKIP) is of a similar magnitude to that of the Trie. This is to be expected, as the top-level of Hybrid(SKIP) stores keys drawn from the same distribution as the trie. However, Hybrid(ARRAY) stores more precisely spaced values from the current set and shows sharper variations from one value of n to the next.

	16K	32K	64K	128K	256K	512K	1M	2M	4M
Hybrid(ARRAY)	17.5	10.4	6.0	3.8	2.7	2.2	2.0	7.4	17.1
Hybrid(SKIP)	5.7	6.1	4.8	3.2	2.3	2.3	2.9	4.0	5.4
Trie	160.5	233.4	321.1	357.3	284.1	179.0	126.3	121.9	NA

It is not clear how to explain point (4). The difference in average bucket sizes (about 24 for the splay tree vs. about 64 for the skip list) is a factor, but as searches are logarithmic this should not matter much. We believe that a splay tree is slow for large input sizes largely because of its poor use of the memory hierarchy; a small splay tree may not suffer to the same extent.

Figure 3 shows that Hybrid(ARRAY) maintains its performance advantage for biased-bit keys with uniform access as well as for uniform keys with pseudo-zipf access. In Figure 3(b) it can be seen that Hybrid(SPLAY) is competitive with Hybrid(RB); this is probably because the splay trees adjust to the non-uniform access patterns. Since the zipf distribution is widely regarded as a good

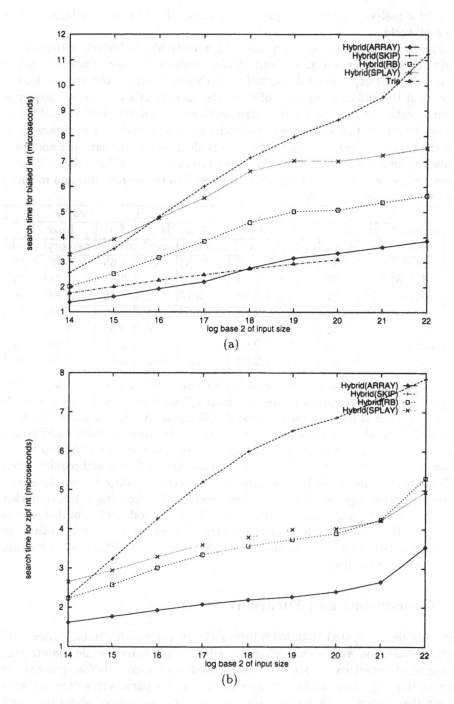

Fig. 3. Non-uniform data: (a) biased-bit integers with uniform access (b) uniform integers with pseudo-zipf access. The trie exceeded available memory at $n = 2^{22}$ in (a).

model for real-world access frequencies, Hybrid(SPLAY) may perform well in the real world.

Finally, we compare the lazy and eager methods of bucket maintenance, using the test described at the end of the previous chapter. The table below summarises the experimental results from this test. It shows the average bucket fullness at the end of a sequence of n inserts, as well as after a mixed sequence of $8n$ inserts and deletes. For the lazy method, for bucket sizes 48 and 64, it shows the search times and memory used by the top-level and auxiliary data. For the eager method, the bucket fullness is the same as the lazy method after n inserts, and is not shown. We do show, however, the bucket fullness after a mixed sequence of $8n$ inserts and deletes, as well as the search time and memory used for a bucket size of 48.

input	lazy						eager		
	fullness		bsize 48		bsize 64		full	bsize 48	
size	insert	mixed	memory	search	memory	search	mixed	memory	search
16384	0.66	0.52	19.08	1.59	15.35	1.46	0.65	16.80	1.34
32768	0.66	0.52	10.60	1.75	10.08	1.71	0.65	10.19	1.83
65536	0.66	0.52	6.27	2.20	5.73	2.08	0.65	5.84	1.98
131072	0.66	0.52	4.11	2.18	3.58	2.18	0.65	3.68	2.14
262144	0.66	0.52	3.02	2.53	2.49	2.56	0.65	2.59	2.47
524288	0.66	0.52	2.49	2.62	1.96	2.62	0.65	2.06	2.76
1048576	0.66	0.52	2.45	2.90	1.68	3.03	0.65	1.81	2.91

We note that a sequence of inserts leads to buckets being about 2/3 full. The lazy method only manages a fullness of about 1/2 with the mixed sequence, while the eager method has a fullness of about 2/3 again. We do not show this here, but the lazy method also processes the mixed update sequence more rapidly than the eager method (by virtue of performing fewer operations on the top level). However, it is more wasteful of memory—in the ArrayBC it would require about 33% more memory in the bucket collection. One could reduce the top-level and auxiliary space requirements of the lazy method by choosing a larger bucket size. In the table, we demonstrate that the lazy method with a bucket size of 64 essentially reduces top-level and auxiliary memory, as well as search time, to the level of the eager method. Hence, in a SplayBC or RedBlackBC one would prefer the lazy method.

5   Conclusions and Future Work

We have demonstrated that hybrid data structures which combine tries with buckets based on arrays can deliver excellent performance while maintaining a degree of genericity. A lot more work needs to be done before one can be prescriptive regarding the best choices of the various parameters. Here we have shown that parameter choices set using worst-case assumptions about the input distributions give surprisingly good performance. We believe that this, rather than tuning based on various distributions, may be the best approach towards parameter selection for very general-purpose data structures.

References

1. A. Acharya, H. Zhu and K. Shen. Adaptive algorithms for cache-efficient trie search. Presented at *ALENEX '99*.
2. J.-I. Aoe and K. Morimoto. An efficient implementation of trie structures. *Software Practice and Experience*, **22** (1992), pp 695–721.
3. A. Brodnik, S. Carlsson, M. Degermark and S. Pink. Small Forwarding tables for fast routing lookups. In *Proc. ACM SIGCOMM Conference: Applications, Technologies, Architectures, and Protocols for Computer Communication (SIGCOMM-97)*, pp. 3–14, ACM Press, 1997.
4. M. R. Brown and R. E. Tarjan. Design and analysis of a data structure for representing sorted lists. *SIAM Journal on Computing*, **9** (1980), pp. 594–614.
5. T. H. Cormen, C. E. Leiserson and R. Rivest. *Introduction to Algorithms*. MIT Press, 1990.
6. J. J. Darragh, J. G. Cleary and I. H. Witten. Bonsai: A compact representation of trees. *Software Practice and Experience*, **23** (1993), pp 277–291.
7. P. F. Dietz. Maintaining order in a linked list. In *Proc. 14th ACM STOC*, pages 122–127, May 1982.
8. P. F. Dietz and D. D. Sleator. Two algorithms for maintaining order in a list. In *Proc. 19th ACM STOC*, pages 365–372, 1987.
9. M. L. Fredman and D. E. Willard. Surpassing the information theoretic bound with fusion trees. *J. Comput. System Sci.*, **47** (1993), pp. 424–436.
10. T. Hagerup. Sorting and Searching on the Word RAM. In *STACS: Annual Symposium on Theoretical Aspects of Computer Science*, LNCS 1373, 1998.
11. J. L Hennessy and D. A. Patterson, *Computer Architecture: A Quantitative Approach, 2nd ed.* Morgan Kaufmann, 1996.
12. S. Huddleston and K. Mehlhorn. A new data structure for representing sorted lists. *Acta Informatica*, **17** (1982), pp. 157–184.
13. D. E. Knuth. *The Art of Computer Programming. Volume 3: Sorting and Searching, 3rd ed.* Addison-Wesley, 1997.
14. C. Levcopolous and M. H. Overmars. A balanced search tree with O(1) worst-case update time. *Acta Informatica*, 26:269–278, 1988.
15. David B. Lomet. A simple bounded disorder file organisation with good performance. *ACM Trans. Database Systems* **13** (1988), pp. 525–551.
16. U. Manber and S. Wu. GLIMPSE: a tool to search through entire file systems. In *Proc Usenix Winter 1994 Technical Conference*, San Francisco, 1994, pp. 23–32.
17. K. Mehlhorn and St. Näher. *LEDA. Cambridge University Press*, 1998.
18. S. Nilsson and M. Tikkanen. Implementing a dynamic compressed trie. In *2nd Workshop on Algorithm Engineering, WAE '98—Proceedings*, K. Mehlhorn, Ed., Max Planck Institute Research Report MPI-I-98-1-019, 1998, pp. 25–36.
19. M. H. Overmars. *The design of dynamic data structures*, volume 156 of *Lecture Notes in Computer Science*. Springer-Verlag, Berlin, 1983.
20. S. Sahni. *Data Structures, Algorithms and Applications in C++*. McGraw-Hill, 1998.
21. B. Stroustrup. *The C++ programming language, 3rd Ed.* Addison-Wesley, 1997.
22. A. K. Tsakalidis. Maintaining order in a generalized linked list. *Acta Informatica*, 21(1):101–112, 1984.
23. P. van Emde Boas. Preserving order in a forest in less than logarithmic time and linear space. *Information Processing Letters*, 6(3):80–82, 1977.
24. D. E. Willard New trie data structures which support very fast search operations *Journal of Computer and System Sciences*, 28(3), pp. 379–394, 1984.

LEDA-SM
Extending LEDA to Secondary Memory

Andreas Crauser and Kurt Mehlhorn*

Max-Planck-Institut für Informatik Saarbrücken, Germany
crauser,mehlhorn@mpi-sb.mpg.de

Abstract. During the last years, many software libraries for *in-core* computation have been developed. Most internal memory algorithms perform very badly when used in an *external memory* setting. We introduce LEDA-SM that extends the LEDA-library [22] towards secondary memory computation. LEDA-SM uses I/O-efficient algorithms and data structures that do not suffer from the so called *I/O bottleneck*. LEDA is used for in-core computation. We explain the design of LEDA-SM and report on performance results.

1 Introduction

Current computers have large main memories, but some applications need to manipulate data sets that are too large to fit into main memory. Very large data sets arise, for example, in geographic information systems [3], text indexing [5], WWW-search, and scientific computing. In these applications, *secondary memory* (mostly disks) [24] provides the required workspace. It has two features that distinguish it from main memory:

- Access to secondary memory is slow. An access to a data item in external memory takes much longer than an access to the same item in main memory; the relative speed of a fast internal cache and a slow external memory is close to one million.
- Secondary memory rewards locality of reference. Main memory and secondary memory exchange data in *blocks*. The transfer of a block between main memory and secondary memory is called an *I/O-operation* (short I/O).

Standard data structures and algorithms are not designed for locality of reference and hence frequently suffer an intolerable slowdown when used in conjunction with external memory. They simply use the virtual memory (provided by the operating system) and address this memory as if they would operate in internal memory, thus performing huge amounts of I/Os. In recent years the algorithms community has addressed this issue and has developed I/O-efficient algorithms for many data structure, graph-theoretic, and geometric problems [28, 6, 9, 19, 18]. Implementations and experimental work are lacking behind.

---

* research partially supported by the EU ESPRIT LTR Project Nr. 20244 (ALCOM-IT)

J.S. Vitter and C.D. Zaroliagis (Eds.): WAE'99, LNCS 1668, pp. 228–242, 1999.
© Springer-Verlag Berlin Heidelberg 1999

External memory algorithms move data in the memory hierarchy and process data in main memory. A platform for external memory computation therefore has to address two issues: movement of data and co-operation with internal memory algorithms.

We propose *LEDA-SM* (LEDA secondary memory) as a platform for external memory computation. It extends LEDA [22, 23] to secondary memory computation and is therefore directly connected to an efficient internal-memory library of data structures and algorithms. LEDA-SM is portable, easy to use and efficient. It consists of:

- a kernel that gives an abstract view of secondary memory and provides a convenient infrastructure for implementing external memory algorithms and data structures. We view secondary memory as a collection of disks and each disk as a collection of blocks. There are currently four implementations of the kernel, namely by standard I/O (`stdio`), system call I/O (`syscall`), memory mapped I/O (`mmapio`) and memory disks (`memory`). All four implementations are portable across Unix-based (also Linux) platforms, and `stdio` and `syscall` are also for Microsoft operating systems.
- a collection of external memory data structures. An external memory data structure offers an interface that is akin to the interface of the corresponding internal memory data structures (of LEDA), uses only a small amount of internal memory, and offers efficient access to secondary memory. For example, an *external stack* offers the stack operations *push* and *pop*, requires only slightly more than two blocks of internal memory, and needs only $O(1/B)$ I/O-operations per push and pop, where B is the number of data items that fit into a block.
- algorithms operating on these data structures. This includes basic algorithms like sorting as well as matrix multiplication, text indexing and simple graph algorithms.
- a precise and readable specification for all data structures and algorithms. The specifications are short and abstract so as to hide all details of the implementation.

The external memory data structures and algorithms of LEDA-SM (items (2) and (3)) are based on the kernel; however, their use requires little knowledge of the kernel. LEDA-SM[1] supports fast prototyping of secondary memory algorithms and therefore can be used to experimentally analyze new data structures and algorithms in a secondary memory setting.

The database community has a long tradition of dealing with external memory. Efficient index structures, e.g, B-trees [7] and extendible hashing [16], have been designed and highly optimized implementations are available. "General purpose" external memory computation has never been a major concern for the database community.

In the algorithms community implementation work is still in its infancy. There are implementations of particular data structures [10, 20, 4] and there is

---

[1] LEDA-SM can be downloaded from `www.mpi-sb.mpg.de/~crauser/leda-sm.html`

TPIE [27], the transparent I/O-environment. The former work aimed at investigating the relative merits of different data structures, but not at external memory computation at-large. TPIE provides some *external programming paradigms* like scanning, sorting and merging. It does not offer external memory data structures and it has no support for internal memory computation. Both features were missed by users of TPIE [20], it is planned to add both features to TPIE (L. Arge, personal communication, February 99). Another different approach is ViC* [11], the Virtual Memory C^* compiler. The ViC* system consists of a compiler and a run-time system. The compiler translates C* programs with shapes declared `outofcore`, which describe parallel data stored on disk. The compiler output is a program in standard C with I/O and library calls added to efficiently access out-of-core parallel data. At the moment, most of the work is focussed on out-of-core fast Fourier transform, BMMC permutations and sorting.

This paper is structures as follows. In Section 2 we explain the software architecture of LEDA-SM and discuss the main layers of LEDA-SM (kernel, data structures and algorithms). We describe the kernel and give an overview of the currently available data structures and algorithms. In Section 3 we give two examples to show (i) how the kernel is used for implementing an external data structure and (ii) the ease of use of LEDA-SM and its with LEDA. Some experimental results are given Section 4. We close with a discussion of future developments.

2   LEDA-SM

LEDA-SM is a C++ class library that uses LEDA for internal memory computations. LEDA-SM is designed in a modular way and consists of 4 layers:

Layer	Major Classes
algorithms	sorting, graph algorithms, ...
data structures	*ext_stack*, *ext_array*, ...
abstract kernel	*block<E>*, *B_ID*, *U_ID*
concrete kernel	*ext_memory_manager*, *ext_disk*, *ext_free_list*, *name_server*

We use application layer for the upper two layers and kernel layer for the lower two layers.

The *kernel layer* of LEDA-SM manages secondary memory and movement of data between secondary memory and main memory. It is divided into the *abstract* and the *concrete* kernel. The abstract kernel consists of 3 C++ classes that give an abstract view of secondary memory. Two classes model disk block locations and users of disk blocks while the third class is a container class that is able to transfer elements of type E to and from secondary memory in a simple way. This class provides a typed view of data stored in secondary memory, data on disk is always untyped (type *void*).

The concrete kernel is responsible for performing I/Os and managing disk space in secondary memory and consists of 4 C++ classes. LEDA-SM provides several realizations of the concrete kernel; the user can choose one of them at run-time. The concrete kernel defines functions for performing I/Os and managing disk blocks, e.g. read/write a block, read/write k consecutive blocks, allocate/free a disk block etc. These functions are used by the abstract kernel or directly by data structures and algorithms.

The *application layer* consists of data structures and algorithms. LEDA is used to implement the in-core part of the applications and the kernel of LEDA-SM is used to perform the I/Os. The physical I/O calls are completely hidden in the concrete kernel, the applications use the container class of the abstract kernel to transport data to and from secondary memory. This makes the data structures simple, easy to use and still efficient.

We now explain in more detail the kernel layer. Examples for the application layer are given in Section 2.2.

2.1   The Kernel

The abstract kernel models secondary memory as files of the underlying file system. Secondary memory consists of *NUM_OF_DISKS* logical disks (files of the file system), *max_blocks[d]* blocks can reside on the d-th disk, $0 \leq d <NUM\_OF\_BLOCKS$. The blocks on any disk are numbered consecutively starting at zero. A *block identifier* is a pair (d, num) of integers. A block identifier is called *valid* if $0 \leq d <NUM\_OF\_DISKS$ and $0 \leq num< max\_blocks[d]$ and it is called *active* if it is valid and the block denoted by it was written to. The class *B_ID* realizes block identifiers. Observe that block identifiers refer to physical objects, namely, to regions of storage on disk. In the remainder of this section there is the need to distinguish between blocks as physical objects (= a region of storage on disk) and blocks as logical objects (= a bit pattern of a particular size). We will use the word *disk block* for the physical object and reserve the word *block* for the logical object. The disk blocks are managed by the *external memory manager* (class *ext_memory_manager*). There is only one instance of this class. The external memory manager can be asked to allocate disk blocks, to free disk blocks, and to transfer blocks between main memory and external memory. The allocation of a disk block is either on a disk chosen by the user or on a disk chosen by the system (if no disk is specified in the allocation request). The return value of an allocation request is a block identifier which can later be used in read- and write-operations. An allocated disk block is always owned by a particular user. Only the owner of a disk block can write the disk block. A user is identified by an *user identification* (= an integer) of class *U_ID* and user identifiers are managed by class *name_server*. We use user identifiers for memory protection. Every instance of a data structure is a different user of the kernel and hence data structures are protected against one-another. There are different ways how these user checks can be performed: a conservative approach checks during read-and write accesses to ensure that even no false data is read, while the standard approach only checks user-ids during write access.

The parameterized type *block<E>* is used to store logical blocks in internal memory. An instance *B* of type *block<E>* can store one logical block and gives a typed view of logical blocks. A logical block is viewed as two arrays: an array of links to disk blocks and an array of variables of type *E*. A link is of type *block identifier*; the number *num_of_bids* of links is fixed when a block is created. The number of variables of type *E* is denoted by *blk_sz* and is equal to $(BLK\_SZ * sizeof(GenPtr) - num\_of\_bids * sizeof(B\_ID))/sizeof(E)$ where *GenPtr* is the generic pointer type of the machine (usually *void**) and *BLK_SZ* is the physical disk block size in bytes. Both arrays are indexed starting at zero.

Every block has an associated user identifier and an associated block identifier. The user identifier designates the owner of the block and the block identifier designates the disk block to which the logical block is bound. The container class *block<E>* is directly connected to functions of the concrete kernel (by use of the class *ext_memory_manager*), i.e. function *write* of class *block<E>* uses the write-function of the concrete kernel to initiate the physical I/Os.

The concrete kernel is responsible for performing I/Os and managing disk space in secondary memory and consists of classes *ext_memory_manager*, *ext_disk*, *ext_free_list*, *name_server*. The class *ext_disk* is responsible for file I/O and models an external disk device. There are several choices for file I/O: system call I/O, standard file I/O, memory mapped I/O and memory I/O (this is an in-memory disk simulation). Each of these methods has different advantages and disadvantages. As we model disk locations by an own class (B_ID), we do not ignore the ordering of disk blocks and are able to manage sequential disk accesses. Furthermore, there also exists the possibility to switch to the raw disk drive[2] (underlying the file system). By this, we simply drop the overhead introduced by the file system layer and handle the problem of non-contiguous disk blocks of files, on the other hand we also loose the caching effects that the file system performs. This feature is not portable across all platforms and is therefore not directly implemented in the library[3]. The class *ext_freelist* is responsible for managing allocated and free disk blocks in external memory. Disk blocks can be free or in-use by a dedicated user, the system must manage this. This is done by a specific data structure which is currently implemented in four different ways.

We can now summarize the software layout of LEDA-SM. The concrete kernel consists of class *ext_memory_manager*, class *name_server* and of the two kernel implementation classes for disk access and for disk block management. The abstract kernel consists of classes *block*, *B_ID* and *U_ID*. All data structures and algorithms of LEDA-SM are implemented using only the kernel classes and data

---

[2] This is the very reason why we model disk positions by an own class.

[3] In Solaris systems, this is just a change of a few code lines, but it also requires super-user rights.

structures or algorithms available in LEDA. In the next subsection we will give an overview of the currently available data structures and algorithms.

2.2   Data Structures and Algorithms

We survey the data structures and algorithms currently available in LEDA-SM. Theoretical I/O-bounds are given in the classical external memory model of [28], where M is the size of the main memory, B is the size of a disk block, and N is the input size measured in disk blocks. For sake of simplicity we assume that D,the number of disks, is one.

Stacks and Queues: External stacks and queues are simply the secondary memory counterpart to the corresponding internal memory data structures. Operations *push*, *pop* and *append* are implemented in optimal $O(1/B)$ amortized I/Os.

Priority Queues: Secondary memory priority queues can be used for large-scale event simulation, in graph algorithms or for online-sorting. Three different implementations are available. Buffer trees [2] achieve optimal $O(1/B \log_{M/B}(N/B))$ amortized I/Os for operations *insert* and *del_min*. Radix heaps are an extension of [1] towards secondary memory. This integer-based priority queue achieves $O(1/B)$ I/Os for *insert* and $O(1/B \log_{M/B}(C))$ I/Os for *del_min* where C is the maximum allowed difference between the last deleted minimum key and the actual keys in the queue. Array heaps [8, 14, 25] achieve $O(1/B \log_{M/B}(N/B))$ amortized I/Os for *insert* and $O(1/B)$ amortized I/Os for *del_min*.

Arrays: Arrays are a widely used data structure in internal memory. The main drawback of internal-memory arrays is the fact that when used in secondary memory, it is not possible to control the paging. Our external array data structure consists of a consecutive collection of disk blocks and an internal-memory data structure of fixed size that implements a multi-segmented cache. The caching allows to control the internal-memory usage of the external array data structure. As the cache is multi-segmented, it is possible to index different regions of the external array and use a different segment of the cache for each region. Several page-replacement strategies are supported like LRU, random, fixed, etc. The user can also implement his/her own page-replacement strategy.

Sorting: Sorting is implemented by multiway-mergesort. Internal sorting during the run-creation phase is realized by LEDA-quicksort which is a fast and generic code-inlined template sorting routine. Sorting N items takes optimal $O(N/B \log_{M/B}(N/B))$ I/Os.

B-Trees: B-Trees [7] are the classical secondary memory online search trees. We use a B^+-implementation and support operations *insert*, *delete*, *delete_min* and *search* in optimal $O(\log_B(N))$ I/Os.

Suffix arrays and strings: Suffix arrays [21] are a full-text indexing data structure for large text strings. We provide several different construction algorithms [13] for suffix arrays as well as exact searching, 1- and 2-mismatch searching, and 1- and 2-edit searching.

Matrix operations: We provide matrices with entry type `double`. The operations $+$, $-$, and $*$ for dense matrices are realized with optimal I/O-bounds [26].

Graphs and graph algorithms: We provide a data type external graph and simple graph algorithms like depth-first search, topological sorting and Dijkstra's shortest path computation. External graphs are static, i.e. graph updates are expensive.

2.3   Further Features

Secondary-memory data structures and algorithms also use internal memory. LEDA-SM allows the user to control the amount of memory that each data structure and/or algorithm uses. The amount of memory is either specified at the time of construction of the data structure or it is an additional parameter of a function call. If we look at our stack example of Section 3 we see that the constructor of data type *ext_stack* has a parameter a, the number of blocks of size *blk_size* that are held in internal memory. We therefore immediately know that the internal memory space occupancy is $a \cdot blz\_size + O(1)$ bytes.

The second feature of LEDA-SM is accounting of I/Os. The kernel supports counting read- and write operations. Some of the reads may be logical and can be served by the buffer of the operating system. It is also possible to count consecutive I/Os (also introduced by [17, 13] as bulk I/Os). This allows the user to experimentally classify the I/O-structure of algorithms and in this way to compare algorithms with the same asymptotic I/O-complexity (see also [13]).

3   Examples

We discuss the implementation of secondary memory stacks and secondary memory graph search. The first example shows how the kernel of LEDA-SM is used to implement data structures and algorithms. The second example shows that secondary memory algorithms can be coded in a natural way in LEDA-SM and LEDA. It also shows the interplay between LEDA and LEDA-SM.

3.1   External Memory Stacks

We discuss the implementation of external memory stacks. It is simple, conceptually and programming-wise. The point of this section is to show how easy it is to translate an algorithmic idea into a program.

A external memory stack S for elements of type E (*ext_stack<E>*) is realized by an array (a LEDA data structure) A of $2a$ blocks of type *block<E>* and a linear list of disk blocks. Each block in A can hold *blk_sz* elements, i.e, A can hold up to $2a \cdot blk\_sz$ elements. We may view A as a one-dimensional array of elements of type E. The slots 0 to *top* of this array are used to store elements of S with *top* designating the top of the stack. The older elements of S, i.e., the ones that do not fit into A, reside on disk. We use *bid* to store the block identifier of the elements moved to disk most recently. Each disk block has one link; it is used to point to the block below. The number of elements stored on disk is always a multiple of $a \cdot blk\_sz$.

⟨ext_stack⟩≡

```
template <class E>
class ext_stack
{
  array< block<E> > A;
  int top_cnt, a_sz, s_sz, blk_sz;
  B_ID bid;

public:

ext_stack(int a = 1);
void push(E x);
E pop();
E top();
int size() { return s_sz; };
void clear();
~ext_stack();
};
```

We next discuss the implementation of the operations *push* and *pop*. We denote by $a\_sz = 2a$ the size of array A. A push operation $S.push(x)$ writes x into the location $top + 1$ of A except if A is full. If A is full ($top\_cnt == a\_sz * blk\_sz - 1$), the first half of A is moved to disk, the second half of A is moved to the first half, and x is written to the first slot in the second half.

⟨ext_stack⟩+≡

```
template<class E>
void ext_stack<E>::push(E x)
{
  int i;
  if (top_cnt ==  a_sz*blk_sz - 1)
  {
    A[0].bid(0) = bid;
    bid = ext_mem_mgr.new_blocks(myid,a_sz/2);
    block<E>::write_array(A,bid,a_sz/2);
    for(i=0;i<a_sz/2;i++)
      A[i] = A[i+a_sz/2];
    top_cnt = (a_sz/2)*blk_sz-1;
  }
  top_cnt++;
  A[top_cnt/blk_sz][top_cnt%blk_sz] = x;
  s_sz++;
}
```

The interesting case of *push* is the one where we have to write the first half of A to disk. In this step we have to do the following: we must reserve $a = a\_sz/2$

disk blocks on disk and we must add the first a blocks of array A to the linked list of blocks on disk. The blocks are linked by using the array (of length one) of type B_ID of class $block$ (see Section 2.1) and the block least recently written is identified by block identifier bid. The commands $A[0](0) = bid$ creates a backwards linked list of disk blocks which we use during pop-operations later. We then use the kernel to allocate a consecutive free disk blocks by the command $ext\_mem\_mgr.new\_blocks$. The return value is the first allocated block identifier. The first half of array A is written to disk by calling $write\_array$ of class $block$ which tells the kernel to initiate the necessary physical I/Os. In the next step we copy the last a blocks of A to the first a blocks and reset top_cnt. Now the normal push can continue by copying element x to its correct location inside A.

A pop operation $S.pop(\ )$ is also trivial to implement. We read out the element in slot top except if A is empty. If A is empty and there are elements residing on disk we move $a \cdot blk\_sz$ elements from disk into the left half of A.

⟨ext_stack⟩+≡

```
template<class E>
E ext_stack<E>::pop()
{
  if (top_cnt == -1 && s_sz > 0)
  {
    B_ID oldbid = bid;
    block<E>::read_array(A,oldbid,a_sz/2);
    bid = A[0].bid(0);
    top_cnt = (a_sz/2)*blk_sz - 1;
    ext_mem_mgr.free_blocks(oldbid,myid,a_sz/2);
  }
  s_sz--;
  top_cnt--;
  return
    A[(top_cnt+1)/blk_sz][(top_cnt+1)%blk_sz];
}
```

If array A is empty ($top\_cnt = -1$) we load a blocks from disk into the first a array positions of A by calling $read\_array$. These disk blocks are identified by bid. We then restore the invariant that block identifier bid stores the block identifier of the blocks least recently written to disk. As the disk blocks are linked backwards, we can retrieve this block identifier from the first entry of the array of block identifiers of the first loaded disk block ($A[0](0)$). Array A now stores $a \cdot blk\_sz$ data items of type E. Variable top_cnt is reset to this value. The just loaded disk blocks are now stored internally, therefore there is no reason to keep them again on disk. These disk blocks are freed by calling the kernel routine $ext\_mem\_mgr.free\_blocks$. Return-value of operation pop is the top-most element of A.

Operations $push$ and pop move a blocks at a time. As the read and write requests for the a blocks always target consecutive disk locations, we can choose

a in such a way that it maximizes disk-to-host throughput rate. After the movement A is half-full and hence there are no I/Os for the next $a \cdot blk\_sz$ stack operations. Thus the amortized number of I/Os per operations is $1/blk\_sz$ which is optimal. Stacks with fewer than $2a \cdot blk\_sz$ elements reside in-core.

3.2   Graph Search

We give a simple example that shows how to use both, LEDA data structures and LEDA-SM data structures and how they interact. Our example computes the number of nodes of a graph G reachable from a source node v by using graph search. We assume that a bit vector for all nodes of graph G can be stored in internal memory. Internal-memory graph data types heavily rely on large linked lists or large arrays to implement the adjacency list representation of large graphs. As most of the graph algorithms like depth-first search, Dijkstra's shortest path etc. access the graph in an unstructured way, they are slowed down tremendously by the "uncontrolled" paging activity of the operating system. Our external graph data type uses external arrays to control the paging.

```
⟨graph search⟩≡
template<class T>
long graph_search( T& G, ext_node<T> v,
                    int_set Visited )
{
  ext_stack< ext_node<T> > S;
  ext_edge<T> e;
  long i = 1;

  Visited.insert(v);
  S.push(v);

  while( S.size())
  {
    v = S.pop();

    forall_out_edges(e,v,G)
    if( !Visited.member( G.target(e)) )
    {
      Visited.insert( G.target(e) );
      S.push( G.target(e) );
      i++;
    }
  }
  return i;
}


ext_graph<int,char,int,char> G;
int_set visited(1,G.number_of_nodes());
ext_node<ext_graph<int,char,int,char> > v;
int reachable = graph_search( G, v, visited );
```

The LEDA-SM graph data type *ext_graph* is parameterized so that algorithm-dependent information can be associated with nodes and edges. For efficiency, it is important that node and edge labels are stored directly in the nodes and edges and are not accessed through pointers or indices (as this would imply an I/O-operation for each label). Different algorithms need different labels and hence it is crucial that the space allocated for labels can be reused. We have chosen the following design: each node (and similarly edge) stores an information of an arbitrary type X and an array of characters; X and the size of the array are fixed when the graph is defined. We use the array of characters to store arbitrary information of fixed length (through appropriate casting) and use the field of type X for "particularly important" information; we could also do without it. In our example, each node and edge has an *int* and a single *char* associated with it. Edges and nodes are parameterized with the graph data type to which they belong.

The example shows the interaction between LEDA and LEDA-SM. Data structures from both libraries are used. The bit vector *Visited* is implemented by LEDA data type *int_set*. Reached nodes are directly inserted into the *int_set*. The necessary type conversion from type *ext_node* to type *int* is performed by LEDA-SM, i.e. type *ext_node* can be converted to type *int*, because the nodes of an external graph are consecutively numbered. We have implemented *graph_search* in a recursion-free way by using a stack (type *ext_stack*). We have chosen an external stack with minimum internal space requirement.

The example shows that LEDA-SM and LEDA support a natural programming style and allow the user to freely combine the required data structures and algorithms. The I/O operations of LEDA-SM are completely hidden in the data and not visible in the algorithms layer.

We conclude this section by explaining how one can eliminate the use of an internal bit-field with size equal to the number of nodes of G. If the main memory cannot store a bit-field, we use external depth-first search that proceeds in rounds (see [9]). In each round, we store visited nodes in an internal dictionary (LEDA data type *dictionary*) of a fixed size. If the dictionary is full (no space left in main memory), we compact the edges of the external graph by deleting (marking) those edges that point to nodes that were already visited. We then delete the dictionary and start the next round.

4   Performance Results

We report on the performance of LEDA-SM. In our tests, we compare secondary memory data structures and algorithms of LEDA-SM to their internal memory counterparts (of LEDA). For more detailed results on the performance of LEDA-SM see [12]. All tests were performed on a SUN UltraSparc1/143 using 64 Mbytes main memory and a single SCSI-disk connected to a fast-wide SCSI controller. All tests used a disk block size of 8 kbytes. We note that this is not the optimal disk block size for the disk according to data throughput versus

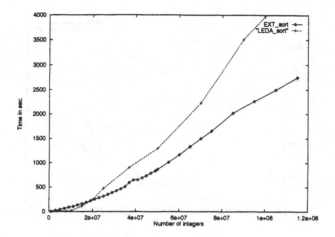

Fig. 1. *Comparison of Multiway-Mergesort of LEDA-SM and quicksort of LEDA.*

service time. However, this disk block size allows us to compare the secondary memory algorithms in a fair way to internal memory algorithms in swap space because the page size of the machine is also 8 kbytes. All external memory algorithms and data structures get faster if we use the optimal (according to data throughput) disk block size of 64 or 128 kbytes.

Figure 1 compares external multiway mergesort and LEDA quicksort. External multiway mergesort uses LEDA quicksort to partition the data into sorted runs before merging proceeds. The sorted runs are merged using a priority queue. The external sorting routine uses approx. 16 Mbytes of main memory. As soon as the input to be sorted reaches the size of the main memory, the external sorting routine is faster than the internal sorting routine. The sharp bend in the curve of external multiway mergesort occurs because we change the priority queue implementation inside the merging routine.

Figure 2 compares internal and external matrix multiplication of dense matrices of type *double*. The external matrix multiplication code uses tiling and matrix reordering (see also [27]). External matrix multiplication is faster than internal matrix multiplication even if the total input size is smaller than the main memory. This effect is due to the better cache behavior of the external matrix code (although it uses the same internal matrix multiplication code as LEDA).

Figure 3 compares the performance of operation *insert* of B-Trees and 2-4 trees. Both data structures are classical online search trees. The internal data structure is slowed down dramatically if it is running in swap space. The 2-4 tree does not exhibit locality of reference and its use of pointers leads to many page faults.

Figure 4 compares the performance of operation *insert* of external radix heaps [14] to Fibonacci heaps and radix heaps [1] of LEDA. We see that the internal priority queues fail if the main memory size is exceeded.

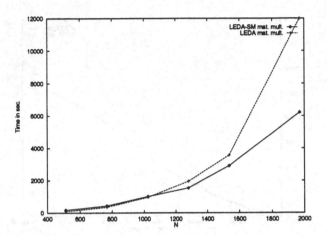

Fig. 2. *Comparison of LEDA-SM matrix multiplication against LEDA matrix multiplication.*

5 Conclusions

We proposed LEDA-SM, an extension of LEDA towards secondary memory. The library follows LEDA's main features: portability, ease-of-use and efficiency. The performance results of LEDA-SM are promising. Although we use a high-level implementation for the library, we are orders of magnitudes faster than the corresponding internal data structures. The speedup increases for larger disk block sizes. The performance of many secondary memory data structures is determined by the speed of their internally used data structures; recall that the goal of algorithm design for external memory is to overcome the I/0-bottleneck. If successful, external memory algorithms are compute-bound. LEDA-SM profits from the efficient in-core data structures and algorithms of LEDA[4]. LEDA-SM is still growing. Future directions of research cover geometric computation [15] as well as graph applications. Important practical experiments should be done for parallel disks (RAID arrays) as well as low-level disk device access. We plan to do both in the near future.

Acknowledgments

The first author wants to thank Stefan Schirra for his help on C++ language problems and the Rechnerbetriebsgruppe of the Max-Planck-Institute for providing the necessary test equipment.

---

[4] LEDA's quicksort routine was recently improved (as a consequence of our tests of LEDA-SM-mergesort). For user-defined data types this led to a speed-up of almost five. LEDA-SM's multiway-mergesort profited immediately; its running time was reduced by a factor of two.

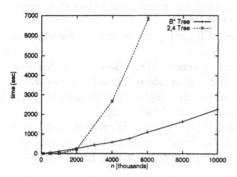

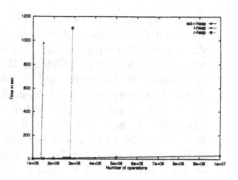

Fig. 3. *Performance of insert-operation for LEDA-SM B-Trees and 2-4-trees of LEDA.*

Fig. 4. *Comparison of LEDA-SM radix heaps against LEDA's radix-heaps and Fibonacci-heaps.*

References

1. R. Ahuja, K. Mehlhorn, J. Orlin, and R. Tarjan. Faster algorithms for the shortest path problem. *Journal of the ACM*, 37(2):213–223, 1990.
2. L. Arge. *Efficient external-memory data structures and applications*. PhD thesis, University of Aarhus, 1996.
3. L. Arge. External-memory algorithms with applications in geographic information systems. CS Department, Duke Univeristy, technical report, 1996.
4. L. Arge, K. H. Hinrichs, J. Vahrenhold, and J. S. Vitter. Efficient bulk operations on dynamic R-trees. In C. C. McGeoch and M. T. Goodrich, editors, *Proceedings of the 1st Workshop on Algorithm Engineering and Experimentation (ALENEX '99)*, volume (to appear) of *Lecture Notes in Computer Science*. Springer, 1999.
5. R. Baeza-Yates, E. Barbosa, and N. Ziviani. Hierarchies of indices for text searching. *Information Systems*, 21(6):497–514, 1996.
6. R. D. Barve, E. F. Grove, and J. S. Vitter. Simple randomized mergesort on parallel disks. In *8th Annual ACM Symposium on Parallel Algorithms and Architectures (SPAA '96)*, pages 109–118. ACM, June 1996.
7. R. Bayer and E. McCreight. Organization and maintenance of large ordered indizes. *Acta Informatica*, 1:173–189, 1972.
8. G. S. Brodal and J. Katajainen. Worst-case efficient external-memory priority queues. In S. Arnborg and L. Ivansson, editors, *Proceedings 6th Scandinavian Workshop on Algorithm Theory*, volume 1432 of *Lecture Notes in Computer Science*, pages 107–118, Stockholm, Sweden, July 1998. Springer.
9. Y.-F. Chian, M. T. Goodrich, E. Grove, R. Tamassia, D. E. Vengroff, and J. S. Vitter. External-memory graph algorithms. In *6th Annual ACM-SIAM Symposium on Discrete Algorithms (SODA95)*, pages 139–149, New York, 1995. acm-Press.
10. Y.-J. Chiang. *Dynamic and I/O-Efficient Algorithms for Computational Geometry and Graph Algorithms*. PhD thesis, Brown University, 1995.
11. A. Colvin and T. Cormen. Vic*: A compiler for virtual-memory c*. *3rd International Workshop on High-Level Parallel Programming Models and Supportive Environments*, pages 23–33, 1998.
12. Crauser, Mehlhorn, Althaus, Brengel, Buchheit, Keller, Krone, Lambert, Schulte, Thiel, Westphal, and Wirth. On the performance of LEDA-SM. Technical Report MPI-I-98-1-028, Max-Planck Institut für Informatik, 1998.

13. A. Crauser and P. Ferragina. On constructing suffix arrays in external memory. *European Symposium on Algorithms (ESA '99)*, 1999.

14. A. Crauser, P. Ferragina, and U. Meyer. Efficient priority queues for external memory. Working Paper, Oktober 1997.

15. A. Crauser, K. Mehlhorn, and U. Meyer. Kürzeste-Wege-Berechnung bei sehr großen Datenmengen. In O. Spaniol, editor, *Promotion tut not : Innovationsmotor "Graduiertenkolleg" (Aachener Beiträge zur Informatik; Band 21)*, pages 113–132S., Aachen, 1997. Verlag der Augustinus Buchhandlung.

16. R. Fagin, J. Nievergelt, N. Pippenger, and H. R. Strong. Extendible hashing — A fast access method for dynamic files. *ACM Transactions on Database Systems*, 4(3):315–344, Sept. 1979. Also published in/as: IBM, Research Report RJ2305, Jul. 1978.

17. M. Farach, P. Ferragina, and S. Muthukrishnan. Overcoming the memory bottleneck in suffix tree construction. *IEEE Symposium on Foundations of Computer Science*, 1998.

18. P. Ferragina and R. Grossi. A fully-dynamic data structure for external substring search. In *Proceedings of the 27th Annual ACM Symposium on the Theory of Computing (STOC '95)*, pages 693–702. ACM, May 1995.

19. M. T. Goodrich, J.-J. Tsay, D. E. Vengroff, and J. S. Vitter. External-memory computational geometry. In IEEE, editor, *Proceedings of the 34th Annual Symposium on Foundations of Comptuer Science*, pages 714–723, Palo Alto, CA, November 1993. IEEE Computer Society Press.

20. D. Hutchinson, A. Makeshwari, J.-R. Sack, and R. Velicescu. Early experiences in implementing the buffer tree. *Workshop on Algorithmic Engineering*, 1997.

21. U. Manber and G. Myers. Suffix arrays: a new method for on-line string searches. *SIAM Journal of Computing*, 22(5):935–948, 1993.

22. K. Mehlhorn and S. Näher. The LEDA Platform for Combinatorial and Geometric Computing. Cambridge University Press, 1999. Draft versions of some chapters are available at http://www.mpi-sb.mpg.de/~mehlhorn.

23. K. Mehlhorn and S. Näher. LEDA, a platform for combinatorial and geometric computing. *Communications of the ACM*, 38:96–102, 1995.

24. C. Ruemmler and J. Wilkes. An introduction to disk drive modeling. *IEEE Computer*, pages 17–28, 1994.

25. P. Sanders. Fast priority queues for cached memory. In *Workshop in Algorithmic Engineering and Experimentation (ALENEX)*, Lecture Notes in Computer Science. Springer, 1999.

26. Ullman and Yannakakis. The Input/Output Complexity of Transitive Closure. *Annals of Mathematics and Artificial Intelligence*, 3:331–360, 1991.

27. D. E. Vengroff and J. S. Vitter. Supporting I/O-efficient scientific computation in TPIE. In *Symposium on Parallel and Distributed Processing (SPDP '95)*, pages 74–77. IEEE Computer Society Press, October 1995.

28. J. Vitter and E. Shriver. Optimal algorithms for parallel memory I:two-level memories. *Algorithmica*, 12(2-3):110–147, 1994.

A Priority Queue Transform

Michael L. Fredman*

Rutgers University, New Brunswick, NJ 08903
fredman@cs.rutgers.edu

Abstract. A transformation, referred to as *depletion*, is defined for comparison-based data structures that implement priority queue operations. The depletion transform yields a representation of the data structure as a forest of heap-ordered trees. Under certain circumstances this transform can result in a useful alternative to the original data structure. As an application, we introduce a new variation of skew-heaps that efficiently implements decrease-key operations. Additionally, we construct a new version of the pairing heap data structure that experimentally exhibits improved efficiency.

1 Introduction

The focus of this paper centers on comparison-based priority queue data structures. A priority queue is considered comparison-based provided that the elements that it stores can belong to an arbitrarily chosen linearly ordered universe. In particular, the only operations that can be performed on these elements (by the data structure) are simple comparisons. A *mergeable* priority queue is one that efficiently supports the merge operation. Among such data structures can be found those that are represented as a forest consisting of one or more heap-ordered trees, and moreover, have the property that comparisons take place only among tree roots in the forest as operations are performed. Examples include binomial queues [7], Fibonacci heaps [3] and the various forms of pairing heaps [2,6]. We shall refer to such data structures as *forest-based* priority queues. Although the context of our discussion concerns mergeable priority queues, this is not an essential aspect.

Let Q consist of a collection of individual priority queues, $Q_1, Q_2, \cdots$. Assume that the priority queue operations being supported consist of merge(Q_i,Q_j), which replaces Q_i with the merged result (removing Q_j in the process), deletemin(Q_i), and the operation make-queue(Q_k,x), which creates a priority queue Q_k containing the single value x. Insertion is defined in terms of make-queue and merge. The *depletion transform* of Q, defined next, yields a forest-based representation of Q. As priority queue operations take place, the depletion transformation maintains a collection S of corresponding *shadow* structures, $S_1, S_2, \cdots$, where S_i is referred to as the shadow of Q_i and consists of a forest of heap-ordered trees that store the values contained in Q_i. Associated with

---

* Supported in part by NSF grant CCR-9732689

J.S. Vitter and C.D. Zaroliagis (Eds.): WAE'99, LNCS 1668, pp. 243–257, 1999.
© Springer-Verlag Berlin Heidelberg 1999

each S_i is a queue C_i consisting of comparison operations, with each comparison in C_i having an associated time-stamp. The comparisons in C_i are arranged in order by their time-stamps, with the back of C_i containing the comparison having the most recent time-stamp. The role of the comparison queues will be described shortly. As operations take place with Q, the depletion transform manipulates S and the associated comparison queues C_i as follows. When executing merge(Q_i, Q_j), the forests of the respective shadows S_i and S_j are combined and replace S_i. Moreover, the comparison queues C_i and C_j are merged (respecting time-stamp order) and replace C_i. When executing make-queue(Q_k, x), a forest S_k consisting of one node that stores the value x is created, and C_k is set to be the empty queue. When deletemin(Q_i) is performed, the corresponding tree root in S_i is deleted, and the subtrees of this root become new trees in the forest of S_i. As comparisons are performed involving values in Q_i, each such comparison is inserted into C_i with its time-stamp set to the current time. We refer to mergings, deletemins, and individual comparisons as *events*.

We now describe the manner in which linkings are established in the shadow S. After each event takes place, a given queue C_i is processed as follows. Assume that their exists in C_i a comparison that involves two tree roots in S_i, and let c be the one having the oldest time-stamp (closest to the front of C_i). Then the corresponding trees in S_i are linked to reflect the comparison c (the root which "loses" the comparison becomes the leftmost child of the root which "wins" the comparison) and c is removed from C_i. The process is repeated until C_i no longer contains any applicable comparisons. Informally, the depletion transform defers a comparison until the two items being compared each become tree roots, at which point the comparison get executed (provided that there are no earlier comparisons free to be executed). As will become clear, our definition of depletion is strictly conceptual, and will not in itself constitute a performance issue.

The following figures illustrate depletion, highlighting the deletemin operation. Figure 1 show a priority queue Q, its shadow S and associated comparison queue C; and figure 2 shows the result of executing deletemin. Referring to figure 1, upon executing the deletemin operation, treating Q as a skew heap, we find that the comparisons $(7 : 4)$ and $(7 : 10)$ get executed, and these comparisons get appended to the back of C. Now with respect to S, the removal of its root permits execution of the linkings $(5 : 4)$ and $(7 : 8)$ from C, as well as the linking corresponding to the newly appended comparison $(7 : 4)$, since the nodes that hold these values have become tree roots (and remain so at the respective moments that these linkings take place). The two remaining comparisons $(5 : 9)$ and $(7 : 10)$ in C are still deferred. The structures are thus modified as shown in figure 2.

The depletion transform typically loses information; some of the comparisons in the C_i's never get executed. When this happens the shadow S is not a viable representation of Q as an independent data structure. However, under appropriate circumstances no such loss occurs, and moreover, the manipulations of S, as induced by the operations of Q, can be inferred without maintaining any structure beyond that represented by the S_i's.

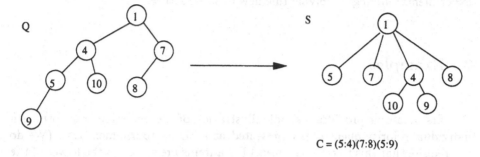

C = (5:4)(7:8)(5:9)

Fig. 1. A priority queue and corresponding shadow structure

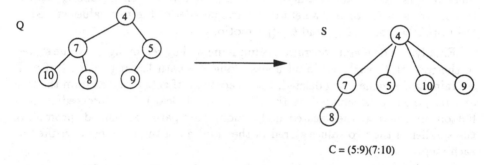

C = (5:9)(7:10)

Fig. 2. Corresponding structures following deletemin execution

When this is the case, we refer to Q as being *faithfully depletable*, and its shadow structure S can enjoy an independent existence. The depletion transformation, in such instances, results in a forest-based data structure enjoying the property that its amortized costs (measured in comparisons) do not exceed those of the original data structure Q. Moreover, decrease-key operations are easily implementable in the shadow data structure. (Recall that the decrease-key operation replaces the value of the referenced item with a smaller value.) Exploiting the fact that the resulting data structure is forest-based, a decrease-key operation is implemented by positioning the subtree of the node affected by the operation as a new tree in its forest (possibly just momentarily), assuming that it is not already a tree root; the same technique used for Fibonacci heaps [3] and pairing heaps [2].

In the following section we consider two examples of depletion. The first provides a simple illustration of the depletion transform, and the second provides an application of depletion to the skew-heap data structure [5], which is shown to be faithfully depletable. In section 3 we construct a new variation of pairing heaps that enjoys improved efficiency, and note that our construction can be motivated in terms of considerations involving depletion. In section 4, we report experimental findings involving this new data structure.

2   Examples

Our first example provides a simple illustration of the depletion transform. An individual priority queue Q is represented as a binary tournament tree. (We do not require that this tree be balanced.) Data items are stored in the leaves of the tree and the winners of comparisons are promoted upwards (copied rather than moved) through the internal nodes of the tree, so that the minimum data item is found in the root of the tree. For the purpose of this example, we implement the merge operation as follows. Let Q_1 and Q_2 be the tournament trees being merged. First, we establish a new root node r and link both Q_1 and Q_2 as the two subtrees of r. Second, we store in r the smaller of the two values stored in the respective roots of Q_1 and Q_2 (promotion).

Finally, the deletemin operation is implemented by removing the value stored in the root of T (say) and in all nodes along the path leading to the leaf of T containing the value being deleted. (The vacated leaf gets removed from the tree and its parent gets replaced by the sibling of this leaf.) Then proceeding in a bottom-up manner, the vacated nodes along this path are refilled, promoting the smaller of the two values stored in the children of the node being refilled at each step.

Now we consider the depletion transform applied to the above data structure. In this instance the resulting shadow structure is similar to a pairing heap [2], and there are no deferred comparisons, so that the comparison queues associated with the shadow structures are empty. (Proofs pertaining to this discussion are left to the reader.) An individual priority queue in the shadow structure is represented as a single heap-ordered tree, and the priority queue operations are performed as follows. Given two heap-ordered trees, a *pairing operation* on these trees combines them by comparing the values in their respective roots, and making the root which loses the comparison the leftmost child of the other root. The merge operation is performed by pairing the respective individual trees.

The deletemin operation is performed by removing the root of the tree and then pairing the remaining subtrees in order from right-to-left; each subtree is paired with the result of the linkings of the subtrees to its right. We refer to this right-to-left sequence of node pairings as a *right-to-left incremental pairing pass*. (In contrast with the pairing heap data structure [2], the deletemin operation omits the first (left-to-right) pairing pass encompassed by the pairing heap deletemin operation, performing only the second pass. Below, we will need to recall that the first, omitted pass, proceeds by pairing the first and second subtrees, then the third and fourth subtrees, etc.)

2.1   Application to Skew-Heaps

We turn next to our second example, skew-heaps [5]. First, we describe the particular variant of the skew-heap data structure whose depletion will prove to be particularly attractive. The skew-heap consists of a binary heap-ordered tree. The deletemin operation is performed by merging the two subtrees of the root (the root having been first removed).

The merge operation proceeds by merging the right spines of the two respective trees. Suppose that the right spine of one tree consists of $x_1, x_2, \cdots, x_r$, that the right spine of the other tree consists of $y_1, y_2, \cdots, y_s$ (in order, starting with the respective tree roots), and that $v_1, v_2, \cdots, v_t$ $(t = r + s)$ gives the result of merging the two spines. Suppose that (say) the sequence of x_i's is exhausted prior to the sequence of y_i's in this merged result. Let $v_h = x_r$, $h < t$ be the point at which the x_i's are exhausted. Then in the resulting merged tree the v_i's comprise a path starting at the root; the links joining $v_1, v_2, \cdots, v_{h+1}$ are all left links, and the remaining links (those joining $v_{h+1}, \cdots, v_t$) all consist of right links. In other words, as merging takes place in top-down order, subtrees are swapped, but not below the level of the first node beyond the exhausted sublist. An alternative, recursive formulation of the merge operation is given as follows. Let u and v be the two trees being merged and let merge(u,v) denote the result of the merging. If v is empty, then merge(u,v) is given by u. Otherwise, assume that the root r of u wins its comparison with the root of v and let u_L and u_R be the left and right subtrees of u, respectively. Then merge(u,v) has r as its root, u_L as its right subtree, and merge(u_R,v) as its left subtree.

Next, we describe the shadow data structure S given by the depletion transform of the skew-heap. This data structure bears striking similarity to the pairing heap; only the sequence of pairings performed in the course of a deletemin operation differ. An individual priority queue is represented as a single heap-ordered tree. Considering the deletemin operation, let $y_1, y_2, \cdots, y_m$ be the children of the tree root being deleted, in left-to-right order. The deletemin operation proceeds to combine these nodes as follows. First, a right-to-left incremental pairing pass is performed among the nodes in odd numbered positions (the nodes $y_1, y_3, y_5, \cdots$). Let S_{odd} denote the resulting tree. Second, a right-to-left incremental pairing pass is performed among the nodes in even numbered positions (the nodes $y_2, y_4, y_6 \cdots$). Let S_{even} denote the resulting tree. Finally, S_{odd} and S_{even} are paired. (If S_{even} is empty, then the final result is given by S_{odd}.) We refer to this shadow data structure as a *skew-pairing heap*.

Theorem 1. *The depletion transform, when applied to the skew-heap data structure, induces the skew-pairing heap data structure.*

(Proof omitted due to space limitation.)

There are two arguable advantages of the skew-pairing heap relative to the skew-heap. First, we avoid the necessity of swapping subtrees as the operations take place. Second, and more fundamentally, the skew-pairing heap provides for efficient implementation of decrease-key operations, as described next.

Because the skew-pairing heap can be viewed as a new variant of pairing heap, decrease-key operations can be implemented, as is the case with the usual pairing heap, by removing the subtree of the affected node (assuming that it is not the root) and then linking it with the root.

It is readily shown that the $O(\log n)$ amortized bounds derived for pairing heaps extend to this data structure. Indeed, the same potential function used in the original analysis of pairing heaps [2] applies here. Moreover, in practice we can expect that the decrease-key operation would be considerably more efficient than suggested by this bound, although constant amortized cost is precluded as a consequence of a lower bound established for a large family data structures that generalize the pairing heap [1]. (Roughly speaking, this bound shows that decrease-key operations have an intrinsic cost of $\Omega(\log\log n)$, $n =$ heap size, for any self-adjusting variation of the pairing heap that restricts comparisons to take place among root nodes of trees in the forest.)

3   Improved Pairing Heaps

The perspective gained by consideration of the depletion transform motivates certain practical improvements for pairing heaps, which we proceed to describe. Returning to the example of the tournament tree, discussed at the beginning of the preceding section, two observations come to mind. First, an alternative implementation of insertion for this data structure can be entertained, and proceeds as follows. We choose a particular leaf of the tree and expand it, replacing it with an internal node having two leaves as children. In one of the new leaves we store the new value being inserted, and in the other we store the value found in the original leaf. We then proceed in a bottom-up manner, restoring the conditions that define a tournament tree. The advantage of this implementation relative to that based on merging (as described at the beginning of the previous section) is that the length of just one path from the root increases, instead of *all* paths having their lengths increase.

Now which leaf should we choose? Keeping in mind that we want our data structure to be faithfully depletable, one possibility is suggested by considering a particular implementation of changemin(x), which replaces the minimum priority queue value with x. Let ℓ be the leaf of the tournament tree that stores the minimum value. A natural implementation of changemin(x) first replaces the value in ℓ with the value x, and then reestablishes appropriate values on the path leading to the root. Viewed in terms of depletion, this operation translates as follows. We remove the root of the shadow tree, and then proceed with a right-to-left incremental pairing pass, but first position a new node containing x to the right of the existing children of the root.

This suggests that to implement insertion in the shadow data structure, we add a new singleton node tree to the forest. When executing a deletemin operation, we treat the inserted element, if it is not minimal, as a rightmost child of the root. As a concept, this also seems sensible in the context of the usual pairing heap data structure, particularly when combined with the *auxiliary area* method of Stasko and Vitter [6]. As originally described [6], the auxiliary area method works as follows.

A deletemin operation is always completed with a single tree remaining, referred to as the *main* tree. Between successive deletemin operations, the nodes and subtrees resulting from insertions and decrease-key operations are stored in what is referred to as the *auxiliary area*. When the next deletemin operation takes place, the trees in the auxiliary area are coalesced into a single tree using what is referred to as the *multipass method* [2] (to be described shortly). This tree is then combined with the main tree using a linking operation. The root of the resulting tree is then removed, and its subtrees are then coalesced in the same manner as for the usual two-pass pairing heap

The multipass method for coalescing a list of trees begins by linking the trees in pairs. This is now repeated for the list of trees resulting from this first pass (of which there are half as many as initially present), and then repeated again, as necessary, until a single tree remains. (An alternative and preferable implementation places the result of each linking at the end of a queue consisting of the trees being linked, proceeding with the linkings in a round robin manner until a single tree remains.)

Now we modify the above by maintaining two auxiliary forests: one for the subtrees that result from decrease-key operations, and the other for newly inserted items. When executing a deletemin operation, we apply the multipass method to both auxiliary areas, obtaining two trees, a *decrease-key tree* and an *insertion tree*. Second, we pair the decrease-key tree with the main tree. Call this the *augmented main tree*. Third, we *compare* the roots of the insertion tree with that of the augmented main tree. If the root of the insertion tree wins the comparison, we link the two trees and proceed with the root removal and subsequent pairing passes in the usual way. On the other hand, if the root of the insertion tree loses the comparison, then we do not link the two trees together. Instead, we remove the main tree root and proceed with the pairing passes, treating the insertion tree as though it is the rightmost subtree of the root. In the sequel we refer to this modification as the *insertion heuristic*.

Next, we turn to our second observation. Viewed in terms of a tournament tree, the first pass of the usual (two-pass) deletemin algorithm for pairing heaps restructures the tournament tree by performing a rotation on every second link along the path leading to the root, thereby "halving" this path. In the long run, this restructuring keeps the tree reasonably balanced. However, for a tree that happens to be highly balanced this path halving operation is somewhat harsh and may even worsen the overall balance of the tree. This motivates our second modification. Only when the number of children of the root (in the pairing heap) is relatively large should the path halving pass take place. Experimentally, a good dividing point seems to be when the number of children exceeds $1.2 \cdot \lg n$, where n is the number of items in the priority queue. (Note that the quantity $1.2 \cdot \lg n$ can be maintained with minimal overhead as priority queue operations take place.) In the sequel we refer to this modification as the *second-pass heuristic*.

4 Experimental Findings

Our experiments fall into two basic categories. With respect to the first category, the priority queue is utilized to sort data inserted into an initially empty priority queue, accomplished by repeated execution of the deletemin operation. With respect to the second category, after building a priority queue of a specified size, the remaining operations are subdivided into multiple *rounds*, where each round consists of a specified number of decrease-key operations, a single insertion, and a single deletemin operation. A round of operations leaves the priority queue size unchanged. All of our experimental results are reported in terms of average number of comparisons per deletemin operation.

As noted by Jones [4], with respect to the second category of experiments, the initial "shape" of the priority queue (upon reaching its steady-state size) can influence subsequent performance. Most of our category 2 experiments (those reflected in figures 3(a), 3(b), 5, 6, 7, and 8) build the initial priority queue configuration by repeated execution of the pattern: insert, insert, deletemin; until the steady-state size is reached. (We return to this issue below.)

Our experiments involve both numerical and adversarial data. For simplicity, we implement a decrease-key operation so that no change in the affected data value takes place. Our numerical experiments (considered first) adhere to the negative exponential distribution model outlined in Jones [4]; the next data value to be inserted is chosen to be the previously deleted value plus an offset $\delta = -\ln(\text{rand})$, where rand is a random variable uniformly distributed over the interval $(0,1)$. In the steady-state the next data item to be inserted has the same distribution as the items remaining in the priority queue [4]. The items chosen for decrease-key operations are randomly selected from the priority queue.

For the sorting experiments it suffices to utilize random numerical data values uniformly distributed over the interval $(0,1)$ since this induces a uniform distribution on the corresponding order permutations. We note that for comparison-based sorting algorithms, the induced distribution on the order permutations is the *only* determinant of performance, when measured in comparisons.

The choice of the coefficient 1.2, appearing in the threshold $1.2 \lg n$ for the first-pass heuristic, can be a matter of fine tuning, and in making this choice we use test runs that utilize numerical rather than adversarial data. Figure 3(a) typifies the considerations involved. Case A plots average deletemin cost versus coefficient choice for steady-state priority queue size 60000, with 0 decrease-key operations per round. Case B represents steady-state priority queue size 10000, with 2 decrease-key operations per round.

We choose the "two-pass auxiliary area" variation of pairing heap [6] (described above) as our starting point. In the sequel this is referred to as the *basic* data structure. We refer to the data structure described in the preceding section, employing the first-pass and insertion heuristics, as the *augmented* data structure. Our experiments are primarily designed to compare the performance of the basic data structure with that of the augmented data structure.

Our experiments involving adversarial data utilize an adversary defined by the rule that when a comparison takes place between two tree roots, the root

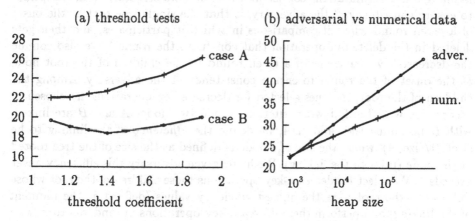

(a) threshold tests

(b) adversarial vs numerical data

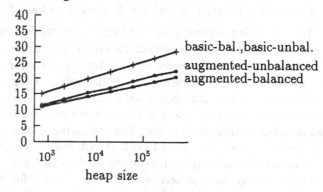

(c) initial configuration effect - numerical data

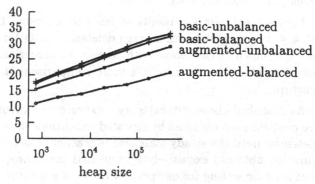

(d) initial configuration effect - adversarial data

Fig. 3. Initial considerations

of the larger tree wins the comparison. This outcome yields the least amount of information for a given comparison, in effect maximizing the number of remaining order permutations consistent with the data structure. (An exception, to ensure consistency of the adversary, is that the single tree root at the onset of a given round wins all comparisons in which it participates, and then gets deleted in the deletemin operation that concludes the round.) We also confine the decrease-key operations of a given round to the children of the root node at the onset of the round to ensure consistency of the adversary. Among the children of the root, the ones selected for decrease-key operations are chosen in accordance with the following criterion. When two nodes A and B are linked, with B becoming the child of A, we define the *efficiency* of the linking to be size(B)/size(A), where the size of a node is defined as the size of the tree rooted at the node (prior to the linking). With our given adversary this efficiency never exceeds 1. We select for decrease-key operations those children of the root whose linkings to the root have the highest efficiency values (defined at the moment of the linking) and perform these decrease-key operations in random order. The number of children chosen in a given round is a parameter of the experiment. (This selection criterion is designed to maximize loss of information [1].)

In contrasting experiments based on numerical data with those based on the use of our adversary, we note that the latter is intended to elicit poor behavior, whereas the former is intended to elicit typical behavior. With respect to experiments involving the adversary, the nodes selected for decrease-key operations reflect the data structure being tested, whereas for the numerical experiments, these choices are independent of the data structure. Figure 3(b) contrasts these two forms of testing. Both of the plots appearing in this figure reflect 8 decrease-key operations per round, and both reflect the augmented data structure. For this and all of remaining figures (except for figures 4 and 7), our plots reflect runs involving steady-state priority queue sizes $n = 3^k$, for $6 \leq k \leq 12$ ($3^6 = 729$, $3^{12} = 531441$), and average deletemin cost is computed over $3n$ deletemin operations ($3n$ rounds per run).

Figure 4 exhibits the results of our sorting runs. Data sets of size 3^k for $6 \leq k \leq 12$ are sorted, and average deletemin costs for the respective data set sizes are shown in the plots. With respect to numerical data, we find that the basic data structure requires 20% more comparisons than the augmented data structure.

As described above, our category 2 experiments involve an initial data structure configuration obtained by repeated executions of the pattern: insert, insert, deletemin; until the steady-state size is reached. Compared with an initial configuration obtained exclusively by repeated insertions, this seems to present a more realistic setting for our priority queue algorithms; the alternative configuration has extremely good initial *balance*, perhaps likely to result in unusually good subsequent performance. Figures 3(c) and (d) exhibit this effect. The plots labeled "balanced" concerns runs that are based on an initial phase consisting exclusively of insertions. The plots labeled "unbalanced" involves runs based on the (insert, insert, deletemin) building phase, employed in our subsequent experiments. (None of the runs depicted in these two figures include decrease-key operations.) Considering the runs involving numerical data, using the basic data structure, the results are indistinguishable with respect to choice of initial configuration. In contrast, the runs involving the augmented data structure are more sensitive to the initial configuration, particularly in the presence of adversarial data.

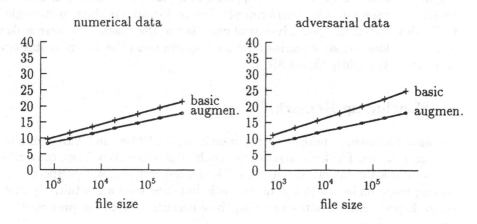

Fig. 4. Sorting data sets

The results of our category 2 experiments are shown in figures 5 and 6, which involve, respectively, numerical and adversarial data. Our category 2 numerical experiments reveal that the basic data structure performs 27% more comparisons than the augmented data structure when no decrease key operations occur. In the presence of decrease-key operations, the two data structures exhibit similar performance. This raises the following question. *Might it be that after a prolonged phase characterized by a high frequency of decrease-key operations, the augmented data structure permanently enters a state from which it no longer enjoys a performance advantage relative to the basic data structure, even if there are no subsequent decrease-key operations?* Fortunately, Figure 7 suggests that this is *not* the case. The figure concerns a run involving numerical data and a steady-state priority queue size of 30000. After the initial building phase, 270000 rounds are executed. The middle 90000 rounds contain 8 decrease-key operations per round; the first and last 90000 rounds contain no decrease-key operations. The 270000 rounds are subdivided into intervals of 1000 rounds each, and the average deletemin costs over the respective intervals are computed and plotted. As seen from the plot, after completion of the middle phase of the computation, the data structure quickly recovers its prior level of performance. This is similarly the case with respect to adversarial data (not shown).

The category 2 experiments exhibit a larger performance gap between the augmented data structure and the basic data structure as compared with the category 1 (sorting) experiments (27% versus 20%), reflecting the fact that the insertion heuristic does not come into play in the category 1 experiments.

A final issue we address concerns the skew-pairing mechanism. With respect to the preceding experiments the augmented data structure employs the strategy that upon removing the minimal root in the course of a deletemin operation, the children of that root are then combined using the two-pass method (provided that their number exceeds $1.2 \lg n$). An alternative would be to combine these children in the manner of the skew-pairing heap, as described above. Unfortunately, this rarely improves performance. Figure 8 contrasts the two strategies. Only when operating upon adversarial data and in the absence of decrease-key operations does the skew-pairing strategy improve upon the two-pass strategy, and then only slightly (figure 8(b)).

5   Concluding Remarks

We have presented a priority queue transform, depletion, and have exhibited two applications. The first extends the capabilities of the skew-heap, converting it into a new variant of pairing heap. The second introduces a perspective on pairing heaps under which certain heuristic improvements are intuitively motivated. Experimental results concerning these heuristics have been presented and appear promising.

Our transform effects a reordering of atomic operations, comparisons in this instance. Other interesting applications of this, and other analogous transforms may exist.

As is the case with the other variations of pairing heaps, an asymptotic determination of the amortized complexities of the skew-pairing heap is an open problem.

References

1. M. L. Fredman, *On the efficiency of pairing heaps and related data structures*, to appear in Journal of the ACM.
2. M. L. Fredman, R. Sedgewick, D. D. Sleator, and R. E. Tarjan, *The pairing heap: a new form of self-adjusting heap*, Algorithmica 1,1 (1986), pp. 111–129.
3. M. L. Fredman and R. E. Tarjan, *Fibonacci heaps and their uses in improved network optimization problems*, Journal of the ACM 34,3 (1987), pp. 596–615.
4. D. W. Jones, *An empirical comparison of priority-queue and event-set implementations*, Communications of the ACM 29, 4 (1986), pp. 300–311.
5. D. D. Sleator and R. E. Tarjan, *Self-adjusting heaps*. SIAM Journal on Computing 15,1 (1986), pp. 52–69.
6. J. T. Stasko and J. S. Vitter, *Pairing heaps: experiments and analysis*, Communications of the ACM 30,3 (1987), pp. 234–249.
7. J. Vuillemin, *A data structure for manipulating priority queues*, Communications of the ACM 21 (1978) pp. 309–214.

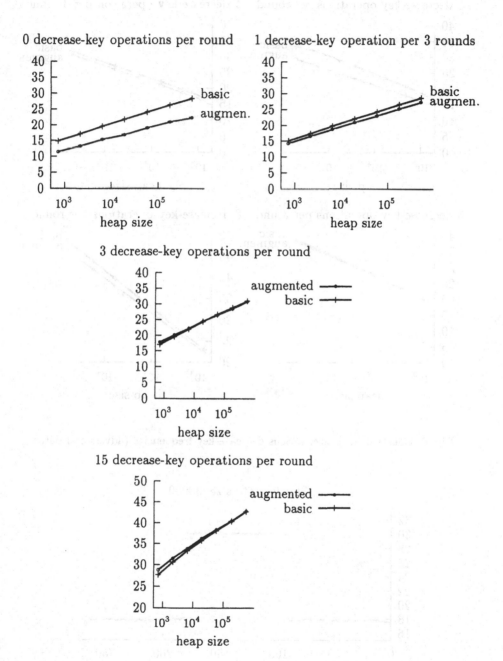

Fig. 5. Constant heap size; various decrease-key frequencies (numerical data)

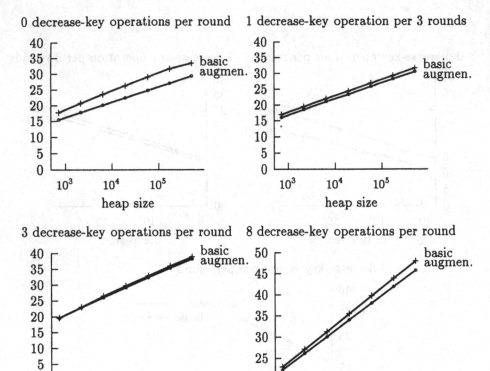

Fig. 6. Constant heap size; various decrease-key frequencies (adversarial data)

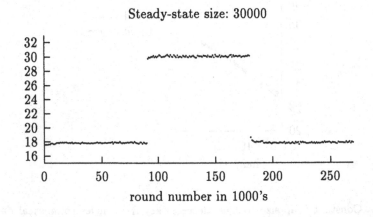

Fig. 7. Middle phase with 8 decrease-key operations per round.

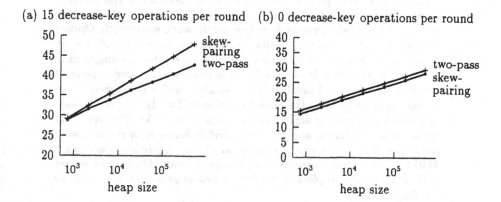

Fig. 8. skew-pairing versus two-pass: (a) numerical data, (b) adversarial data

Implementation Issues and Experimental Study of a Wavelength Routing Algorithm for Irregular All–Optical Networks*

Athanasios Bouganis, Ioannis Caragiannis**, and Christos Kaklamanis***

Computer Technology Institute and
Dept. of Computer Engineering and Informatics
University of Patras, 26500 Rio, Greece

Abstract. We study the problem of allocating optical bandwidth to sets of communication requests in all–optical networks that utilize Wavelength Division Multiplexing (WDM). WDM technology establishes communication between pairs of network nodes by establishing transmitter–receiver paths and assigning wavelengths to each path so that no two paths going through the same fiber link use the same wavelength. Optical bandwidth is the number of distinct wavelengths. Since state–of–the–art technology allows for a limited number of wavelengths, the engineering problem to be solved is to establish communication between pairs of nodes so that the total number of wavelengths used is minimized.

In this paper we describe the implementation and study the performance of a wavelength routing algorithm for irregular networks. The algorithm proposed by Raghavan and Upfal [17] and is based on a random walk technique. We also describe a variation of this algorithm based on a Markov chain technique which is experimentally proved to have improved performance when applied to random networks generated according to the $G_{n,p}$ model.

1 Introduction

Optical fiber transmission is rapidly becoming the standard transmission medium for networks and can provide the required data rate, error rate, and delay performance for future networks. A single optical wavelength supports rates of gigabits-per-second (which in turn support multiple channels of voice, data, and video [18]). Multiple laser beams that are propagated over the same fiber on distinct optical wavelengths can increase this capacity much further; this is achieved through WDM (Wavelength Division Multiplexing). However, data rates are limited in opto–electronic networks by the need to convert the optical signals on the fiber to electronic signals in order to process them at the network nodes.

---

* This work has been partially supported by EU Esprit Project 20244 ALCOM–IT, the Greek General Secretariat of Research and Technology, and the Greek PTT.

** e-mail: caragian@ceid.upatras.gr, http://www.ceid.upatras.gr/~caragian

*** e-mail: kakl@cti.gr

J.S. Vitter and C.D. Zaroliagis (Eds.): WAE'99, LNCS 1668, pp. 258–270, 1999.
© Springer-Verlag Berlin Heidelberg 1999

Networks using optical transmission and maintaining optical data paths through the nodes are called all–optical networks.

In such networks, we consider communication requests as ordered transmitter–receiver pairs of network nodes. WDM technology establishes connectivity by finding transmitter–receiver paths, and assigning a wavelength to each path, so that no two paths going through the same fiber link use the same wavelength. Optical bandwidth is the number of available wavelengths. As state–of–the–art optics technology allows for a limited number of wavelengths (even in the laboratory) the important engineering problem to be solved is to establish communication between pairs of nodes so that the total number of wavelengths used is minimized.

Theoretical work on optical networks mainly focuses on the performance of wavelength routing algorithms on regular networks or arbitrary networks using oblivious (predefined) routing schemes. We point out the pioneering work of Pankaj [15] who considered shuffle exchange, De Bruijn, and hypercubic networks. Aggarwal et al. [1] consider oblivious wavelength routing schemes for several networks. Raghavan and Upfal in [17] consider mesh–like networks. Aumann and Rabani [2] improve the bounded of Raghavan and Upfal for mesh networks and also give tight results for hypercubic networks. Rabani in [16] gives almost optimal results for the wavelength routing problem on meshes.

These topologies reflect architectures of optical computers rather than wide–area networks. For fundamental practical reasons, the telecommunication industry does not deploy massive regular architectures: backbone networks need to reflect irregularity of geography, non–uniform clustering of users and traffic, hierarchy of services, dynamic growth, etc. In this direction, Raghavan and Upfal [17], Aumann and Rabani [2], and Bermond et al. [4], among other results, focus on bounded–degree networks and give upper and lower bounds in terms of the network expansion. The wavelength routing problem in tree–shaped networks has also received much attention. Erlebach and Jansen [7, 8] prove that several versions of the problem are NP–hard, while extensive study of a class of algorithms for such networks has been made in a series of papers [13, 10, 12, 11].

In this paper we describe the implementation and experimentally study the performance of a wavelength routing algorithm for irregular (arbitrary) sparse networks. This algorithm proposed by Raghavan and Upfal in [17] and theoretical results were obtained for networks of bounded degree. The model of bounded degree networks reflects the irregularity property of real communication networks. We also describe a variation of this algorithm (algorithm MC) that is experimentally proved to have improved performance.

The paper is structured as follows. We give basic definitions on the optical model we follow in section 2. The description of algorithm RW and issues concerning its implementation are presented in section 3. Algorithm MC is described in section 4. We present experimental results in section 5. Some extensions of this work currently in progress are briefly discussed in section 6.

2   The Optical Model

We follow the notation proposed in [3]. A network is modeled as a graph $G = (V(G), E(G))$ where $V(G)$ is the set of nodes and $E(G)$ the set of fiber links. We denote by $P(x, y)$ a *path* in G that consists of consecutive edges beginning in node x and ending to node y. A *request* is an ordered pair of nodes (x, y) in G. An *instance* I is a collection of requests. Note that a given request may appear more than once in an instance.

A *routing* R for an instance I on network G is a set of paths $R = \{P(x, y) | (x, y) \in I\}$. The *conflict graph* associated to a routing R of instance I on network G is a graph $G_R = (V(G_R), E(G_R))$ such that each node in $V(G_R)$ corresponds to a path in R and two nodes x, y in $V(G_R)$ are adjacent (there exist an edge $(x, y) \in E(G_R)$) if and only if the corresponding paths in R share a fiber link of network G.

Let G be a network and I an instance of requests. The *wavelength routing problem* consists of finding a routing R for instance I and assigning each request $(x, y) \in I$ a wavelength, in such way that no two paths of R sharing a fiber link of G have the same wavelength. Intuitively, we can think the wavelengths as colors and the wavelength routing problem as a node–coloring problem of the corresponding conflict graph.

Early models of optical networks assumed that optical transmission through fibers is performed bidirectionally, so undirected graphs were used to model optical networks. It has since become apparent that directed graphs are essential to model state–of–the–art technology, so recent theoretical work on WDM optical networks focuses to the directed model. In this model, the network is a symmetric digraph (unless otherwise specified), i.e. between two adjacent nodes, there exist two opposite directed fiber links. Note that the definitions above apply to both models.

Given a routing R for an instance I on network G, we denote by $w(G, I, R)$ the minimum number of wavelengths used for a valid wavelength assignment to paths in R. Obviously, $w(G, I, R)$ is the chromatic number of the corresponding conflict graph G_R. We denote by $w(G, I)$ the minimum $w(G, I, R)$ over all routings R of instance I. The notation for the undirected model is $w(G, I, R)$ and $w(G, I)$, respectively.

Given a routing R for an instance I on network G, the *load* of a fiber link $\alpha \in E(G)$, denoted by $\pi(G, I, R, \alpha)$, is the number of paths in R that contain α. We denote by $\pi(G, I, R)$ the maximum load of any fiber link in $E(G)$, and by $\pi(G, I)$ the minimum $\pi(G, I, R)$ over all routings R of an instance I. The notation for the undirected model is $\pi(G, I, R, \alpha)$, $\pi(G, I, R)$ and $\pi(G, I)$, respectively. Obviously, $\pi(G, I, R)$ is a lower bound for $w(G, I, R)$, so $\pi(G, I)$ is a lower bound for $w(G, I)$.

Given a routing R for an instance I on network G, the *length* of a path $P(x, y) \in R$, denoted by $\delta(G, I, R, P(x, y))$, is the number of consecutive fiber links of $E(G)$ contained in $P(x, y)$. The *dilation* of a routing R for instance I on network G, denoted by $\delta(G, I, R)$ is the maximum length of any path in R. The average length of paths in R, denoted by $\tilde{\delta}(G, I, R)$ is defined in the obvious way. We can also define $\delta(G, I)$ as the minimum $\delta(G, I, R)$ over all routings R. In general computing $w(G, I)$, $w(G, I, R)$ and $\pi(G, I)$ is NP–hard for most networks [9, 7, 8], while computing $\delta(G, I)$ can be solved in polynomial time [6].

An important instance that has received much attention is the k–*relation*. A directed k–relation is defined as an instance I_k in which each node appears as a source or as a destination in no more than k requests. An undirected k–relation is defined as an instance I_k in which each node appears in no more than k requests (either as source or as destination). In the following sections, we consider k–relations that have the maximum number of requests, i.e. in a directed k–relation, each node appears as source or designation in exactly k requests, while in an undirected k–relation, each node appears (either as source or as destination) in exactly k requests. Note that under this definition of k–relations, given a network G, the number of requests in a directed k–relation is twice the number of requests in an undirected k–relation. A 1–relation I_1 is called permutation and has received much attention in the literature (not only within the context of wavelength routing).

3 The Algorithm of Raghavan and Upfal

In this section we describe the implementation of the algorithm proposed by Raghavan and Upfal [17] for approximating $w(G, I_k)$ on bounded–degree all–optical networks (algorithm RW).

Given a k–relation I_k on a network G, the algorithm uses a random walk technique for finding a routing R for I on G and properly assigns colors to paths of R. The algorithm considers requests one by one, finds a path for each one of them and assigns a proper color to it. When a path has been established for a request of I and colored, it is never reestablished or recolored again.

We define a *random walk* on a graph $G = (V(G), E(G))$ (directed or not) as a Markov chain $\{X_t\} \subseteq V$ associated to a particle that moves from vertex to vertex according to the following rule: the probability of a transition from vertex v_i, of degree d_i, to vertex v_j is $1/d_i$ if $(i, j) \in E$, and 0 otherwise. The *transition matrix* of the random walk denoted by P is such that element P_{ij} (or P_{v_i, v_j}) is the probability that the particle moves from vertex v_i to vertex v_j. The *stationary distribution* of the random walk is a vector ρ such that $\rho = P\rho$. We define a *trajectory* W of length τ as a sequence of vertices $[w_0, w_1, \ldots, w_\tau]$ such that $(w_t, w_{t+1}) \in E(G)$ for $0 \leq t < \tau$.

Algorithm RW

1. Compute the transition matrix P and the stationary distribution ρ.
2. Compute a sufficient length L for trajectories.
3. For each request $(a_i, b_i) \in R$ do:
   (a) Choose a node $r_i \in V$ according to the stationary distribution.
   (b) Choose a trajectory W_i' (resp. W_i'') of length L from a_i to r_i (resp. from b_i to r_i) according to the distribution on trajectories, conditioned on the endpoints being a_i and r_i (resp. b_i and r_i).

(c) Connect a_i to b_i by the path $P(a_i, b_i)$, defined by W'_{a_i, r_i} followed by W''_{r_i, b_i}.

(d) Eliminate cycles of the path $P(a_i, b_i)$.

(e) Assign a proper color to $P(a_i, b_i)$.

Computing the length of trajectories. It is known [14] that since a random walk is an ergodic Markov chain, P^t converges to a unique equilibrium as $t \to \infty$. The length L of trajectories must be such that the power P^L is very close to $\lim_{t \to \infty} P^t$. Although this computation in [17] is $L = -3\frac{\log kn}{\log \lambda}$ where λ is the second largest eigenvalue of the transition matrix P in absolute value, we observed that P^t is very close to $\lim_{t \to \infty} P^t$ for smaller values of t. The computation $L = -\frac{1.5 \log n}{\log \lambda}$ from [14] suffices.

Random choice in step 3a. A node $r_i \in V$ is chosen according to the stationary distribution by simulating the casting an n–faced die with probabilities $\rho_i, i = 1, ..., n$ associated with the n faces.

Finding trajectories. The method proposed in [5] which is also implied in [17] for computing a random trajectory $W = [u = w_1, w_2, ..., w_t = v]$ of length t from node u to node v is to:

1. Choose a node w according to the following rule: let w be a neighbor of v. Then

$$Pr[w_{t-1} = w | w_t = v] = \frac{P_{u,w}^{(t-1)} P_{w,v}}{P_{u,v}^{(t)}}.$$

   where $P^{(t)}$ denotes the t–th power of the transition matrix P.

2. Recursively, choose a random trajectory of length $t - 1$ from u to w.

Eliminating cycles. This phase is not analyzed in [17]. We observed that the paths defined by two trajectories contain cycles, almost always. Cycles in a path $P(a_i, b_i)$ created by two trajectories W'_i and W''_i are emilinated by finding a shortest path between nodes a_i and b_i in the subgraph H_i defined by the two trajectories (i.e. $H_i = (V(H_i), E(H_i))$ s.t. $V(H_i) = \{v \in V(G) | v \in W'_i$ or $v \in W''_i\}$ and $E(H_i) = \{(v, u) \in E(G) | v, u \in W'_i$ or $v, u \in W''_i\}$).

Coloring paths. The algorithm we use for coloring paths is the obvious (greedy) one. We use a palette of colors $\chi_1, \chi_2, ..., \chi_{kn}$ and assign to a path $P(x, y)$ the color χ_c with minimum c s.t. χ_c has not been assigned to any established path that shares a fiber link with $P(x, y)$. The total number of colors (wavelengths) actually used is the maximum index of colors assigned to paths.

4   The Algorithm MC

The algorithm MC is a variation of the algorithm RW presented in the previous section. The main structure of algorithm MC remains the same with the one of algorithm RW. Also algorithm MC maintains the main characteristics of algorithm RW. Given a k–relation I_k on a network G, the algorithm uses a Markov chain technique for finding a routing R for I on G and properly assigns colors to paths of R. The algorithm considers requests one by one, finds a path for each one of them and assigns a proper color to it, so that once a path has been established for a request of I and colored, it is never reestablished or recolored again.

The main difference is that algorithm MC, while computing the path for a request of instance I, it considers a Markov chain $\{Z_t\} \subseteq V$ associated to a particle that moves from node to node. The definition of the Markov chain (i.e. the transition probabilities) is such that it takes into account the paths that have been already established in previous steps. This means that the transition matrix, the stationary distribution and the sufficient length of trajectories is recomputed for each request given as input to the algorithm. There are many different ways to define the Markov chain $\{Z_t\}$. In the following we present algorithm MC together with a simple strategy for computing the transition probabilities of the Markov chain $\{Z_t\}$. This strategy is the one used for performing the experiments presented in section 5.

Algorithm MC

1. For each request $(a_i, b_i) \in R$ do:
   (a) Recompute the transition matrix P' and the stationary distribution ρ'.
   (b) Recompute a sufficient length L for trajectories.
   (c) Choose a node $r_i \in V$ according to the stationary distribution.
   (d) Choose a trajectory W'_i (resp. W''_i) of length L from a_i to r_i (resp. from b_i to r_i) according to the distribution on trajectories, conditioned on the endpoints being a_i and r_i (resp. b_i and r_i).
   (e) Connect a_i to b_i by the path $P(a_i, b_i)$, defined by W'_i followed by W''_i.
   (f) Eliminate cycles of the path $P(a_i, b_i)$.
   (g) Assign a proper color to $P(a_i, b_i)$.

Recomputing the transition matrix. Consider a step of the algorithm. Assume that paths for $s < |I|$ requests have been already been established and colored. Let I_s denote the subset of instance I which contains the s already considered by the algorithm MC requests, and R_s the set of paths established for requests of I_s. We denote by $\pi(G, I_s, R_s, \alpha)$ the load of a fiber link $\alpha \in E(G)$ at the current step.

Given a network $G = (V(G), E(G))$, we define a Markov chain $\{Y_t\} \subseteq V(G)$ associated to a particle that moves from node to node. The transition probability between non–adjacent nodes is $P''_{ij} = 0$. The definition of the transition probabilities from a node u_i to its neighbors, distinguishes between two cases:

1. Some of the fiber links adjacent to u_i are not used by any path of R_s. Let d be the degree of node u_i and m be the number of nodes adjacent to $u_i, u_1, ..., u_m$ such that the fiber links $(u_i, u_j), j = 1, ..., m$ have $\pi(G, I_s, R_s, (u_i, u_j)) = 0$, for $1 \le j \le m$. It is $\pi(G, I_s, R_s, (u_i, u_j)) > 0$, for $m + 1 \le j \le d$. Then, the

transition probability from node u_i to a neighbor node u_j is $P''_{ij} = 1/m$ if $j \leq m$, $P''_{ij} = 0$ otherwise.

2. All fiber links adjacent to u_i are used by paths in R_s. The transition probabilities P''_{ij} from node u_i to two neighbors u_j and u_l satisfy

$$\frac{P''_{ij}}{P''_{i,l}} = \frac{\pi(G, I_s, R_s, (u_i, u_l))}{\pi(G, I_s, R_s, (u_i, u_j))} \quad \text{s.t.} \quad \sum_{(u_i, u_j) \in E(G)} P''_{ij} = 1.$$

This gives

$$P''^{-1}_{ij} = \pi(G, I_s, R_s, (u_i, u_j)) \sum_{(u_i, u_k) \in E(G)} \frac{1}{\pi(G, I_s, R_s, (u_i, u_j))}.$$

It is clear that, the Markov chain $\{Y_t\}$ constructed at a step of the algorithm, has the following main property: *for a transition from a node, it prefers the less loaded fiber links.* The algorithm MC uses a parameter $f \in (0, 1]$ and assigns to each transition of Markov chain $\{Z_t\}$ a probability

$$P'_{ij} = f P_{ij} + (1 - f) P''_{ij}.$$

Intuitively, $\{Z_t\}$ maintains the ergodicity feature of the random walk, while it encapsulates the main property of $\{Y_t\}$. Note that for $f = 1$, algorithm MC is identical to algorithm RW, since $\{Z_t\}$ is a random walk. Values close to $f = 0$ are not valid since the Markov chain $\{Y_t\}$ may not be ergodic.

Note that the transition probabilities of P' that must be recomputed at each step of the algorithm MC are only those associated to nodes of the path established in the previous step.

5   Experiments and Results

The algorithms have been tested using as input random networks and random k-relations. Random networks are constructed according to the $G_{n,p}$ model. For maintaining connectivity, a network given as input to the algorithms has an arbitrary Hamilton circuit. The parameters that effect the results of the wavelength routing algorithms are:

- Parameter n is the number of nodes of the random network. All the experiments presented in this paper have been made on networks with 200 nodes.
- Parameter c indicates that the probability that a fiber link (except those in the Hamilton path) exists in the network is $p = c/n$. This parameter gives a measurement for the density of the network.
- Parameter k indicates that the instance given to the algorithms as input is chosen randomly among all k-relations on a network of n nodes. For the experiments presented in this paper, we used small values for k (up to 4). Larger values increase the running time of the algorithms dramatically.

We first measure the number of wavelengths assigned to a k relation by the algorithm RW as a function of the density of the network. Note that, while the standard algorithm of Raghavan and Upfal as it was presented in [17] considers undirected paths only, the description given in section 3 applies to both the directed and the undirected model. We consider three types of k relations: (1) a random k-relation of undirected paths, (2) the corresponding symmetric k-relation of directed paths (i.e. for each request $(u,v) \in I_k$, $(v,u) \in I_k$ as well) and (3) a random k-relation of directed paths (non–symmetric). Note that the total number of requests in a undirected k-relation is half the number of requests of a directed k-relation. Moreover, the two versions of the wavelength routing problem in the directed and undirected model have important inherent differences. In particular, there exist a permutation I_1 on a n node graph which, although it can be routed with load $\pi(G, I_1) = 2$, it requires $w(G, I_1) = n$ wavelengths [1]. No such case has been observed in the directed model, and it has been conjectured that $w(G, I) = O(\pi(G, I))$ for every network G and instance I [3]. Surprisingly, algorithm RW has the same performance when applied to random undirected or directed requests on random (sparse) networks. The results for permutations, 2–relation and 4–relation are depicted in table 1.

Density	Undirected			Symmetric			Non–Symmetric		
3	19	29	55	19	31	54	18	30	55
5	14	23	36	13	22	36	13	23	36
10	9	14	21	9	14	22	8	14	21
15	5	10	14	6	10	15	6	10	15
20	5	8	11	5	7	12	6	7	11

Table 1. Routing undirected and directed requests on random networks of 200 nodes. The table contains the number of wavelengths used for random undirected, the corresponding symmetric directed, and random directed permutations, 2–relations, and 4–relations on 5 random networks of different density.

In the following we concentrate to directed instances. The results for the performance of algorithm RW on random permutations, 2–relations, and 4–relations according to the data contained in the three rightmost columns of table 1 are graphically represented in figure 1.

Furthermore, data of table 2 show that the number of wavelengths used by algorithm RW in each case is slightly greater than the load of the routings and is much smaller than the degree of the conflict graph. Also, for sparse networks (density 3 or 5), the dilation is smaller than the theoretical length of paths, meaning that all paths produced by connecting two trajectories contained cycles. This is not true for dense networks (density 10 or 20). For any case, most paths produced by connecting trajectories contained cycles, so the average length of paths is much smaller than the dilation. The difference between average length

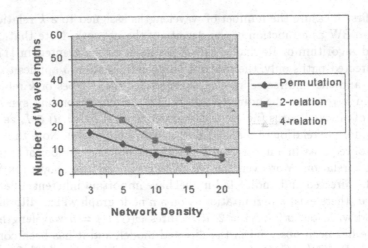

Fig. 1. Routing directed permutations, 2–relations, and 4–relations on random networks with 200 nodes using algorithm RW.

of paths and dilation (and the theoretical length of paths which is twice the length of trajectories) increases as the network density decreases.

Density	#Wav.	Load	Av. Length	Dilation	Traj. Length	CG Degree
3	20	11	17.73	35	22	76
5	15	9	17.04	27	15	54
10	8	6	12.53	18	9	24
20	7	5	10.60	12	6	13

Table 2. Routing a permutation on random networks with 200 nodes using algorithm RW. The columns of the tables contain results for the number of wavelengths used, the load, the average length of paths, the dilation, the length of trajectories, and the degree of the conflict graph of the produced routing for 4 random networks of different density.

We observed that algorithm MC improves algorithm RW concerning the number of wavelengths assigned to requests. This improvement can be about 20% of the performance of RW in some cases and is significant especially for instances with many requests (4–relations) on very sparse networks (density 3 or 5).

This is mainly due to the fact that the routings produced by algorithm MC are qualitatively better than those produced by algorithm RW. We performed experiments by running algorithm MC with parameter $f = 0.5$ for the same permutation on random networks of density 3, 5, 10, and 20. The conflict graph produced has smaller degree in any case; as a result, the number of wavelengths used is decreased related to the number of wavelengths used by algorithm RW.

A somewhat surprising result is that, for sparse networks, the average length of paths produced by algorithm MC is slightly smaller than the average length of paths produced by algorithm RW. Table 3 summarizes data produced by algorithm MC with $f = 0.5$.

Density	#Wav.	Load	Av. Length	Dilation	CG Degree
3	15	11	15.63	33	64
5	13	9	15.31	29	42
10	6	5	12.90	18	15
20	4	3	10.75	12	9

Table 3. Routing a permutation on random networks with 200 nodes using algorithm MC with parameter f set to 0.5. The columns of the tables contain results for the number of wavelengths used, the load, the average length of paths, the dilation, and the degree of the conflict graph of the produced routing for 4 random networks of different density. Corresponding rows of tables 2 and 3 represent results of experiments on the same network. Data in both tables were obtained for the same permutation.

Concerning the number of wavelengths used, which is the main performance metric we use for comparing our algorithms, the performance of algorithm MC is better in most cases as the parameter f decreases. Algorithm MC was executed with several values of parameter f (included $f = 1.0$, a value for which algorithm MC is identical to algorithm RW) on permutations and 4–relations on networks of different density. The results that correspond to these experiments are depicted in tables 4 and 5 and are graphically represented in figures 2 and 3, respectively.

Density	$f = 0.2$	$f = 0.4$	$f = 0.6$	$f = 0.8$	RW
3	14	16	16	17	20
5	13	12	13	12	14
10	5	6	7	7	8
20	4	4	4	5	5

Table 4. Routing permutations on networks with 200 nodes using algorithm MC. Data represent the number of wavelengths used, for several values of parameter f.

6   Extensions

The algorithms RW and MC were implemented in the Matlab v5.0 environment and the experiments presented in this paper were conducted on a Pentium

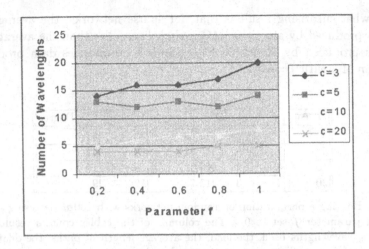

Fig. 2. Routing permutations on networks with 200 nodes using algorithm MC.

Density	$f = 0.2$	$f = 0.4$	$f = 0.6$	$f = 0.8$	RW
3	44	47	48	50	53
5	30	32	32	35	36
10	16	17	18	19	19
20	10	11	10	10	11

Table 5. Routing 4–relations on networks with 200 nodes using algorithm MC. Data represent the number of wavelengths used for several values of parameter f. Corresponding rows of tables 4 and 5 represent results of experiments on the same network.

PC/200MHz. Recently, we completed the implementation and testing of the algorithms in C; we currently perform several experiments on a powerful Pentium III/500MHz running Solaris 7.

In experiments not included here, we experimentally studied the performance of modified MC algorithms that use more complex functions to define the transition probabilities of the Markov chain. In some cases, we observed that results can be further improved.

In addition to networks generated according to the $G_{n,p}$ model, we currently perform experiments on random regular networks with link faults; a model that also reflects irregularity of real networks and is closer to the theoretical model of bounded–degree graphs assumed in [17]. We plan to include all these results in the full version of the paper.

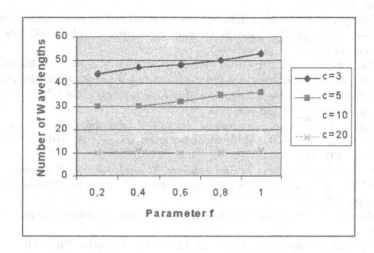

Fig. 3. Routing 4–relations on networks with 200 nodes using algorithm MC.

References

1. A. Aggarwal, A. Bar–Noy, D. Coppersmith, R. Ramaswami, B. Schieber, M. Sudan. Efficient Routing and Scheduling Algorithms for Optical Networks. In *Proc. of the 5th Annual ACM–SIAM Symposium on Discrete Algorithms (SODA 94)*, pp. 412–423, 1994.
2. Y. Aumann, Y. Rabani. Improved Bounds for All Optical Routing. In *Proc. of the 6th Annual ACM–SIAM Symposium on Discrete Algorithms (SODA 95)*, pp. 567–576, 1995.
3. B. Beauquier, J.–C. Bermond, L. Gargano, P. Hell, S. Perennes, U. Vaccaro. Graph Problems arising from Wavelength Routing in All–Optical Networks. *2nd Workshop on Optics and Computer Science (WOCS 97)*, 1997.
4. J.–C. Bermond, L. Gargano, S. Perennes, A. Rescigno, U. Vaccaro. Efficient Collective Communication in Optical Networks. In *Proc. of the 22nd International Colloquium on Automata, Languages, and Programming (ICALP 96)*, LNCS 1099, Springer–Verlag, pp. 574–585, 1996.
5. A.Z. Broder, A. Frieze, and E. Upfal. Existence and Construction of Edge Disjoint Paths on Expander Graphs. In *Proc. of the 24th Annual ACM Symposium on the Theory of Computing (STOC 92)*, pp. 140–149, 1992.
6. E. Dijkstra. A Note on Two Problems in Connection with Graphs. Num. Math., 1:269–271, 1959.
7. T. Erlebach, K. Jansen. Scheduling of Virtual Connections in Fast Networks. In *Proc. of the 4th Workshop on Parallel Systems and Algorithms (PASA 96)*, pp. 13–32, 1996.
8. T. Erlebach, K. Jansen. Call Scheduling in Trees, Rings, and Meshes. In *Proc. of the Hawaii Int. Conf. on Syst. Sciences (HICSS 97)*, 1997.

9. M.R. Garey, D.S. Johnson, G.L. Miller, C.H. Papadimitriou. The Complexity of Coloring Circular Arcs and Chords. *SIAM Journ. Alg. Disc. Meth.*, vol. 1, no. 2, (1980), pp. 216–227.

10. C. Kaklamanis, P. Persiano. Efficient Wavelength Routing on Directed Fiber Trees. In *Proc. of the 4th European Symposium on Algorithms (ESA 96)*, LNCS 1136, Springer Verlag, 1997, pp. 460–470.

11. C. Kaklamanis, P. Persiano, T. Erlebach, K. Jansen. Constrained Bipartite Edge Coloring with Applications to Wavelength Routing. In *Proc. of the 24th Internation Colloquium on Automata, Languages, and Programming (ICALP 97)*, LNCS 1256, Springer Verlag, 1997, pp. 493–504.

12. V. Kumar, E. Schwabe. Improved Access to Optical Bandwidth in Trees. In *Proc. of the 8th Annual ACM-SIAM Symposium on Discrete Algorithms (SODA 97)*, 1997, pp. 437–444.

13. M. Mihail, C. Kaklamanis, S. Rao. Efficient Access to Optical Bandwidth. In *Proc. of the 36th Annual Symposium on Foundations of Computer Science (FOCS 95)*, 1995, pp. 548–557.

14. R. Motwani, P. Raghavan. Randomized Algorithms. *Cambridge University Press*, 1995.

15. R.K. Pankaj. Architectures for Linear Lightwave Networks. PhD. Thesis, Dept. of EECS, MIT, 1992.

16. Y. Rabani. Path Coloring on the Mesh. In *Proc. of the 37th Annual Symposium on Foundations of Computer Science (FOCS 96)*, 1996.

17. P. Raghavan, E. Upfal. Efficient Routing in All-Optical Networks. In *Proc. of the 26th Annual Symposium on Theory of Computing (STOC 94)*, 1994, pp. 133–143.

18. R. Ramaswami, K. Sivarajan. Optical Networks. *Morgan Kaufmann*, 1998.

Estimating Large Distances in Phylogenetic Reconstruction

Daniel H. Huson[1], Kelly Ann Smith[2], and Tandy J. Warnow[3]

[1] Applied and Computational Mathematics
Princeton University, Princeton NJ 08544-1000
huson@math.princeton.edu
[2] Department of Computer and Information Science
University of Pennsylvania, Philadelphia PA 19104
kellyann@gradient.cis.upenn.edu
[3] Department of Computer Science
University of Arizona, Tucson AZ 85721
tandy@cs.arizona.edu

Abstract. A major computational problem in biology is the reconstruction of evolutionary (a.k.a. "phylogenetic") trees from biomolecular sequences. Most polynomial time phylogenetic reconstruction methods are *distance-based*, and take as input an estimation of the evolutionary distance between every pair of biomolecular sequences in the dataset. The estimation of evolutionary distances is standardized except when the set of biomolecular sequences is "saturated", which means it contains a pair of sequences which are no more similar than two random sequences. In this case, the standard statistical techniques for estimating evolutionary distances cannot be used. In this study we explore the performance of three important distance-based phylogenetic reconstruction methods under the various techniques that have been proposed for estimating evolutionary distances when the dataset is saturated.

1 Introduction

Evolutionary trees (also called "phylogenetic trees") are rooted leaf-labeled trees whose branching order represents the order of speciation events (or gene duplication events) that make up the evolutionary events leading to the given species or gene sequences. Much biological research (including multiple alignment, gene finding, protein structure and function prediction, etc.), begins with the estimation of an evolutionary tree, and the accuracy of the research thus depends (sometimes very significantly) upon the accurate reconstruction of the branching order of the unknown evolutionary tree. There are many methods for reconstructing trees. While some of these methods are based upon NP-hard optimization criteria such as Maximum Parsimony (MP) and Maximum Likelihood (ML), other methods are polynomial time, and these most typically are "distance-based".

Distance-based phylogenetic reconstruction methods require an estimation of the evolutionary distances between the sequences in the dataset. These distances are usually obtained by applying standardized statistical techniques to aligned biomolecular sequences. However, these standard techniques cannot be

J.S. Vitter and C.D. Zaroliagis (Eds.): WAE'99, LNCS 1668, pp. 271–285, 1999.
© Springer-Verlag Berlin Heidelberg 1999

used when the dataset is saturated (because some distances are then undefined) , which means that it contains a pair of sequences whose sequence similarity is at or below the threshold given by chance. Alternative techniques have been developed for estimating distances in the presence of saturation. For example, a technique that is recommended by authorities in the field [21] and much in favor in practice is the *replacement by large value* technique, where all undefined distances are given the same large value, and this large value is typically selected to be arbitrarily larger than any other value in the distance matrix. A less-favored technique is to use normalized Hamming distances, which are also known as *uncorrected distances* or *p*-distances.

Since the accuracy of the tree obtained using a distance method will depend upon the input given to the distance method, it is possible that the particular technique used to estimate distances could have a dramatic effect upon the accuracy of the tree reconstructed by a given distance method. However, to our knowledge, no studies have been published in the scientific literature on this question. The purpose of this paper is to explore the consequences of the choice of technique for distance estimation upon the accuracy of the trees reconstructed by the major distance-based phylogenetic reconstruction methods.

Our study examines this question experimentally by simulating sequence evolution on different model trees, and with respect to the most popular distance based method, Neighbor-Joining (NJ) [18], and two of its variants, BIO-NJ (BJ) [7] and Weighbor (WJ) [1]. We make two surprising observations: first, the particular choice of the "large value" in the large value substitution technique is very important, so that the typical ways of instantiating that technique are not to be recommended at all. Secondly, we find that the less favored technique of using uncorrected distances is sometimes superior to the substitution by large value technique, even when the large value is chosen optimally. These findings are significant and could yield improved estimations of phylogenetic trees. We also show that the choice of distance calculation technique has serious implications for experimental performance studies.

Acknowledgment We are grateful to Lisa Vawter for helping us to find suitable biological model trees.

2   Distance Methods and Distance Estimations

Distance-Based Tree Reconstruction Many techniques exist for reconstructing phylogenetic trees from DNA sequences, and distance-based techniques are an important class of such phylogenetic tree reconstruction methods. Distance-based tree reconstruction methods are motivated by the observation that each phylogenetic tree T is determined (up to the location of the root) by the matrix whose i, j^{th} entry is the number of mutations of a random site on the path in the tree between the leaves i and j. This is an example of a more general phenomenon called *additivity*, which we now define.

Definition 1. *An $n \times n$ matrix D is said to be **additive** if there exists a tree T with leaves labeled by $1, 2, \ldots, n$, and a positive edge-weighting $w : E \to \mathbb{R}^+$, so that for each pair of leaves i, j, we have $d_{ij}^T = \sum_{e \in P_{i,j}} w(e) = D_{ij}$, where $P_{i,j}$ is the path between i and j in the tree T.*

Given an additive matrix D, the unique tree T and its edge weighting w can be constructed in polynomial time using a number of different distance-based methods [3, 18, 22].

If we let λ denote the additive matrix associated to the phylogenetic tree T, then the application of a phylogenetic distance method proceeds in two steps:

1. an approximation d to the matrix λ is computed, and then,
2. the matrix d is mapped, using some distance method ϕ, to a (nearby) additive matrix $\phi(d) = D$.

If D and λ define the same unrooted leaf-labeled tree, then the method is said to be *accurate*, even if they assign different weights to the edges of the tree. While complete topological accuracy is the objective, partial accuracy is the rule. In Section 3 we discuss a standard technique for measuring partial accuracy.

The Jukes-Cantor Model of Evolution T. Jukes and C. Cantor [13] introduced a very simple model of DNA sequence evolution. While the questions we address apply quite generally, we will explore our techniques under the Jukes-Cantor model; however our techniques can be naturally extended to other models of evolution. (Please see the excellent review articles by Felsenstein [6] and Swofford *et al.* [21] for more information about statistical aspects of phylogenetic inference.)

Definition 2. *Let T be a fixed rooted tree with leaves labeled $1 \ldots n$. The **Jukes-Cantor** model of evolution has several basic assumptions:*

1. *The sites (positions within the sequence) evolve identically and independently (i.i.d.) down the tree from the root.*
2. *The possible states for each site are A, C, T, G (denoting the four possible nucleotides), and for each site, the state at the root is drawn from a distribution (typically uniform).*
3. *For each edge $e \in E(T)$, if the state of a site changes on e, then the probability of change to each of the 3 remaining states are equal.*

Some of these assumptions are clearly over-simplistic, but the study of how well phylogenetic methods perform under the Jukes-Cantor model is standard within the phylogenetics literature, and no methods are known to perform well under models that are much more realistic.

The Jukes-Cantor Distance Estimation We now show how to calculate λ_{ij}, the expected number of mutations of a random site on the path in a Jukes-Cantor tree between the sequences at leaves i and j.

Definition 3. *Let $H(i,j)$ denote the **Hamming** distance between the sequences at leaves i and j (where the Hamming distance is the number of indices for which the two sequences differ), and let k be the sequence length. We define $h_{ij} := \frac{H(i,j)}{k}$, and call this the **normalized Hamming distance**, or **p-distance**.*

Asymptotically, $h_{i,j}$ converges to the probability that a random site will have a different value at i than at j.

Definition 4. *The **maximum likelihood distance** between a pair of sequences i, j (that is, the most likely λ_{ij} to have generated the observed sequences) is given by the following formula:*

$$d_{ij} = -\frac{3}{4}\ln\left(1 - \frac{4}{3}h_{i,j}\right).$$

*This is called the **standard Jukes-Cantor distance estimation** technique, and the distances are called **standard JC-distances**.*

The Saturation Problem and Techniques for Overcoming It The standard JC-distance will be undefined on pairs i, j of sequences for which $h_{i,j} \geq \frac{3}{4}$, as the logarithm of a non-positive number is undefined. This occurs in practice when the dataset of taxa includes pairs that are only very distantly related. We call such a pair of taxa *saturated*, as their corresponding DNA sequences are essentially randomized with respect to each other, and refer to this as the *saturation problem*.

In [21], Swofford *et al.* discuss the saturation problem and make several basic suggestions. They recommend eliminating a few sequences from the dataset, if their removal can eliminate all the saturation (so that all remaining distances can be calculated using the standard technique). When this is not possible, they suggest replacing all undefined values by an "arbitrarily large value". This is the standard technique used in systematic biology:

Definition 5. *The **substitution by large value** technique takes a parameter C (the "large value"), and computes distances d_{ij} from p-distances h_{ij} as follows. If $h_{i,j} < \frac{3}{4}$, then the standard JC-distance estimation technique is used, but if $h_{i,j} \geq \frac{3}{4}$, then $d_{i,j}$ is set to C. In other words,*

$$d_{ij} := \begin{cases} C, & \text{if } h_{ij} \geq \frac{3}{4}, \\ -\frac{3}{4}\ln\left(1 - \frac{4}{3}\cdot h_{ij}\right), & \text{else.} \end{cases}$$

Swofford *et al.* suggest that C should be twice that of the largest defined distance in the matrix. We formalize and generalize their suggested technique as a particular *fix-factor correction*, as follows:

Definition 6. *The **fix-factor correction** takes a parameter F. Let*

$$MAX = \max\left\{-\frac{3}{4}\ln\left(1 - \frac{4}{3}\cdot h_{ij}\right) \text{ such that } h_{ij} < \frac{3}{4}\right\}.$$

The calculation of the fix-factor distance matrix has two steps:

1. For each i, j, we set d_{ij} using the standard Jukes-Cantor estimation technique, and leave as undefined those d_{ij} for which $h_{ij} \geq 3/4$.

2. For each i, j, if d_{ij} is undefined or greater than $F \times MAX$, then d_{ij} is reset to $F \times MAX$.

Using this terminology, Swofford *et al.* suggested using a fix-factor correction with $F = 2$. Note that if $F \geq 1$, then this is a large-value substitution, but the fix-factor distances are defined even if $F < 1$. However, when $F < 1$, more values are reset besides those that are undefined using the standard JC distance estimation technique. Thus, when $F < 1$, this technique is not an instantiation of the substitution by large value technique. It is worth noting that the large-value corrections do not generally preserve the order in the p-distances, since all large p-distances are mapped onto a single value C.

Uncorrected distances (also called p-distances, i.e. setting $d_{ij} := h_{ij}$) are also used when the dataset is saturated, but this is not as popular as the substitution by large value technique.

To our knowledge, there has been no published study on the relative merits of different techniques for estimating distances for saturated datasets.

3 Experimental Setup

The Simulation of DNA Evolution The performance of phylogenetic methods is usually investigated experimentally by simulating sequence evolution on different model trees and applying the methods to the simulated data sets (see [8, 17, 14, 20, 19, 11, 10, 9, 12] for a small sample of such studies). Such experiments generally follow the same basic structure. The evolution of biomolecular sequences is simulated by "evolving" an initial (usually random) root sequence down a given *model* or *true* tree T according to a specified model of evolution. The sequences obtained at the leaves of the tree form the simulated data set S. Each tree-reconstruction method M under study is then applied to S and the resulting tree $M(S)$ is compared to the model tree.

Comparing the Performance of Tree-Reconstruction Methods Biologists usually quantify degrees of accuracy by comparing the true tree and the estimated tree in terms of their shared bipartitions: Given a tree T leaf-labeled by S and given an internal edge $e \in E(T)$, the *bipartition* induced by e is the bipartition on the leaves of T obtained by deleting e from T. We denote this bipartition by Π_e. It is clear that every such tree T is defined uniquely by the set $C(T) = \{\Pi_e \mid e \in E_{int}(T)\}$, where $E_{int}(T)$ denotes the internal edges of T (i.e., those edges not incident to leaves). This is called the *character encoding* of T.

Definition 7. *If T is the model tree and T' an estimation of T, then the errors in T' can be classified into two types:*

- **false positives** *(FP)* $= C(T') - C(T)$, *i.e. edges in T' but not in T, and*
- **false negatives** *(FN)* $= C(T) - C(T')$, *i.e. edges in T missing from T'.*

The **false positive rate** *is obtained by dividing the number of false positives by the number of internal edges in the estimated tree, while the* **false negative rate** *is obtained by dividing the number of false negatives by the number of internal edges in the model tree.*

For the case where both the model and inferred trees are binary, the false positive rate and false negative rate will be identical. In our experimental performance study, our model trees and inferred trees will be binary, by design. Consequently, we will report our error rate without identifying that it is the FN rate or FP rate, as these will be equal. Note also that when the trees are constrained to be binary, the number of internal edges is $n - 3$, where n is the number of leaves.

Choice of Model Trees In our experiments, we used binary model trees, some of which were randomly generated model trees (with random topologies and branch lengths) as well as some "biologically based" model trees. Each basic model tree included lengths on the branches, indicating the expected number of mutations of a random site on that edge. We modified the rate of evolution up and down by multiplying the branch lengths by a "scale" factor. The larger the scale factor, the greater the proportion of entries of the normalized Hamming distance matrix that were at or above saturation (for JC evolution, saturation occurs when the normalized Hamming distance reaches $\frac{3}{4}$). We looked at several different scalings on each tree, to ensure that we generated datasets that were below saturation, at saturation, and at various degrees above saturation.

Randomly Generated Model Trees We used a number of random model trees having up to 300 leaves. These random trees were generated by first constructing a leaf-labeled tree with a topology drawn at random from the uniform distribution, and then assigning branch lengths to the tree. Branch lengths were either set uniformly to 1, or drawn from the uniform distribution on $1\ldots15$ or $1\ldots30$.

Biological Model Trees We also used a number of basic model trees whose topologies and branch (i.e. edge) lengths were obtained from published phylogenies of biological datasets (however, if the published tree had an edge of 0 length, we modified the length to be half the smallest non-zero edge length in the tree). These basic model trees are as follows: (a) *rbcL-93*: a 93-taxon tree inferred on a subset of the *rbcL* (ribulose 1,5-biphosphate carboxylase large subunit) dataset, which was originally used in an attempt to infer seed plant phylogeny [16]. (b) *rbcL-35*: a 35-taxon tree inferred on the *Lamiidae sensu stricto* clade of the *rbcL* dataset. (c) *crayfish-72*: a 72-taxon tree based on 16S mtDNA sequence data for crayfish [4]. (d) *rRNA-111*: a 111-taxon tree inferred on 5.8S nuclear ribosomal RNA genes [5]. (e) *basidio-152*: a 152-taxon tree inferred on a set of basidiomycetous fungi, using DNA from the central part of the mitochondrial large subunit [2]. As with the randomly generated model trees, the rates of evolution on these basic model trees were then scaled upwards and downwards.

Choice of Distance Methods and Techniques for Handling Saturation
Our study examined neighbor joining (NJ) (the most popular distance based method) and two modifications of it, Bio-NJ (BJ), and Weighbor (WJ). We now briefly describe NJ. Given a matrix d of estimated leaf-to-leaf distances,

1. Compute the matrix Q as follows: $Q_{ij} = (n-2)d_{ij} - \sum_k (d_{ik} + d_{jk})$, where k ranges from 1 up to n, the total number of leaves.
2. **WHILE** there are at least 4 leaves remaining **DO:**
   - Find i, j minimizing Q_{ij}, and make i and j siblings.
   - Replace i and j by a new node, x, and delete the row and column associated with i and j from the matrix d.
   - Compute $d_{xk} = \frac{1}{2}(d_{ik} + d_{jk} - d_{ij})$ for each $k \notin \{i, j\}$.
   - Add row and column for x to the matrix d.
   - Recompute Q_{ab} for every a, b.
3. If three or fewer leaves remain, make them a sibling set.
4. Given the sequence of siblinghood operations, the tree is then reconstructed by reversing the mergers.

NJ, BJ, and WJ only compute binary trees. BJ differs from NJ in how it defines the distances d_{xk} for newly created nodes x, and WJ differs from both NJ and BJ in that it uses statistical techniques to downplay the importance and impact of large distances.

4   Description and Discussion of Experiments

Overview Our experimental study simulated sequence evolution on the Jukes-Cantor model trees described in Section 3. Distance matrices were computed for the sequences at the leaves of the tree, using uncorrected distances, substitution by large-value with $C = 10,000$, and some fix-factor corrections. Trees were computed for each of these distance matrices by the distance methods we study, NJ, BJ, and WJ. These trees were then evaluated for topological accuracy with respect to the model trees. Since our model trees and inferred trees are by constraint always binary, the FN and FP rates are always identical. Consequently, we will simply report error rates without specifying the type of error.

For each model tree (topology and edge length settings), we generated either 5 or 10 sets of leaf sequences; however, due to time constraints, only 2 of the random 300-taxon tree settings were able to run to completion. We used seqgen [15] to generate sequences under the Jukes-Cantor model of evolution, and then truncated the sequences to sequence lengths of 100, 200, 400, 800, and 1600 for every dataset.

Our experimental study consists of two parts. In the first part, we compared different fix-factor techniques to the substitution by large value technique with $C = 10,000$. We found that $C = 10,000$ is *much worse* than many fix-factor corrections, and that the best fix-factor are based upon small values for F; this contradicts the common practice in the systematics community. In the second part, we compared a few good fix-factor corrections to uncorrected distances;

for reference, we also included the data for the substitution by large value technique with $C = 10,000$. Surprisingly, we found no direct domination between the uncorrected distance and small fix-factor techniques: sometimes uncorrected distances were better than small fix-factors, and sometimes they were worse. The good performance of NJ under uncorrected distances goes against the common wisdom, and suggests that some mathematical theory needs to be established to explain this phenomenon.

Throughout the study we observed that NJ and BJ had almost identical performance on every model tree and sequence length, but that WJ differed from NJ and BJ, sometimes in interesting ways. Thus, we omit the data on BJ, and focus most of our analysis and discussion on NJ. When space permits and there is an interesting difference, we also present data on WJ.

Exploring the Fix-Factor Correction In the first part of our experimental study, we explored how to select F on our model trees under a range of scalings. The results we obtained here were very consistent across all trees and all scalings.

In Figure 1 we present the results of our study on datasets generated by a 35 taxon random tree with branch lengths drawn from $1 \ldots 15$, with the rates of evolution scaled so that approximately 20% of the normalized Hamming distances are above the correctable threshold. We generated sequences with lengths from 100 up to 1600 nucleotides.

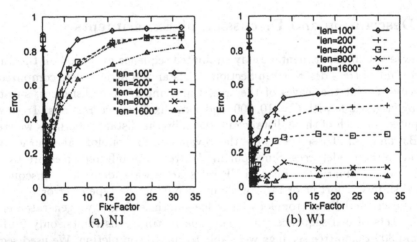

Fig. 1. We report the error rate of NJ (a) and WJ (b) as a function of the fix-factor used in the Jukes-Cantor correction, for a number of different sequence lengths. Each data-point represents the average of 10 independent simulations. This experiment was performed on a random 35-taxon tree with edge length ratio 15:1 and a scale factor of 0.5. The performance by BJ was quite similar to that of NJ and is omitted.

We show the false negative rate obtained by NJ and WJ for a wide range of fix-factor calculations. Notice that the choice of fix-factor correction greatly impacts the error rate for all methods at all sequence lengths, although for NJ

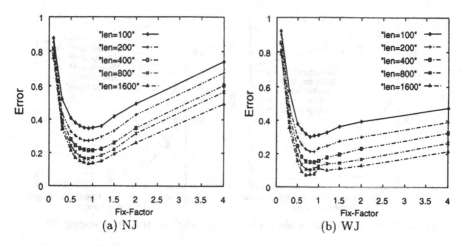

Fig. 2. Here we report the error rate obtained by averaging over all our random trees at one scaling per tree, selected so that a number of the normalized Hamming distances were above the threshold. We show the results for NJ (a) and WJ (b). The results for BJ are very similar to those of NJ so they aren't shown here.

the effect is greater than for WJ. Also note that the error rate depends upon the sequence length and the fix-factor, with lower error rates for longer sequences. The effect of longer sequence lengths upon error rates is well-established in the literature, and is to be expected.

In Figure 2 we explored the optimal choice for the fix-factor correction, and it is clearly seen to be located in the range $[.5, 1]$ for both NJ and WJ. It is interesting that WJ's performance is less affected by increasing fix-factors than NJ. We suspect this is true because WJ is designed to downplay the effects of large distances in the input matrix. (Since the use of a large fix-factor results in distance matrices with large values, WJ's design strategy would lessen the impact of large fix-factors by comparison to NJ.)

These experiments show that all three distance methods have better performance when using fix-factors in the range $[.5, 1]$ than when using fix-factors significantly greater than 1. The study also suggests the recommendation in [21] of using $F = 2$ probably would produce worse phylogenetic reconstructions than when using $F = 1$, and the common practice of using *very* large values for the substitution by large value technique would probably produce very bad results.

To test this hypothesis, we explored the accuracy of the three phylogenetic methods when used with four fix-factor corrections (.7, .9, 1 and 2) and the large-constant correction in which $C = 10,000$. In Figure 3 we show the results of this study for NJ on two biologically based model trees with fairly high rates of evolution. Note that while NJ had relatively good accuracy using each of the three smallest fix-factor corrections, it had approximately twice the error rate when using $F = 2$, and its error rate under the large value correction with $C = 10,000$ was very much worse. (Both BJ and WJ also showed the same relative performance on all trees and scalings we explored.)

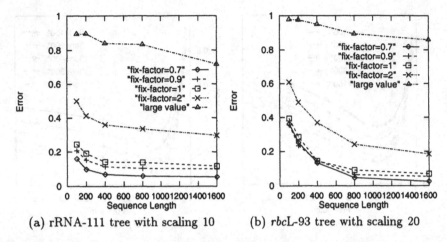

(a) rRNA-111 tree with scaling 10   (b) *rbcL*-93 tree with scaling 20

Fig. 3. Here we report the error rate for NJ, obtained for two biological trees at different fix-factor corrections as well as the large value substitution with $C = 10,000$. The rRNA-111 tree is shown at a scaling of 10 times the original evolution rate (a) and the *rbcL*-93 tree is shown at a scaling of 20 times the original evolution rate (b).

Comparing Fix-Factor and Uncorrected Distances In the second part of this study we compared the performance of NJ using the fix-factor correction with small fix-factors and using uncorrected distances. For reference, we also include the data for NJ under the substitution by large value technique with $C = 10,000$. (BJ, WJ, and NJ are quite similar with respect to their *relative* performance under these different distance estimation techniques.)

In many of our experiments we observed NJ having worse performance on uncorrected distances than on our best fix-factors, but this was not always the case. For example, see Figure 4. Here we show the performance of NJ on the rRNA-111 tree at two scalings, using the uncorrected distance technique, the large value substitution technique with $C = 10,000$, and several fix-factor techniques where F ranges from .7 up to 1. On the rRNA-111 tree scaled up to 5 times its original rate, we see that NJ using uncorrected distances outperforms NJ using these fix-factor corrections at all sequence lengths 100 up to 1600. However, on the same basic model tree with a higher rate of evolution (scaling 10), the best performer at all sequence lengths is the smallest fix-factor we examine ($F = .7$).

Since the relative performance between uncorrected and good fix-factors changed with the scaling factor, we explicitly studied the performance of NJ under uncorrected and the fix-factor correction with $F = 1$ while varying the rate of evolution and keeping the sequence length fixed. For reference we also include the large value substitution with $C = 10,000$. In Figure 5(a) we show the result of this experiment on the rRNA-111 tree for sequence length 800. Note that NJ performs better using uncorrected distances than using fix-factor correction with $F = 1$, but only in a relatively small range; once the rate of evolution is high enough, NJ does worse with uncorrected distances than with

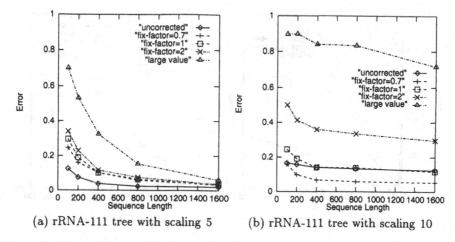

(a) rRNA-111 tree with scaling 5   (b) rRNA-111 tree with scaling 10

Fig. 4. The error rate for NJ on the model tree rRNA-111, using uncorrected distances, fix-factors of 0.7, 1 and 2, and the large-value substitution with $C = 10,000$. The results are shown for various sequence lengths at a scaling of 5 (a) and 10 (b).

the fix-factor correction with $F = 1$. In Figure 5(b) we show the results for the same experiment on the rbcL-35 tree. On this tree, the opposite is true. For most rates of evolution sufficient to produce saturation, we find NJ doing better with fix-factor corrections with $F = 1$, but for *very* high rates of evolution it does very slightly better with uncorrected distances (though at this point, the error rates are about 60% or higher, and the improvement is probably irrelevant).

Our experiments thus have not indicated that any particular correction outperforms the others. For example, sometimes uncorrected is the best technique, and sometimes some of the small fixfactors outperforms uncorrected. However, we cannot even recommend any particular fix-factor as being the best, as no particular fix-factor correction outperforms the other fix-factor corrections on all datasets we examined. We do not yet understand the conditions under which a given correction is the best, and there is clearly a need to select the fix-factor correction based upon the input; we leave this to future research.

What remains quite puzzling, however, is the good performance of uncorrected distances, as this goes against the common wisdom in the community. We do not have an explanation for this phenomenon, and our study is too preliminary for us to draw much of a conclusion about how often we will find NJ's performance better under uncorrected distances than under good fix-factor corrections. We also do not have a reasonable conjecture about the conditions under which the use of uncorrected distances will be preferable to good fix-factor corrections. This remains probably the most interesting and mysterious open research problem that this study has uncovered.

Conclusions Clearly, when using a "substitution by large value" technique (of which the fix-factor correction is an example), it is important to choose the large value *carefully* rather than *arbitrarily*. In particular, our study suggests

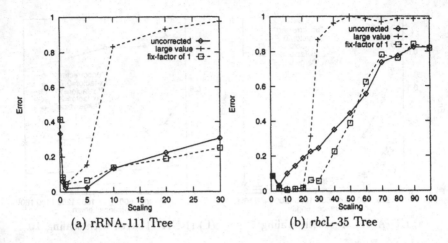

(a) rRNA-111 Tree   (b) rbcL-35 Tree

Fig. 5. The error rate for NJ on the model trees (a) *rRNA-111* and (b) *rbcL-35*, using uncorrected distances, substitution by large value with $C = 10,000$, and fix-factor with $F = 1$. The results are shown for various scalings at sequence length $k = 800$.

that popular ways of selecting the large constant (either twice the largest defined distance in the matrix, as suggested by Swofford *et al.*, or a very large value such as 10,000 or even the largest floating point number supported by a given software system) are definitely inferior to the more modest technique of replacing all undefined values by the largest value in the matrix (i.e. the fix-factor technique with $F = 1$). We also found that fix-factor techniques with F slightly less than 1 often outperformed the fix-factor technique with $F = 1$; consequently, greater improvements may be obtained by even smaller fix-factor techniques. Finally, and surprisingly, the use of uncorrected distances has quite good performance when compared to many other techniques for handling saturated datasets, and on some trees it outperforms the best techniques. We do not understand the conditions under which each technique will outperform the other, and it is clear that additional research is needed before a definitive recommendation can be made.

5   Consequences for Other Studies

The use of experimental techniques involving simulations of sequence evolution is well established in phylogenetic performance studies. These studies have revealed a great deal about the performance of different phylogenetic estimation techniques on small trees (fewer than 15 taxa) and somewhat less about their performance on moderate sized trees (up to 30 or so taxa).

Only very recently have studies examined the performance of phylogenetic reconstruction methods on large trees. Perhaps the first published study is [8], where Hillis showed that on a single large biologically inspired model tree with 228 taxa, a popular hill-climbing heuristic for maximum parsimony (MP) and the popular polynomial time method Neighbor-Joining (NJ) performed comparably well, obtaining highly accurate reconstructions of the true tree from sequences of 5,000 nucleotides. This study suggested the polynomial time methods would be competitive with more computationally expensive methods on large complex trees. In another study, [12], Huson *et al.* found differently. Their study looked also at biologically inspired model trees, where they found that for low evolutionary rates the two methods (MP and NJ) were comparable in accuracy, but for high evolutionary rates, MP had better accuracy than NJ. (These two conclusions seem to be in conflict, but the model tree used by Hillis had low evolutionary rates, and hence the predictions made by Huson *et al.* were consistent with the Hillis study.) Huson *et al.* concluded that for large evolutionary trees with high rates of evolution, it was safer to use MP or other computationally intensive methods (such as ML) rather than to use NJ and other distance methods.

We believe that our current study indicates that the Huson *et al.* study needs to be redone, because their study used a very large fix-factor distance estimation technique, and as we have seen, the large fix-factor corrections are not appropriate choices for NJ. The fix-factor technique they used set $F = n$, where n is the number of taxa in the dataset. Since $F = 2$ produces worse results than $F = 1$, and $n >> 2$ in their study, it is possible that the bad performance of NJ in that study may have been due to the poor distance correction.

We repeated the NJ analysis of the datasets used in that paper using a better fix-factor, $F = .9$, and we observed a great difference in performance; see Figure 6. Indeed, as Huson *et al.* observed, when $F = n$ and there is high divergence,

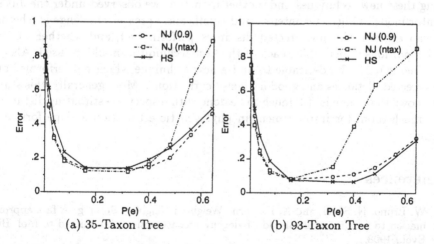

Fig. 6. We compare the performance of NJ and a simple version of heuristic search Maximum Parsimony (HS) on the *rbcL*-35 (a) and *rbcL*-93 (b) model trees. For both trees, at different settings of $p(e)$ (the maximal probability of change on an edge, we show the false negative rates for NJ using a fix-factor of $ntax = 35$ (a) and $ntax = 93$ (b), for HS (data for both plots was taken from the previous study) and for NJ using a fix-factor of 0.9. Each data point represents the average result of 20 independent experiments using sequence length 200.

then NJ has worse performance than MP, but under the same conditions, if $F = .9$, then NJ has *about the same* performance as MP. Thus, simply changing the fix-factor correction from $F = n$ to $F = .9$ changed the relative performance of NJ and MP.

Nevertheless, the conclusions drawn in that study [12] may be valid, for the following reason. MP is an NP-hard optimization problem, and the accuracy of the optimization problem is hard to study experimentally on large datasets, except through the use of (inexact) heuristics. The particular heuristic for MP that Huson *et al.* used in [12] was a very simple one, and hence is likely to have worse performance than the better heuristics used in practice (in particular, they used only one random starting point for the MP heuristic search, rather than multiple starting points). Thus, what Huson *et al.* observed was that NJ with a bad distance estimation technique is worse than MP with a bad heuristic, and what we have observed is that NJ with a good distance estimation technique is about as good as MP with a bad heuristic. It remains to be seen whether MP with a good heuristic is better than NJ with a good distance estimation technique!

6   Future Work

The techniques we have described are quite general and are easily defined for other models of evolution besides the Jukes-Cantor model. In future work we plan to test these techniques for estimating evolutionary distances under other models of evolution, and we predict that similar improvements will be obtained using these new techniques under other models as we observed under the Jukes-Cantor model. Of greater interest and significantly greater challenge is whether we can explain why uncorrected distances perform well, and whether we can find conditions under which each of the best techniques should be used. Also, if possible, it would be desirable to find a new technique which outperforms both uncorrected distances and good fix-factor corrections. More generally, this study has shown that there is still much to be done with respect to estimating distances, and much room for improvement in phylogenetic estimation and performance studies.

References

1. W. Bruno, N. Socci, and A. Halpern. Weighted Neighbor Joining: A fast approximation to maximum-likelihood phylogeny reconstruction. Submitted to Mol. Bio. Evol., 1998.
2. T. Bruns, T. Szaro, M. Gardes, K. Cullings, J. Pan, D. Taylor, T. Horton, A. Kretzer, M. Garbelotto, and Y. Li. A sequence database for the identification of ectomycorrhizal basidiomycetes by phylogenetic analysis. *Molecular-Ecology*, 7(3), March 1998. Accession Number 199800223328.
3. P. Buneman. The recovery of trees from measures of dissimilarity. In *Mathematics in the Archaeological and Historical Sciences*, pages 387–395. Edinburgh University Press, 1971.

4. K. A. Crandall and J. F. Fitzpatrick. Crayfish molecular systematics: using a combination of procedures to estimate phylogeny. *Systematic Biology*, 45(1):1–26, March 1996.

5. K. Cullings and D. Vogler. A 5.8s nuclear ribosomal rna gene sequence database: Applications to ecology and evolution. *Molecular Ecology*, 7(7):919–923, July 1998. Accession number 199800404928.

6. J. Felsenstein. Phylogenies from molecular sequences: inference and reliability. *Annu. Rev. Genet.*, 22:521–565, 1988.

7. O. Gascuel. BIONJ: An improved version of the NJ algorithm based on a simple model of sequence data. *Mol. Biol. Evol.*, 14:685–695, 1997.

8. D. Hillis. Inferring complex phylogenies. *Nature*, 383:130–131, 12 September 1996.

9. D. Hillis, J. Huelsenbeck, and C. Cunningham. Application and accuracy of molecular phylogenies. *Science*, 264:671–677, 1994.

10. J. Huelsenbeck. Performance of phylogenetic methods in simulation. *Syst. Biol.*, 44:17–48, 1995.

11. J. Huelsenbeck and D. Hillis. Success of phylogenetic methods in the four-taxon case. *Syst. Biol.*, 42:247–264, 1993.

12. D. Huson, S. Nettles, K. Rice, and T. Warnow. Hybrid tree reconstruction methods. In *Proceedings of the "Workshop on Algorithm Engineering"*, Saarbrücken, pages 172–192, 1998.

13. T. Jukes and C. Cantor. Evolution of protein molecules. In H. Munro, editor, *Mammalian Protein Metabolism*, pages 21–132. Academic Press, 1969.

14. M. Kuhner and J. Felsenstein. A simulation comparison of phylogeny algorithms under equal and unequal evolutionary rates. *Mol. Biol. Evol.*, 11:459–468, 1994.

15. A. Rambaut and N. Grassly. An application for the monte carlo simulation of DNA sequence evolution along phylogenetic trees. *Comput. Appl. Biosci.*, 13(3):235–238, 1997.

16. K. Rice, M. Donoghue, and R. Olmstead. Analyzing large data sets: *rbcL* 500 revisited. *Syst. Biol.*, 46(3):554–563, 1997.

17. N. Saitou and T. Imanishi. Relative efficiencies of the Fitch-Margoliash, maximum parsimony, maximum likelihood, minimum evolution, and neighbor-joining methods of phylogenetic tree construction in obtaining the correct tree. *Mol. Biol. Evol.*, 6:514–525, 1989.

18. N. Saitou and M. Nei. The Neighbor-Joining method: a new method, for reconstructing phylogenetic trees. *Mol. Biol. Evol.*, 4:406–425, 1987.

19. M. Schöniger and A. von Haeseler. Performance of maximum likelihood, neighbor-joining, and maximum parsimony methods when sequence sites are not independent. *Syst. Biol.*, 44(4):533–547, 1995.

20. J. Sourdis and M. Nei. Relative efficiencies of the maximum parsimony and distance-matrix methods in obtaining the correct phylogenetic tree. *Mol. Biol. Evol.*, 5(3):293–311, 1996.

21. D. L. Swofford, G. J. Olsen, P. J. Waddell, and D. M. Hillis. Chapter 11: Phylogenetic inference. In D. M. Hillis, C. Moritz, and B. K. Mable, editors, *Molecular Systematics*, pages 407–514. Sinauer Associates, Inc., 2nd edition, 1996.

22. M. Waterman, T. Smith, and W. Beyer. Additive evolutionary trees. *Journal Theoretical Biol.*, 64:199–213, 1977.

The Performance of Concurrent Red-Black Tree Algorithms

Sabine Hanke

Institut für Informatik, Universität Freiburg
Am Flughafen 17, D-79110 Freiburg, Germany
hanke@informatik.uni-freiburg.de, fax: ++49 +761 203 8162

Abstract. Relaxed balancing has become a commonly used concept in the design of concurrent search tree algorithms. Many different relaxed balancing algorithms have been proposed, especially for red-black trees and AVL-trees, but their performance in concurrent environments is not yet well understood. This paper presents an experimental comparison of the strictly balanced red-black tree and three relaxed balancing algorithms for red-black trees using the simulation of a multi-processor machine.

1 Introduction

In some applications—like real-time embedded systems as the switching system of mobile telephones—searching and updating a dictionary must be extremely fast. Dictionaries stored in main-memory are needed which can be queried and modified concurrently by several processes. Standard implementations of main-memory dictionaries are balanced search trees like red-black trees or AVL-trees. If they are implemented in a concurrent environment, there must be a way to prevent simultaneous reading and writing of the same parts of the data structure. A common strategy for concurrency control is to lock the critical parts. Only a small part of the tree should be locked at a time in order to allow a high degree of concurrency. For an efficient solution to the concurrency control problem, the concept of relaxed balancing is used in the design of many recent concurrent search tree algorithms (see [17] for a survey). *Relaxed balancing* means that the rebalancing is uncoupled from the update and may be arbitrarily delayed and interleaved.

With regard to the comparison of relaxed balancing algorithms with other concurrent search tree algorithms and an experimental performance analysis, the B-tree is the only algorithm which has been studied well. There are analytical examinations and simulation experiments which show that the B^{link}-tree has a clear advantage over other concurrent but non-relaxed balanced B-tree algorithms [9, 18]. Furthermore, the efficiency of group-updates in relaxed balanced B-trees is experimentally analyzed [15]. So far, the performance of concurrent relaxed balanced binary search trees has hardly been studied by experiments. None of the experimental research known to us examines relaxed balanced binary trees in concurrent environments.

J.S. Vitter and C.D. Zaroliagis (Eds.): WAE'99, LNCS 1668, pp. 286–300, 1999.
© Springer-Verlag Berlin Heidelberg 1999

Bougé et al. [1] and Gabarró et al. [5] perform some serial experiments in order to study the practical worst-case and average convergence time of their relaxed balancing schemes. Malmi presents sequential implementation experiments applying a periodical rebalancing algorithm to the chromatic tree [11] and the height-valued tree [12]. The efficiency of group-operations is studied in height-valued trees [13]. It is shown that group updates provide a method of considerably speeding up search and update time per key [13].

This paper presents an experimental comparison of the strictly balanced red-black tree and three relaxed balanced red-black trees in a concurrent environment. We examine the algorithm proposed by Sarnak and Tarjan [16] combined with the locking scheme of Ellis [4], the chromatic tree [2], the algorithm that is obtained by applying the general method of making strict balancing schemes relaxed to red-black trees [8], and Larsen's relaxed red-black tree [10]. The three relaxed balancing algorithms are relatively similar and have the same analytical bounds on the number of needed rebalancing transformations. The chromatic tree and Larsen's relaxed red-black tree can both be seen as tuned versions of the algorithm obtained by the general relaxation method. However, it is unclear how much worse the performance of the basic relaxed balancing algorithm really is. Furthermore, from a theoretical point of view, it is assumed that the relaxed red-black tree has an advantage over the chromatic tree, since its set of rebalancing transformations is more grained, so that fewer nodes at a time must be locked. But whether, in practice, the relaxed red-black tree really fits better is unknown.

In order to clarify these points, we implemented the four algorithms using the simulation of a multi-processor machine, and we compared the performance of the algorithm in highly dynamic applications of large data. The results of our experiments indicate that in a concurrent environment, the relaxed balancing algorithms have a significant advantage over the strictly balanced red-black tree. With regard to the average response time of the dictionary operations, the three relaxed balancing algorithms perform almost identically. Regarding the quality of the rebalancing, we find that the chromatic tree has the best performance.

The remainder of this paper is organized as follows. In Section 2 we first give a short review of the red-black tree algorithms; then we present the relief algorithms that are necessary to implement the algorithms in a concurrent environment. In Section 3 we describe the performed experiments and the results.

2   Concurrent Red-Black Tree Algorithms

2.1   The Rebalancing Schemes

In this section we give a short review of the four red-black tree algorithms that we want to compare. For each of them, the number of structural changes after an update is bounded by $O(1)$ and the number of color changes by $O(\log n)$, if n is the size of the tree. The rebalancing is amortized constant [2, 8, 10, 16].

The Standard Red-Black Tree In the following, we consider the variant of red-black trees by Sarnak and Tarjan [16]. Deviating from the original proposal by Guibas and Sedgewick [6], here the nodes, and not the edges, are colored red or black. Since the relaxed balanced red-black trees described below can all be seen as relaxed extensions of the strict rebalancing algorithm by Sarnak and Tarjan, in the following we call it the *standard red-black tree*.

The balance conditions of a standard red-black tree require that:

1. each path from the root to a leaf consists of the same number of black nodes,
2. each red node (except for the root) has a black parent, and
3. all leaves are black.

Inserting a key may introduce two adjacent red nodes on a path. This is called a *red-red conflict*. Immediately after generating a red node p that has a red parent, the insert operation performs the rebalancing. If p's parent has a red sibling, then the red-red conflict is handled by color changes. This transformation may produce a new violation of the red constraint at the grandparent of p. The same transformation is repeated, moving the violation up the tree, until it no longer applies. Finally, if there is still a violation, an appropriate transformation is carried out that restores the balance conditions again.

Deleting a key may remove a black node from a path. This violation of the black constraint is handled immediately by the delete operation. Using only color changes, it is bubbled up in the tree until it can be finally resolved by an appropriate transformation.

The Basic Relaxed Balancing Algorithm The nodes of a *basic relaxed balanced red-black tree* [8] are either red or black. Red nodes may have *up-in requests*, black nodes may have *up-out requests*, and leaves may have *removal requests*. The relaxed balance conditions require that:

1. the sum of the number of black nodes plus the number of up-out requests is the same on each path from the root to a leaf,
2. each red node (except for the root) has either a black parent or an up-in request, and
3. all leaves are black.

In contrast to the standard red-black tree, a red-red conflict is not resolved immediately after the actual insertion. The insert operation only marks a violation of the red constraint by an up-in request. In general, deleting a key leads to a removal request only, where the actual removal is part of the rebalancing. The handling of a removal request may remove a black node from a path. This is marked by an up-out request.

The task of rebalancing a basic relaxed balanced red-black tree involves handling all up-in, removal, and up-out requests. For this the rebalancing operations of the standard red-black tree are used, which are split into small restructuring steps that can be carried out independently and concurrently at different locations in the tree. The only requirement is that no two of them interfere at the same location. This can be achieved very easily by handling requests in a top-down manner if they meet in the same area; otherwise, in any arbitrary order.

Furthermore, in order to keep the synchronization areas as small as possible, two additional rebalancing transformations are defined that apply to situations that cannot occur in a strictly balanced red-black tree.

Chromatic Tree The nodes of a *chromatic tree* [14] are colored red, black, or overweighted. The color of a node p is represented by an integer $w(p)$, the *weight* of p. The weight of a red node is zero, the weight of a black node is one, and the weight of an *overweighted node* is greater than one. The relaxed balance conditions are:

1. all leaves are not red and
2. the sum of weights on each path from the root to a leaf is the same.

Analogous to the basic relaxed red-black tree, the rebalancing is uncoupled from the updates. Here the delete operation performs the actual removal. A deletion may remove a non-red node and increment the weight of a node on the path instead. Thereby, the node may become overweighted. This is called an *overweight conflict*.

The task of rebalancing a chromatic tree involves removing all red-red and overweight conflicts from the tree. For this purpose, Nurmi and Soisalon-Soininen propose a set of rebalancing operations [14]. A more efficient set of transformations is defined by Boyar and Larsen [2,3].

Several red-red conflicts at consecutive nodes must be handled in a top-down manner. Otherwise, the rebalancing transformations can be applied in any arbitrary order.

The basic relaxed balanced red-black tree differs from the chromatic tree mainly in that deletions are accumulated by removal requests. However, an up-out request is equivalent to one unit of overweight. Therefore, a basic relaxed balanced red-black tree always fulfills the balance conditions of chromatic trees. It differs from an arbitrary chromatic tree only in that overweights are bounded by two. Furthermore, a detailed comparison of the rebalancing transformations of both algorithms depicts that the transformations of the chromatic tree can be seen as generalizations of the transformations of the basic relaxed balancing algorithm.

Larsen's Relaxed Red-Black Tree The relaxed balance conditions, the operations insert and delete, and the transformations to handle a red-red conflict of *Larsen's relaxed red-black tree* [10] are exactly the same as those of chromatic trees. The set of rebalancing transformations to handle an overweight conflict is reduced to a smaller collection of operations consisting of only six operations instead of nine. Five of them, which perform at most one rotation or double rotation, are identical with operations of the chromatic tree. The sixth transformation, *weight-temp*, performs one rotation but does not resolve an overweight conflict. If we assume that weight-temp is followed immediately by one of the other transformations, then this combined transformation is almost identical to one of the remaining operations of the chromatic tree. But weight-temp can also be carried out independently and interleaved freely with other operations. This has the advantage that fewer nodes must be locked at a time. On the other hand, in order to prevent weight-temp from generating its symmetric case, it introduces a restriction in which order weight-balancing transformations can be carried out.

2.2 Relief Algorithms

In this section we consider the additional algorithms that are necessary to implement the relaxed balancing algorithms in a concurrent environment. This includes the administration of the rebalancing requests and a suitable mechanism for concurrency control.

Administration of the Rebalancing Requests In the literature, there are three different suggestions for the way in which rebalancing processes can locate rebalancing requests in the tree. One idea is to traverse the data structure randomly in order to search for rebalancing requests [14]. Another idea is to mark all nodes on the path from the root to a leaf where an update occurs, so that a rebalancing process can restrict its search to the marked subtrees [11]. The third suggestion is to use a problem queue [3], which has the advantage that rebalancing processes can search purposefully for imbalance situations. Here it is necessary for each node to have a parent pointer in addition to the left and right child pointers. Since the problem queue is mentioned only very vaguely [3], in the following we propose how to maintain a queue to administrate the rebalancing requests.

In addition to the color field of a node, each rebalancing request is implemented by a pointer to the node, the *request pointer*. Whenever a rebalancing request is generated, the corresponding request pointer is placed in a problem queue.

Since it is possible to remove nodes from the tree which have rebalancing requests, the problem queue may contain request pointers that are no longer valid. In order to avoid side effects, every node should be represented by only one pointer in a queue. This can be guaranteed by using a control bit for each node that is set if a request pointer is placed in a problem queue.

If a node for which a request pointer exists is deleted from the tree, then after the removal the node must be marked explicitly as removed. A rebalancing process following the link of the request pointer is now able to notice that the node is no longer in the tree. Freeing the memory allocated for the node is delayed and becomes part of the rebalancing task.

Concurrency Control A suitable mechanism for concurrency control in strictly balanced search trees is the locking protocol by Ellis [4]. Nurmi and Soisalon-Soininen propose a modification of Ellis's locking scheme for relaxed balanced search trees [14]. In the following, we only mention the features of both locking schemes which are the most important and which are relevant to this paper.

Both protocols use three different kinds of locks: *r-locks*, *w-locks*, and *x-locks*. Several processes can *r*-lock a node at the same time. Only one process can *w*-lock or *x*-lock a node. Furthermore, a node can be both *w*-locked by one process and *r*-locked by several other processes, but an *x*-locked node cannot be *r*-locked or *w*-locked.

A process performing a search operation uses *r*-lock coupling from the root. *Lock coupling* means that a process on its way from the root down to a leaf locks the child to be visited next before it releases the lock on the currently visited node.

Using the locking protocol by Ellis, a process that performs an update operation *w*-locks the whole path. This is necessary in order to perform the rebalancing immediately after the update. In the locking protocol for relaxed balanced search trees, it is sufficient only to use *w*-lock coupling during the search phase of the update operation.

A rebalancing process uses *w*-locks while checking whether a transformation applies. Just before the transformation, it converts the *w*-locks of all nodes, the contents of which will be changed, to *x*-locks. Structural changes are implemented by exchanging the contents of nodes.

Nurmi and Soisalon-Soininen suggest that rebalancing processes locate rebalancing requests by traversing the tree nondeterministically, so that rebalancing processes can always *w*-lock the nodes incipient with the top-most one. This guarantees that the locking scheme is dead-lock free [14].

If we use a problem queue in order to administrate the rebalancing requests, the situation is different. The rebalancing transformations of red-black trees have the property that the parent, or even the grandparent, of a node with a rebalancing request must be considered in order to apply a transformation. Thus, if a rebalancer follows the link of a request pointer when searching for an imbalance situation in the tree, it cannot lock the top-most node involved first. Therefore, we extend the locking scheme for relaxed balanced search trees in the following way.

A rebalancing process always only tries to *w*-lock ancestor nodes. If the parent of a node is still *w*-locked or *x*-locked by another process, then in order to avoid a dead-lock situation, the rebalancer immediately releases all locks it holds. Since all locks that may block the process are taken in top-down direction, dead-lock situations are avoided. The results of our experiments, which we present in the following section, display that this strategy seems to be very suitable—at least in large trees. In 99.99% of all cases, the *w*-lock could be taken successfully.

Normally the *w*-lock of each node, the contents of which will be changed, is converted to an *x*-lock. However, this conversion is not necessary if only the parent pointer of the node is to be updated. This is because rebalancing processes are the only processes that use parent pointers.

Since structural changes are implemented by exchanging the contents of nodes, the address of the root node is never changed by a rotation. The only two cases in which the pointer to the root may be replaced are if a key is inserted into an empty tree or if the last element of the tree is deleted. The replacing of the root pointer can easily be prevented by using a one-element-tree with a removal request at its only leaf as representation of an empty tree instead of a nil pointer. Then it is not necessary to protect the root pointer by a lock.

3   The Experiments

In order to compare the concurrent red-black tree algorithms, we have used the simulation of a multi-processor environment psim [7] that was developed in co-operation with Kenneth Oksanen from the Helsinki University of Technology. In the following, we give a short description of the simulator psim. Then we describe the experiments and briefly present the main results.

3.1   Simulation of a Concurrent Environment

The interpreter psim serves to simulate algorithms on parallel processors [7]. The algorithms are formulated by using psim's own high-level programming language which can be interpreted, executed, and finally statistically analyzed.

psim simulates an abstract model of a multi-processor machine, a CRCW-MIMD machine with shared constant access time memory. The abstract psim-machine can be described as follows. The psim-machine has an arbitrary number of serial processors that share a global memory. All processors can read or write memory locations at the same time. In order to read a value from the global memory into a local register or, respectively, to write a value from a register into the shared memory, each processor needs one unit of *psim-time*. The execution of a single instruction (for instance an arithmetic operation, a compare operation, a logical operation, etc.) also takes one unit of psim-time. The access of global data can be synchronized by using r-locks, w-locks, and x-locks. Several processes waiting for a lock are scheduled by a FIFO queue. The time that is needed to perform a lock operation is constant for each and can be chosen by the user. The overhead of lock contention is not modeled.

3.2   Experiments

Our aim is to study the behavior of the red-black tree algorithms in highly dynamic applications of large data. For this, we implemented the algorithms in the psim programming language and concurrently performed 1 000 000 dictionary operations (insertions, deletions, and searches) on an initially balanced standard red-black tree, which was built up by inserting 1 000 000 randomly chosen keys into an empty tree. As key space we choose the interval $[0, 2\,000\,000\,000]$. During the experiments, we varied the number of processes and the time that is needed to perform a lock operation. In the case of strict balancing, 2, 4, 8, 16, 32, or 64 user processes apply concurrently altogether 1 000 000 random dictionary operations to the tree. In the case of relaxed balancing, either 1, 3, 7, 15, 31, or 63 user processes are used, plus a single rebalancing process, which gets its work from a FIFO problem queue. Analogous to [9], the probabilities that a dictionary operation is an insert, delete, or search operation are $p_i = 0.5$, $p_d = 0.2$, and $p_s = 0.3$. The costs of an arbitrary lock operation—excluding the time waiting to acquire a lock, if it is already taken by another process—are constant: either 1, 10, 20, 30, 40, 50, or 60 units of psim-time, motivated by measurements of lock costs that we performed on real machines.

Comparison of Strict and Relaxed Balancing In the following, we confirm by experiment that relaxed balancing has a significant advantage over strict balancing in a concurrent environment. For this, we compare the performance of the standard red-black tree with that of the chromatic tree.

The search is the only dictionary operation in which the standard red-black tree supplies almost the same results as the chromatic tree (cf. Figure 1.1). On average, a search is only 1.18 times faster in a chromatic tree than in a standard red-black tree. This is because both use r-lock coupling, and no rebalancing is needed after the search.

In the case of the update operations, the chromatic tree is considerably faster than the standard red-black tree (cf. Figure 1.2). The speed-up factor depends on the number of processes and the lock costs. It varies from a minimal factor of 1.21 (in the case of 2 processes and lock costs 60) to a maximal factor of 25.21 (in the case of 32 processes and lock costs 1, see Table 1).

The total time needed on average to perform 1 000 000 dictionary operations is depicted in Figure 2. Only in the case of two concurrent processes does the standard red-black tree terminate earlier than the chromatic tree. But note that, in this case, the dictionary operations are performed serially in the chromatic tree, and the rebalancing process is hardly used (cf. Table 3). If more than two processes work concurrently, then the chromatic tree always terminates earlier than the standard red-black tree. The higher the number of processes, the clearer the differences between both algorithms become. In the case of 64 processes, the chromatic tree is up to 20.22 times faster than the standard red-black tree (see Table 2). As Figure 3 shows, this is because an increase in the number of user processes in a standard red-black tree leads to almost no saving of time.

The utilization of the user processes reflects why the standard red-black tree performs so badly (cf. Figure 4). Since during an update operation, the whole path from the root to a leaf must be w-locked, the root of the tree becomes a bottleneck. Therefore, with an increasing number of concurrent processes, the utilization of the processes decreases in a steep curve towards only 4%. In contrast, the user processes that modify a chromatic tree use w-lock coupling during the search phase of an update operation. The rebalancing process locks only a small number of nodes at a time. Thus, for up to 16 concurrent processes, the

(1)

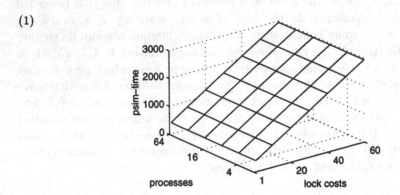

(2)

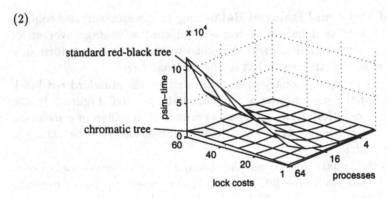

Fig. 1. Average time needed to perform (1) a search operation and (2) an update operation.

	number of processes					
	2	4	8	16	32	64
1	1.84	3.30	6.86	13.71	25.21	23.19
10	1.51	2.61	5.53	11.15	19.53	17.16
20	1.37	2.32	4.97	10.08	17.23	14.98
30	1.30	2.18	4.69	9.54	16.07	13.94
40	1.26	2.09	4.52	9.21	15.38	13.34
50	1.23	2.03	4.40	8.99	14.94	12.93
60	1.21	1.99	4.32	8.83	14.61	12.63

(row labels at left under "lock costs": 1, 10, 20, 30, 40, 50, 60)

Table 1. Factor of speed-up when performing an update operation in a chromatic tree compared with a standard red-black tree.

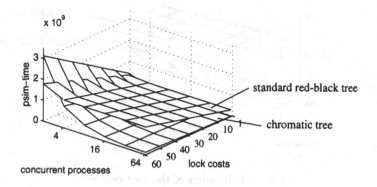

Fig. 2. Total time needed on average to perform 1 000 000 dictionary operations.

	number of processes					
	2	4	8	16	32	64
1	0.81	2.04	4.73	9.97	19.14	20.22
10	0.68	1.63	3.78	7.94	14.68	14.98
20	0.63	1.46	3.39	7.13	12.91	13.07
30	0.61	1.38	3.20	6.73	12.04	12.17
40	0.59	1.33	3.08	6.49	11.51	11.63
50	0.58	1.30	3.01	6.33	11.18	11.28
60	0.57	1.28	2.95	6.21	10.96	11.03

(row labels on left: lock costs)

Table 2. Factor of speed-up when performing 1 000 000 dictionary operations in a chromatic tree compared with a standard red-black tree.

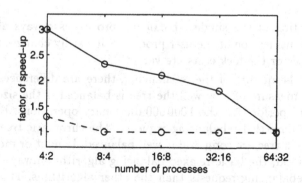

Fig. 3. Gain in time by increasing the number of processes. ——— chromatic tree, – – standard red-black tree

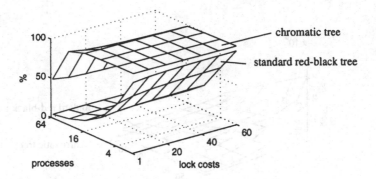

Fig. 4. Utilization of the user processes.

utilization of the user processes is always over 98%. By a further increase of the number of user processes, the utilization decreases clearly, but even in the case of 64 concurrent processes, it is still between 42% and 46% (cf. Figure 4).

Comparison of the Relaxed Balancing Algorithms In this section we experimentally compare the performance of the chromatic tree [2], Larsen's relaxed red-black tree [10], and the basic relaxed balanced red-black tree [8].

Regarding the average response time of the dictionary operations and the utilization of the user processes, the three relaxed balancing algorithms supply nearly identical results. Thus, in all three cases, almost the same time is needed on average to perform 1 000 000 dictionary operations.

The length of a path from the root to a leaf is at most 26, if two or four concurrent processes are used. In the case of eight concurrent processes, the maximum length of a path is 27, and in the case of 16, 32, and 64 processes, it is 28. The average length of a search path varies from 21.7 to 21.9. This increase is caused by the proportion of only one rebalancing process to a large number of user processes.

The utilization of the single rebalancing process is always about 99%. In contrast to the utilization of the user processes, it hardly changes if the number of user processes or the lock costs are varied.

Regarding the quality of the rebalancing, there are differences between the algorithms. A measure of how well the tree is balanced is the size of the problem queue after performing the 1 000 000 dictionary operations. Figure 5 shows the number of unsettled rebalancing requests. The chromatic tree and Larsen's relaxed red-black tree are comparably well balanced. If eight or more concurrent processes are used, the basic relaxed balancing algorithm always leaves considerably more rebalancing requests than the other algorithms. This is caused by the additional work a rebalancer must do in a basic relaxed balanced red-black tree in order to settle removal requests as well. In total, about 20–40% more rebalancing requests are generated than in a chromatic tree.

Next we consider the capacity of the rebalancing process. Figure 6 and Table 3 show how much of the time the rebalancing process spends idling because of an empty problem queue. For 16 or more processes, the rebalancing process always works to capacity. For 8 processes or fewer, the chromatic tree needs the least rebalancing work. The case of two concurrent processes depicts the differences between the algorithms most clearly. With a growing number of concurrent processes, the portion of the time the rebalancing process spends idling increases slightly from 82.66% to 88.46% in the case of the chromatic tree, from 75.47% to 85.11% in the case of the basic relaxed balancing algorithm, and from only 38.62% to 41.47% in the case of Larsen's relaxed red-black tree.

Since Larsen's algorithm does not produce considerably more rebalancing requests than the other algorithms, we do not expect such a low idle time. In order to find the reasons for this, we have measured how often the rebalancing process fails to apply a rebalancing operation. The rate of failure to apply a red-balancing transformation is always insignificant. The rate of failure to apply a weight-balancing transformation shows considerable differences between the algorithms and gives reasons for the relatively low idle time of the rebalancing process in Larsen's algorithm.

In the chromatic tree, the rebalancing process never fails to handle an over-weight conflict. In the basic relaxed balancing algorithm, the number of failures is almost always very low and is at most 2.7% of all overweight handling. Larsen's relaxed red-black tree differs crucially from the other two algorithms. Here, the rate of failure to handle an overweight conflict is up to 99% of all overweight handling. Furthermore, it depends only on the weight-temp operation. In all cases in which the tree is relatively well balanced, most of the time the rebalancing process tries to prepare weight-temp, fails because of an interfering overweight conflict in the vicinity, and appends the request again to the problem queue.

The high rate of failure is caused by the use of a FIFO problem queue. A FIFO queue supports the top-down handling of rebalancing requests, since requests are appended to the queue in the order in which they are created. This is very useful when resolving red-clusters in relaxed red-black trees, because red-red conflicts can be handled only in a top-down manner. On the other hand, weight-temp introduces a bottom-up handling of overweight-conflicts. Therefore, if a rebalancing process gets its work from a FIFO problem queue, it must use an additional strategy for the order in which overweight conflicts can be handled.

We have experimentally studied the effect of trying to handle the interfering overweight conflict next, if weight-temp cannot be applied. In the case of randomly chosen dictionary operations, this strategy leads to a satisfying decrease in the rate of failure.

Let us now compare the performance of the weight-balancing transformations. Since the weight-balancing operations of Larsen's algorithm are more grained than the ones of the chromatic tree or the basic relaxed balanced red-black tree, it could be expected that, on average, they need significantly less time to handle an overweight conflict. However, whenever the rate of failure during an experiment is only minimal, the chromatic tree and Larsen's relaxed red-black

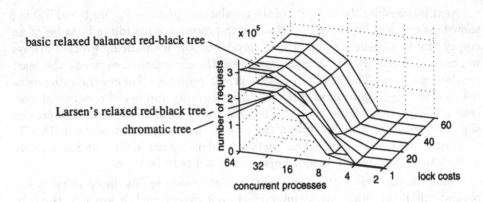

Fig. 5. Number of unsettled rebalancing requests.

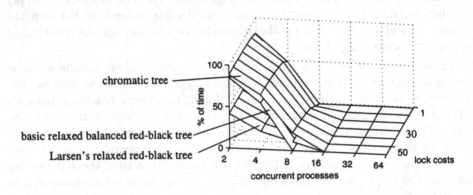

Fig. 6. Portion of the time the rebalancing process spends idling.

	number of processes											
	2	4	8	16	2	4	8	16	2	4	8	16
1	82.66	49.60	0.00	0.00	75.47	29.83	0.00	0.00	38.62	15.74	0.00	0.00
10	85.62	57.82	6.00	0.00	80.48	43.18	0.00	0.00	40.07	18.72	1.88	0.00
20	86.79	61.14	12.82	0.00	82.45	48.63	0.00	0.00	40.65	19.93	5.75	0.00
30	87.51	63.11	16.47	0.00	83.65	51.80	0.00	0.00	41.01	20.63	7.86	0.00
40	87.95	64.32	18.75	0.00	84.36	53.69	0.00	0.00	41.22	21.07	9.17	0.00
50	88.25	65.13	20.27	0.00	84.80	54.90	0.00	0.00	41.37	21.36	10.05	0.00
60	88.46	65.71	21.38	0.00	85.11	55.75	0.00	0.00	41.47	21.57	10.68	0.00
	chromatic tree				*basic relaxed balanced red-black tree*				*Larsen's relaxed red-black tree*			

(lock costs — row labels at left: 1, 10, 20, 30, 40, 50, 60)

Table 3. Portion of the time (in %) the rebalancing process spends idling.

tree perform in a broadly similar fashion. This is caused by the fact that the cases in which the handling of an overweight conflict differ in both algorithms occur very seldom. In less than 1.5% of the cases of overweight handling, one of the larger weight-balancing transformations apply in the chromatic tree. This corresponds exactly to the occurrence of weight-temp during the rebalancing of Larsen's relaxed red-black tree. Therefore, by the results of our experiments, we could not confirm that Larsen's relaxed red-black tree fits better in a concurrent environment.

4 Conclusions

We have experimentally studied the performance of three relaxed balancing algorithms for red-black trees, the chromatic tree [2], the algorithm that is obtained by applying the general method of how relaxing a strict balancing scheme to red-black trees [8], and Larsen's relaxed red-black tree [10], and we compared them with the performance of the strictly balanced red-black tree by Sarnak and Tarjan [16], combined with the locking scheme of Ellis [4]. We find that in a concurrent environment, the relaxed balancing algorithms have a considerably higher performance than the standard algorithm.

A comparison of the relaxed balancing algorithms among themselves shows that, with regard to the average response time of the dictionary operations, the performance of the three relaxed balancing algorithms is comparably good. This gives grounds for the assumption that the general method of relaxing strict balancing schemes [8] generates satisfactory relaxed balancing algorithms if this method is applied to other classes of search trees as well.

Regarding the quality of the rebalancing, there are differences between the three algorithms. We find that the chromatic tree always has the best performance. The rebalancing performance of the algorithm obtained by the general method is considerably lower than the performance of the chromatic tree, but it is still satisfactory. The rebalancing of Larsen's relaxed red-black tree requires the use of additional strategies for handling overweight-conflicts, since a FIFO problem queue and the need for the bottom-up handling of overweight conflicts leads to a considerably high rate of failure. Furthermore, the hoped for increase in the speed of the weight-balancing is insignificant since, during the experiments, one of the larger weight-balancing transformations applies to only less than 1.5% of all overweight handling.

Acknowledgments

I would like to express my thanks to Kenneth Oksanen, who designed and implemented the first prototypes of the interpreter psim and thereby made the basis that enabled the experimental comparison of the concurrent search tree algorithms. I would also like to thank Eljas Soisalon-Soininen and Thomas Ottmann, who have decisively contributed to the origin of this work.

References

1. L. Bougé, J. Gabarró, X. Messeguer, and N. Schabanel. Concurrent rebalancing of AVL trees: A fine-grained approach. In *Proceedings of the 3th Annual European Conference on Parallel Processing*, pages 321–429. LNCS 1300, 1997.
2. J. Boyar, R. Fagerberg, and K. Larsen. Amortization results for chromatic search trees, with an application to priority queues. *Journal of Computer and System Sciences*, 55(3):504–521, 1997.
3. J. Boyar and K. Larsen. Efficient rebalancing of chromatic search trees. *Journal of Computer and System Sciences*, 49:667–682, 1994.
4. C.S. Ellis. Concurrent search in AVL-trees. *IEEE Trans. on Computers*, C-29(29):811–817, 1980.
5. J. Gabarró, X. Messeguer, and D. Riu. Concurrent rebalancing on hyperred-black trees. In *Proceedings of the 17th Intern. Conference of the Chilean Computer Science Society*, pages 93–104. IEEE Computer Society Press, 1997.
6. L.J. Guibas and R. Sedgewick. A dichromatic framework for balanced trees. In *Proc. 19th IEEE Symposium on Foundations of Computer Science*, pages 8–21, 1978.
7. S. Hanke. The performance of concurrent red-black tree algorithms. Technical Report 115, Institut für Informatik, Universität Freiburg, Germany, 1998.
8. S. Hanke, T. Ottmann, and E. Soisalon-Soininen. Relaxed balanced red-black trees. In *Proc. 3rd Italian Conference on Algorithms and Complexity*, pages 193–204. LNCS 1203, 1997.
9. T. Johnson and D. Shasha. The performance of current B-tree algorithms. *ACM Trans. on Database Systems*, 18(1):51–101, March 1993.
10. K. Larsen. Amortized constant relaxed rebalancing using standard rotations. *Acta Informatica*, 35(10):859–874, 1998.
11. L. Malmi. A new method for updating and rebalancing tree-type main memory dictionaries. *Nordic Journal of Computing*, 3:111–130, 1996.
12. L. Malmi. *On Updating and Balancing Relaxed Balanced Search Trees in Main Memory*. PhD thesis, Helsinki University of Technology, 1997.
13. L. Malmi and E. Soisalon-Soininen. Group updates for relaxed height-balanced trees. In *ACM Symposium on the Principles of Database Systems*, June 1999.
14. O. Nurmi and E. Soisalon-Soininen. Chromatic binary search trees: A structure for concurrent rebalancing. *Acta Informatica 33*, pages 547–557, 1996.
15. K. Pollari-Malmi, E. Soisalon-Soininen, and T. Ylönen. Concurrency control in B-trees with batch updates. *Trans. on Knowledge and Data Engineering*, 8(6):975–983, 1996.
16. N. Sarnak and R.E. Tarjan. Planar point location using persistent search trees. *Communications of the ACM*, 29:669–679, 1986.
17. E. Soisalon-Soininen and P. Widmayer. Relaxed balancing in search trees. In D.-Z.Du and K.-I. Ko, editors, *Advances in Algorithms, Languages, and Complexity: Essays in Honor of Ronald V. Book*. Kluwer Academic Publishers, Dordrecht, 1997.
18. V. Srinivasan and M.J. Carey. Performance of B-tree concurrency control algorithms. *Proc. ACM Intern. Conf. on Management of Data*, pages 416–425, 1991.

Performance Engineering Case Study: Heap Construction

Jesper Bojesen[1], Jyrki Katajainen[2], and Maz Spork[2]

[1] UNI•C, Danish Computing Centre for Research and Education,
Technical University of Denmark, Building 304
DK-2800 Lyngby, Denmark
[2] Department of Computing, University of Copenhagen
Universitetsparken 1, DK-2100 Copenhagen East, Denmark
{jyrki|halgrim}@diku.dk
http://www.diku.dk/research-groups/performance_engineering/
http://users/~iekeland/web/welcome.html

Abstract. The behaviour of three methods for constructing a binary heap is studied. The methods considered are the original one proposed by Williams [1964], in which elements are repeatedly inserted into a single heap; the improvement by Floyd [1964], in which small heaps are repeatedly merged to bigger heaps; and a recent method proposed, e.g., by Fadel et al. [1999] in which a heap is built layerwise. Both the worst-case number of instructions and that of cache misses are analysed. It is well-known that Floyd's method has the best instruction count. Let N denote the size of the heap to be constructed, B the number of elements that fit into a cache line, and let c and d be some positive constants. Our analysis shows that, under reasonable assumptions, repeated insertion and layerwise construction both incur at most cN/B cache misses, whereas repeated merging, as programmed by Floyd, can incur more than $(dN \log_2 B)/B$ cache misses. However, for a memory-tuned version of repeated merging the number of cache misses incurred is close to the optimal bound N/B.

1 Introduction

In many research papers on sorting problems, the performance of competing methods is compared by analysing the number of key comparisons, after which it is stated that the number of other operations involved is proportional to that of key comparisons. However, in the case of small keys a key comparison can be as expensive as, for instance, an index comparison. Moreover, to save some key comparisons complicated index manipulation might be necessary, and as the result the methods tuned the number of key comparisons in mind are often impractical (cf., [1]).

Another tradition popularized by Knuth (see, e.g., [9]) is to analyse meticulously the number of all primitive operations performed. However, in current computers with a hierarchical memory the operations involving memory accesses

J.S. Vitter and C.D. Zaroliagis (Eds.): WAE'99, LNCS 1668, pp. 301–315, 1999.
© Springer-Verlag Berlin Heidelberg 1999

can be far more expensive than the operations carried out internally in the processing unit. To predict accurately the execution time of a program, different costs must be assigned to different primitive operations. In particular, a special attention must be laid when assigning the cost for the memory access operations.

Most high-performance computers have several processors and a hierarchical memory, each memory level with its own size and speed characteristics. The different memory levels existent in the vast majority of computers are from the fastest to the slowest: register file, on-chip cache, off-chip cache, main memory, and disk storage. The processors have only access to the lowest memory level, the register file, so the data has to be there before it can be processed. The basic idea is to keep the regions of recently accessed memory as close to the processors as possible. If the memory reference cannot be satisfied by one memory level, i.e., in the case of a *miss* or a *fault*, the data must be fetched from the next level. Not only one datum is fetched, but a *block* of data in the hope that the memory locations in vicinity of the current one will be accessed soon. In real programs a significant proportion of the running time can be spent waiting for the block transfers between the different levels of the memory hierarchy.

In this paper we study, both analytically and experimentally, the performance of programs that construct a binary heap [20] in a hierarchical memory system. Especially, we consider the highest memory level that is too small to fit the whole heap. We call that particular level simply the *cache*. It should be, however, emphasized that our analysis is also valid for the memory levels below this cache, provided that all our assumptions are fulfilled. We let B denote the size of the blocks transferred between the cache and the memory level above it, and M the capacity of the cache, both measured in elements being manipulated.

Recent research papers indicate that a binary heap is not the fastest priority-queue structure (see, e. g., [10, 15]) and heapsort based on a binary heap is not the fastest sorting method (see, e. g., [11, 13]), but we are mainly interested in the performance engineering aspects of the problem. The reason for choosing the heap construction as the topic of this case study was the existence of the elegant Algol programs published by Williams [20] and Floyd [3] in 1964. In our preliminary experiments we translated these old programs to C++ and compared them to the C/C++ programs published in textbooks on algorithms and/or programming. It turned out that none of the newer variants were able to beat the old programs. Hence, the tuning of these old programs seemed to be a challenge.

It is well-known that the worst-case behaviour of Williams' method is bad, but for randomly generated input Williams' program appears to be faster than Floyd's program when the size of the heap exceeds that of the cache. There are two reasons for this:

1. In the average case the number of instructions executed by Williams' program is linear [5], which is guaranteed in the worst case by Floyd's program.
2. Williams' program accesses the memory much more locally than Floyd's program, as shown experimentally by LaMarca and Ladner [10].

Let N denote the size of the heap being constructed. We prove analytically that, if $M \geq rB\lfloor \log_2 N \rfloor$ for some real number $r > 1$, Williams' program incurs never more than $(4r/(r-1))N/B$ cache misses. On the other hand, there exists an input which makes Floyd's program to incur more than $((1/2)N \log_2 B)/B$ cache misses if $M \le (1/4)N$. However, by memory tuning the number of cache misses incurred can be reduced close to the optimal bound N/B if $M \gg 2B^2$.

Also, efficient external-memory algorithms are developed with a multilevel memory model in mind. There the main concern is to keep the number of page transfers between main memory and disk storage as low as possible. Such algorithms can be good sources for cache-efficient programs as observed, e. g., by Sanders [15]. A fast external-memory algorithm for heap construction have been presented by Fadel et al. [2]. Based on their idea a method for constructing a binary heap is obtained that has nice theoretical properties, e. g., it performs well even if the cache can only hold a constant number of blocks, but in practice it could not compete with the reengineered version of Floyd's program.

In this conference version of the paper the proofs of the theorems are omitted. These will be brought out in the full version of the paper. A more thorough experimental evaluation of the methods will also be presented in the full version.

2 Meticulous Analysis

In meticulous analysis, the term coined by Katajainen and Träff [8], the goal is to analyse the running time of a program as exactly as possibly, including the constant factors. This style of analysis is well-known from Knuth's books (see, e. g., [9]). In this section we define the models of computation and cost, under which we carry out a meticulous analysis of heap-construction programs.

2.1 Primitive Operations

We use the standard *word random-access machine* (RAM) (see, e. g., [4]) as our model of computation with the exception that the memory has several levels in addition to the register file. All computations are carried out in the registers while there are special instructions for loading data from memory to registers and for storing data from registers to memory. The number of registers available is unspecified but fixed, and each of the registers can be used both as an index register and a data register.

We assume that the programs executed by the machine are written in C++ [19] and compiled to pure C [8]. Pure C statements are similar in strength to machine instructions on a typical present-day load-store reduced-instruction-set computer (RISC). Let x, y, and z be symbolic names of arbitrary (not necessarily distinct) registers, c a constant, p a symbolic name of an index register (or a pointer), and λ some label. A pure C program is a sequence of possibly labelled statements that are executed sequentially unless the order is not altered by a branch statement. The pure C statements are the following:

1. *Memory-to-register assignments*, that is, load statements: "x = *p".
2. *Register-to-memory assignments*, that is, store statements: "*p = x".

3. *Constant-to-register assignments*: "x = c".
4. *Register-to-register assignments*: "x = y".
5. *Unary arithmetic assignments*: "x = $\ominus$y", where $\ominus \in \{-, \tilde{}, !\}$.
6. *Binary arithmetic assignments*: "x = y $\oplus$ z", where $\oplus \in \{+, -, *, /, \&, |, \char`^,$ <<, >>\}$.
7. *Conditional branches*: "if (x $\triangleleft$ y) goto λ", where $\triangleleft \in \{<, <=, ==, >, >=\}$.
8. *Unconditional branches*: "goto λ".

2.2 Hierarchical Memory Model

Let w denote the *word size* of the word RAM. We assume that the *capacity* of *memory* is at most 2^w words, so that the address of any word can be stored in one word. The memory is arranged into several, say, ℓ levels abiding the principle of inclusion: if some datum is present at level i, it is also present at level $i+1$. The lowest memory level is the register file. Level $i + 1$ is assumed to be larger and slower than level i. The contents of each memory level is divided into disjoint, consecutive *blocks* of fixed uniform size, but the block size at different levels may vary. When the block accessed is not present at some level, i.e., when a *miss* occurs, it is fetched from the next level and one of the existing blocks is removed from the cache.

In general, we are mainly interested in the highest memory level that is too small to fit the input data. We call that particular level simply the *cache*. We let B denote the *size* of *blocks* used in the cache, and M the *capacity* of the *cache*, both measured in elements being manipulated. For the sake of simplicity, we assume that both B and M are powers of 2. Even if many cache-based systems employ set associative caches (for the technical details the reader is referred to the book by Hennessy and Patterson [6]), we assume that the cache is *fully associative*, i.e., when a block is transferred to the cache, it can be placed in any of the M/B cache lines. We assume that the block replacement is controlled by the cache system according to the *least-recently-used* (LRU) policy. In particular, the program has no control over which block is removed when a new block arrives.

2.3 Penalty Cost Model

We use a weighted version of the cost model presented in [8]. Alternatively, one can see the cost model as a simplification of the Fortran-based model discussed in [14].

1. The cost of all pure C operations is assumed to be the same τ.
2. Each load and store operation has an extra penalty τ_i^* if it incurs a miss at level i of the memory hierarchy, $i \in \{1, 2, \ldots, \ell\}$.

Now, if a program executes n pure C operations and incurs m_i misses at memory level i, $i \in \{1, 2, \ldots, \ell\}$, its execution time might be expressed as the sum

$$n \cdot \tau + \sum_{i=1}^{\ell} m_i \cdot \tau_i^* .$$

So, in addition to the normal instruction count, it is important to calculate the number of misses that occur at different memory levels.

This cost model allows computer-independent meticulous analysis of programs, but it is overly simplified. Some computer architectures may offer combinations of the pure C operations as a single primitive operation, e. g., it might not be necessary to translate x = a[i] to the form p = a + i, x = *p; while others are unable to execute some pure C operations as a single operation, e. g., it might only be possible to carry out a comparison to zero so if (y > z) goto λ has to be translated to the form x = y - z, if (x > 0) goto λ. In some computers, x = y * z may be more expensive than x = y + z, et cetera. In addition, the cost model disregards a series of latency sources such as pipeline hazards, branch delays, register spilling, contention on memory ports, and translation look-aside buffer misses.

3   Heap Construction

An array a[1..N] of N elements, each with an associated key, is a *binary maximum heap*, or briefly a *heap*, if the key of $a[\lfloor j/2 \rfloor]$ is larger than or equal to that of $a[j]$ for all $j \in \{2, 3, \ldots, N\}$. We consider the problem of transforming an arbitrary array into a heap. If possible, we would like to carry out the heap construction *in-place*, i. e., by using no more than $O(1)$ extra words of memory.

One can view a heap as a binary tree in which node a[1] is the *root*, the *children* of node a[i] are a[2i] and a[2i + 1] (if they exist), and the *parent* of node a[i] is $a[\lfloor i/2 \rfloor]$ (if this exists). The *depth* or *level* of a *node* in a heap is the length of the path from that node to the root. That is, the root is on level 0, its children on level 1 and so on. The *height* of a *node* in a heap is the length of the longest path from that node to a node with no children. The *height* of a *heap* is the maximum height (or depth) of any of its nodes. That is, the height of a heap of size N is $\lfloor \log_2 N \rfloor$, and this is also the maximum depth of any of the nodes.

For the purpose of our analysis, we assume that the heap is *B-aligned*, i. e., that there is a block boundary between the root a[1] and its first child a[2]. That is, the root is in its own block. Furthermore, we assume that a block can fit at least two elements, i. e., $B \geq 2$, and that a block only keeps whole elements. Recall also that we assumed B to a power of 2. These assumptions guarantee that the children of a node will always be situated in the same block.

In this section we recall and analyse meticulously the heap-construction methods presented by Williams [20], Floyd [3], and Fadel et al. [2]. Both the number of instructions performed and the number of cache misses incurred are analysed. The instruction counts are measured in pure C operations under the assumption that the elements are one-word integers.

3.1   STL Compatibility

We have implemented our programs in C++ in the form compatible with the make_heap routine available in the Standard Template Library (STL). This routine has two overloaded interfaces:

```
template <class random_access_iterator>
void make_heap(random_access_iterator first, random_access_iterator last);

template <class random_access_iterator, class compare>
void make_heap(random_access_iterator first, random_access_iterator last, compare less);
```

The first one uses **operator<** provided for the elements in the comparisons, whereas the second one accepts any function object with **operator()** as its third parameter.

The input sequence is passed by two iterators **first** and **last**, where **first** points to the first element of the sequence and **last** to the first element beyond the sequence. One complication caused by this arrangement is that we only know the type of the iterators, but not the type of the elements referred to by the iterators. To dereference an iterator and manipulate the resulting element, or to operate with the difference of two iterators, we utilize the **iterator_traits** facility [19, Section 19.2.2] provided by the STL.

The **make_heap** routine of the STL (as provided with the GNU **g++** 2.8.1 and Kuch & Associates **kcc** 3.2f compilers) implements Floyd's method (see Section 3.3) with the siftholedown-siftup heuristic [9, Exercise 5.2.3.18], which saves some comparisons [12], but in practice the heap construction becomes slower due to an increase in the number of element moves.

3.2   Repeated Insertion

In Williams' heap-construction method [20] one starts with a heap of size one, and then repeatedly inserts the other elements, one by one, into the heap constructed so far. In Figure 1 a C++ implementation of this method, very much in the spirit of Williams' original program, is given.

In the worst case the inner loop in the **Williams::inheap** routine visits each level of the heap exactly once. Since the height of a heap of size j is $\lfloor \log_2 j \rfloor$, the construction of an N-element heap requires $\sum_{j=2}^{N} \lfloor \log_2 j \rfloor$ key comparisons, which is at most $N \log_2 N - 1.91 \ldots N + O(\log_2 N)$ key comparisons. By simple transformations the inner loop is seen to execute 9 pure C operations, and the outer loop (after inlining **Williams::inheap**) 7 pure C operations. Hence, the pure C operation count is at most $9N \log_2 N - 9N + O(\log_2 N)$.

Hayward and McDiarmid [5] proved that, under the assumption that each initial permutation of the given N elements with distinct keys is equally likely, the number of promotions performed, i.e., assignments **a[i] = a[j]** in the inner loop, is on an average at most 1.3. By using this we get that in the average case the number of pure C operations performed is less than $24N$.

Let us now analyse the cache behaviour of the repeated-insertion heap-construction method. We say that a block is *relevant* if it contains a node of the heap that is on the path from the current node, indicated by variable j in **Williams::make_heap**, to the root. Further, we call a block *active* if it is in the cache, and *inactive* otherwise. When the elements at one level are inserted into the heap, each of the blocks holding a node from any of the previous levels will be relevant only once. Assuming that the capacity of the cache were at least

```
namespace Williams {
  #include <iterator>

  template<class node, class distance, class element, class compare>
  void inheap(node a, distance& n, element in, compare less) {
    iterator_traits<node>::difference_type i, j;
    i = n = n + 1;
  scan:
    if (i > 1) {
      j = i / 2;
      if (less(a[j], in)) {
        a[i] = a[j];
        i = j;
        goto scan;
      }
    }
    a[i] = in;
  }

  template<class node, class compare>
  void make_heap(node first, node last, compare less) {
    iterator_traits<node>::difference_type n = last - first;
    if (n < 2) return;
    node a = first - 1;
    iterator_traits<node>::difference_type j;

    j = 1;
  L:
    inheap(a, j, a[j + 1], less);
    if (j < n) goto L;
  }
}
```

Fig. 1. Williams' repeated-insertion heap-construction method in C++.

$\lfloor \log_2 N \rfloor + 1$ and that we could keep the relevant blocks active all the time, each block holding a node of the heap would be visited at most once.

The ith level of an N-element heap, $i \in \{0, 1, \ldots, \lfloor \log_2 N \rfloor\}$, contains at most 2^i elements which occupy at most $\lceil 2^i/B \rceil + 1$ blocks. Hence, the insertion of these 2^i elements into the heap can incur at most $\sum_{j=0}^{i}(\lceil 2^j/B \rceil + 1)$ cache misses. Summing this over all possible levels gives

$$\sum_{i=0}^{\lfloor \log_2 N \rfloor} \sum_{j=0}^{i}(\lceil 2^j/B \rceil + 1)$$

cache misses, which is at most $4N/B + O\left((\log_2 N)^2\right)$.

The main problem with the above analysis is that we cannot control the replacement of the cache blocks, but this is done by the cache system according to the LRU replacement strategy. In particular, it may be possible that a relevant block is removed from the cache and it has to be reread into the cache if it is needed later on. This complication is formally handled in the proof of the following theorem. The main ingredient of the proof is a result by Sleator and Tarjan [17, Theorem 6] which shows that the LRU strategy cannot be much worse than the optimal replacement strategy if the size of the cache is a bit larger than that used by the optimal algorithm.

Theorem 1. *Let B denote the size of cache blocks and N the size of the heap to be constructed. Assuming that the cache can hold* at least $r\lfloor \log_2 N \rfloor$ *blocks for some real number $r > 1$ and that the block-replacement strategy is LRU, the repeated-insertion heap-construction method incurs at most $(4r/(r-1))N/B + O\left((\log_2 N)^2\right)$ cache misses.*

Our simulations show that the upper bound given by Theorem 1 is pessimistic. On the other hand, even in the best case Williams' method will touch two levels of the heap at each insertion. Based on this observation it is not difficult to show that $(3/2)N/B - M/B$ cache misses will always be necessary, so the number of misses cannot be made arbitrary close to the optimal bound N/B if $N \gg M$.

The worst-case performance of Williams' heap-construction method is dominated by the internal computations. Hence, it is motivated to try to reduce the number of instructions performed. The code tuning of the program should, however, be accomplished carefully since on an average both the inner loop and the outer loop are executed about equally many times. Hence, by moving some code from the inner loop to the outer loop improves the worst-case performance, but the average-case performance remains the same.

In code tuning we used the following principles:

1. All common subexpressions were removed.
2. The function `Williams::inheap` was inlined.
3. Instead of array indexing we used the pointer notation provided by the C++ language. Sometimes inelegant casts were necessary to avoid the extra multiplications and divisions that were generated by a compiler when calculating a difference of two iterators.
4. Before inserting an element into the heap we compared its key to the key of the root. After this the inner loop could be divided into two similar loops that both have only one termination condition: the root reached or an element with a key larger than or equal to the key of the inserted element found.
5. To compensate the extra key comparison the root was kept in registers.
6. By making two copies of both of the two versions of the inner loop the assignment i = j could be avoided as hinted, e. g., in [9, Exercise 5.2.1.33].
7. The loops were finished conditionally as recommended by Sedgewick [16]. This made the program code complicated since now the outer loop is ended in four different places.

Because of space restrictions we cannot give the resulting program here. Note, however, that the foregoing transformations do not change the memory reference pattern of the program.

After code tuning each of the copies of the inner loop executes 5 pure C operations and the copies of the outer loop 8 or 11 operations depending on which of the branches is selected. For an average-case input the longer branch is selected almost always [5], whereas for a worst-case input (a sorted array) the shorter branch is executed at all times. From this we get that the number of pure C operations executed is bounded by $(5 \cdot 1.3 + 11)N + O(\log_2 N) =$

```
namespace Floyd {
  #include <iterator>

  template<class node, class distance, class compare>
  void siftdown(node a, distance i, distance n, compare less) {
    iterator_traits<node>::value_type copy;
    iterator_traits<node>::difference_type  j;

    copy = a[i];
  loop:
    j = 2 * i;
    if (j <= n) {
      if (j < n)
        if (less(a[j], a[j + 1]))
          j = j + 1;
      if (less(copy, a[j])) {
        a[i] = a[j];
        i = j;
        goto loop;
      }
    }
    a[i] = copy;
  }

  template <class node, class compare>
  void make_heap(node first, node last, compare less) {
    node a = first - 1;
    iterator_traits<node>::difference_type n = last - first;
    iterator_traits<node>::difference_type i;

    for (i = n / 2; i > 0; i--)
      siftdown(a, i, n, less);
  }
}
```

Fig. 2. Floyd's repeated-merging heap-construction method in C++.

$17.5N + O(\log_2 N)$ in the average case and by $5N \log_2 N + 8N + O(1)$ in the worst case.

3.3   Repeated Merging

In Floyd's heap-construction method [3] small heaps are merged to form larger heaps. The basic operation (**Floyd::siftdown**) transforms a subtree rooted by a node containing an arbitrary element and the heaps rooted by the children of that node into a heap. In Floyd's scheme the subtrees are processed in order of their heights. Trivially, all subtrees with height 0 form a heap. After this the subtrees with height 1 are processed, then the subtrees with height 2 and so on. Figure 2 gives a C++ program, similar to Floyd's original program, that implements this heap-construction method.

To make the program comparable with the tuned version of Williams' program, we tuned it by using the same principles as earlier with two substantial additions. First, the test of the special case, whether a node has only one child, was moved outside the inner loop. Second, a specialized version of siftdown was written which handles a heap of exactly three elements. Since roughly half of the calls to siftdown are handled by this specialized routine, noticeable savings are achieved. Except for these extra optimizations, we expect that most of our optimizations can be carried out automatically by modern compilers, so the main concern here is to get a realistic estimate for the instruction count of the program. Again due to space restrictions we do not give the resulting program, but note that the transformations do not change the memory reference pattern.

The specialized siftdown is called at least $\lfloor N/4 \rfloor - 1$ times and at most $\lfloor N/4 \rfloor$ times, and it executes at most 12 pure C operations. The control of the outer loop requires 2 additional pure C operations per iteration. It is known that Floyd's method constructs a heap of size N with at most $2N$ key comparisons. Of these at least $N/2 - O(1)$ are performed by the specialized siftdown and the remaining $(3/2)N + O(1)$ by the normal siftdown. Since 2 comparisons are performed in each iteration of the original inner loop, we conclude that the inner loop of the normal siftdown is repeated $(3/4)N + O(1)$ times. All branches of the three copies of the inner loop execute 9 pure C operations, and the outer loop 9 pure C operations. Hence, the total number of pure C operations executed is at most $14(N/4) + 9(3/4)N + 9(N/4) + O(1) = 12.5N + O(1)$.

In Floyd's scheme two consecutive siftdown operations process two disjoint subtrees so temporal locality is absent. The poor cache performance is formalized in the following theorem.

Theorem 2. *Let B denote the size of cache blocks, and assume that the block-replacement strategy is LRU. There exists an input of size $N \geq 2B^2$ that makes the repeated-merging heap-construction method, in the form programmed by Floyd, to incur more than $(1/2)(N \log_2 B)/B$ cache misses if the cache can hold at most $(1/4)N/B$ blocks.*

One could also show that the number of cache misses is upper bounded by $(cN \log_2 B)/B$ for a positive constant c. We leave the proof of this fact as an exercise for the reader.

3.4 Reengineered Repeated Merging

The main drawback in Floyd's method is the order in which the subtrees are considered. The subtrees could be processed in any order, provided that the subtrees of a node form a heap before the node is considered as a root of the larger subtree. Actually, temporal locality can be improved considerably by processing, e. g., the right subtree of a node, then its left subtree, and directly after this the subtree rooted by the node itself. It is not difficult to write a recursive program that implements this idea.

Due to the recursion stack the recursive program does not construct a heap in-place, but the recursion can be removed such that only a constant amount of indices are in use. A reengineered iterative version of the recursive program is given in Figure 3. The iterative version handles the nodes at the level next to the bottommost level in a special manner: if a node has at least one child, the two or three element heap rooted by it is constructed; after this the nested loop follows the path towards the root, calling siftdown for the parent node, as long as a node is found which is not a left child.

```
namespace Reengineered_Floyd {
  #include <iterator>

  template <class node, class compare>
  void make_heap(node first, node last, compare less) {
    node a = first - 1;
    iterator_traits<node>::difference_type n = last - first;
    iterator_traits<node>::difference_type i, j, k, z;

    for (i = 1; i <= n; i *= 2);

    i = i / 2 - 1, j = i / 2, k = n / 2;
    for (; i > j; i--) {
      if (i <= k)
        Floyd::siftdown(a, i, n, less);
      for (z = i; (z & 1) != 0; z /= 2)
        Floyd::siftdown(a, z / 2, n, less);
    }
  }
}
```

Fig. 3. Reengineered version of the repeated-merging heap-construction method in C++.

In our actual implementation the optimizations of siftdown described in Section 3.3 were included. Furthermore, two versions of the outer loop were written to avoid the **if**-test in it. After these optimizations the cost of the outer loop is reduced to 3 pure C operations per call to siftdown. As earlier the pure C cost of the specialized siftdown is 12 and it is executed at most $N/4$ times. The pure C cost of the inner loop of the normal siftdown is 9 and it is executed $(3/4)N + O(1)$ times. The 7 pure C operations in the normal siftdown outside its inner loop are executed at most $N/4$ times. Finally, the computation of the largest power of 2 smaller than or equal to N requires $O(\log_2 N)$ pure C operations. To sum up, the total cost is $13N + O(\log_2 N)$.

In the next theorem we prove that the cache behaviour of the reengineered program is superior to Floyd's original proposal.

Theorem 3. *Let B denote the size of cache blocks and N the size of the heap to be constructed. Assuming that the cache can hold M elements and $M \geq 2B^2$, and that the block-replacement strategy is LRU, the reengineered repeated-merging heap-construction method incurs never more than $N/B + 8(\log_2 M - \log_2 B)N/M$ cache misses, which for $M \gg 2B^2$ is close to N/B.*

3.5 Layerwise Construction

In the layerwise heap-construction method proposed, e. g., by Fadel et al. [2] the elements to be moved to the nodes with height 0, with height 1, et cetera are determined repeatedly. In each phase one half of the remaining elements are excluded from consideration. The process involves the selection of the median element and the (3-way) partitioning around the median element. Figure 4 implements this heap-construction method by using the STL routine **nth_element**, which accomplishes both the selection and the partitioning.

```
namespace Layerwise {
  #include <algorithm>

  template <class node, class compare>
  void make_heap(node first, node last, compare less) {
    while (last - first > 1) {
      node middle = first + (last - first) / 2;
      nth_element(first, middle, last, less);
      last = middle;
    }
  }
}
```

Fig. 4. Layerwise heap-construction method in C++.

We wrote several programs, both deterministic and randomized, for selection and partitioning, but none of our programs were able to beat the library routine in speed. Therefore, we only used our own programs for analytical purposes. We would like to point out that most selection routines are recursive so they need a recursion stack for their operation, i.e., they do not run in-place, but of course a recursion stack with a logarithmic size is acceptable in practice. Moreover, both the selection and partitioning routines should be able to handle multiset data, i.e., the input can contain elements with equal keys. Therefore, not all published methods can be used for our purposes.

We have not made any attempts to analyse the instruction count of the layerwise heap-construction program as done for the other heap-construction programs. It is well-known that the instruction count is high for the deterministic selection routines, even though it is linear in the worst case. The instruction count is moderate for the randomized selection method proposed by Hoare [7], it can handle multisets, and it performs partitioning simultaneously with the selection. The **nth_element** routine of the STL (as provided with the GNU **g++** 2.8.1 and Kuch & Associates **kcc** 3.2f compilers) is an implementation of this randomized method, but it switches to the deterministic method in a clever way if the depth of the recursion gets too large.

Fadel et al. [2] observed the goodness of the layerwise heap-construction method on a paged memory environment. Its cache behaviour is also good, at least in theory, as shown in the next theorem.

Theorem 4. *Let B denote the size of cache blocks and N the size of the heap to be constructed. Assuming that the block-replacement strategy is LRU, the layerwise heap-construction method incurs never more than cN/B cache misses for a positive constant c, even if the cache can only hold a constant number of blocks.*

4   Test Results

The experimental part of this work involved four activities. First, a simple trace-driven cache simulator was build. Second, above this simulator a visualization tool was build which could show the memory reference patterns graphically. Both of these tools are described in detail in the M. Sc. thesis of the third author [18]. Many of the findings in the theoretical part evolved first after we got a deep understanding of the programs with these tools. Third, a micro benchmark, similar to that described in [14], was developed and this was used to extract information about the memory latencies of our computers. Fourth, the actual tests with the programs described in Section 3 were carried out. In this section we give the test results in the case the input is a random permutation of $\{0, 1, \ldots, N - 1\}$ and N ranges from 2^4 to 2^{23}. That is, the elements were 4-byte integers.

Based on micro benchmarking we selected two computers for our tests: IBM RS/6000 397 and IBM 332 MHz F50. In brief, these computers could be characterized as follows: the former has a slow processing unit and fast memory, whereas the latter has a fast processing unit and slow memory. Further characteristics of these computers are given in Table 1.

Table 1. Characteristics of the computers on which the tests were carried out.

Computer	Clock rate	On-chip cache	Off-chip cache
IBM RS/6000 397	160 MHz	128 kbytes, 128 bytes per cache line, 4-way set associative	none
IBM F50	332 MHz	32 kbytes, 32 bytes per cache line, 4-way set associative	256 kbytes, 64 bytes per cache line, 8-way set associative

The C++ programs developed were compiled by the Kuch & Associates kcc compiler (version 3.2f), and the options used were +K3 -O3 -qtune=pwr2 -qarch=pwr2 on IBM RS/6000 and +K3 -O3 -qtune=604 -qarch=ppc on IBM F50. In general, the object code generated by this compiler was about 30 percent faster than the corresponding code produced by the GNU g++ compiler (version 2.8.1).

All the programs described in Section 3 as well as the heap-construction program from the STL (as provided with the GNU g++ 2.8.1 and Kuch & Associates kcc 3.2f compilers) were included in our tests. Their execution times on the two test computers are given in Figures 5 and 6, respectively. For each value of N, the experiment was repeated $2^{24}/N$ times, each time with a new array. Furthermore, the execution trials were carried out on dedicated hardware, so no cache interference from our sources was expected.

The results on IBM RS/6000 indicate clearly the effectiveness of our code tuning. The analytical pure C costs reflect well the speed-ups obtained. The results on IBM F50 show the effectiveness of our memory tuning. In particular, the reengineered version of Floyd's method performs well independently of the problem size. In the runtime curves of the other programs, especially in that of Floyd's original program, two hops can be recognized. These corresponds to the points where the size of on-chip cache and off-chip cache is exceeded. This kind of slowdown, when the size of a problem increases, is a characteristic of programs that are not designed for a computer with a hierarchical memory.

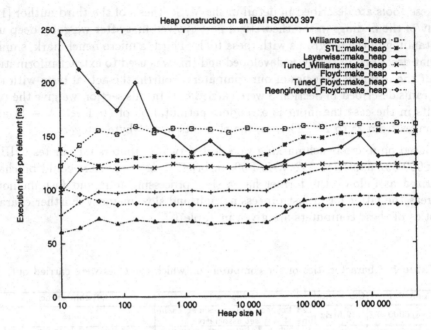

Fig. 5. Execution times of various heap-construction programs on an IBM RS/6000.

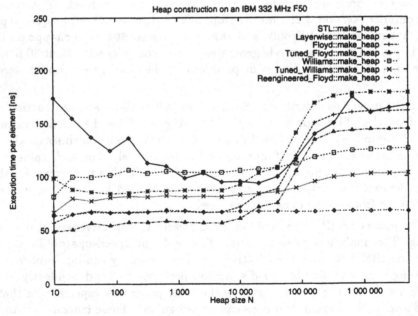

Fig. 6. Execution times of various heap-construction programs on an IBM F50.

References

1. Carlsson, S.: A note on HEAPSORT. *The Computer Journal* **35** (1992), 410–411
2. Fadel, R., Jakobsen, K. V., Katajainen, J., Teuhola, J., Heaps and heapsort on secondary storage, *Theoretical Computer Science* **220** (1999), 345–362
3. Floyd, R. W.: Algorithm 245: TREESORT 3. *Communications of the ACM* **7** (1964), 701
4. Hagerup, T.: Sorting and searching on the word RAM, *Proceedings of the 15th Annual Symposium on Theoretical Aspects of Computer Science*, Lecture Notes in Computer Science **1373**, Springer-Verlag, Berlin/Heidelberg (1998), 366–398
5. Hayward, R., McDiarmid, C.: Average case analysis of heap building by repeated insertion. *Journal of Algorithms* **12** (1991), 126–153
6. Hennessy, J. L., Patterson, D. A.: *Computer Architecture: A Ouantitative Approach*, 2nd Edition. Morgan Kaufmann Publishers, Inc., San Francisco (1996)
7. Hoare, C. A. R.: Algorithm 65: FIND. *Communications of the ACM* **4** (1961), 321–322
8. Katajainen, J., Träff, J. L.: A meticulous analysis of mergesort programs. *Proceedings of the 3rd Italian Conference on Algorithms and Complexity*, Lecture Notes in Computer Science **1203**, Springer-Verlag, Berlin/Heidelberg (1997), 217–228
9. Knuth, D. E.: *The Art of Computer Programming*, Volume 3: *Sorting and Searching*, 2nd Edition. Addison Wesley Longman, Reading (1998)
10. LaMarca, A., Ladner, R. E.: The influence of caches on the performance of heaps. *The ACM Journal of Experimental Algorithmics* **1** (1996), Article 4
11. LaMarca, A., Ladner, R. E.: The influence of caches on the performance of sorting. *Journal of Algorithms* **31** (1999), 66–104
12. McDiarmid, C. J. H., Reed, B. A.: Building heaps fast. *Journal of Algorithms* **10** (1989), 352–365
13. Moret, B. M. E., Shapiro, H. D.: *Algorithms from P to NP*, Volume 1: *Design & Efficiency*. The Benjamin/Cummings Publishing Company, Inc., Redwood City (1991)
14. Saavedra, R. H., Smith, A. J.: Measuring cache and TLB performance and their effect on benchmark runtimes. *IEEE Transactions on Computers* **44** (1995), 1223–1235
15. Sanders, P.: Fast priority queues for cached memory. *Proceedings of the 1st Workshop on Algorithm Engineering and Experimentation*, Lecture Notes in Computer Science **1619**, Springer-Verlag, Berlin/Heidelberg (1999)
16. Sedgewick, R.: Implementing Quicksort programs. *Communications of the ACM* **21** (1978), 847–857. Corrigendum, *ibidem* **22** (1979), 368
17. Sleator, D. D., Tarjan, R. E.: Amortized efficiency of list update and paging rules. *Communications of the ACM* **28** (1985), 202–208
18. Spork, M.: Design and analysis of cache-conscious programs, M. Sc. Thesis, Department of Computing, University of Copenhagen, Copenhagen (1999). Available from http://www.diku.dk/research-groups/performance_engineering/publications/spork99.ps.gz
19. Stroustrup, B.: *The C++ Programming Language*, 3rd Edition. Addison-Wesley, Reading (1997)
20. Williams, J. W. J.: Algorithm 232: HEAPSORT. *Communications of the ACM* **7** (1964), 347–348

A Fast and Simple Local Search for Graph Coloring

Massimiliano Caramia[1] and Paolo Dell'Olmo[2]

[1] Dipartimento di Informatica, Sistemi e Produzione
University of Rome "Tor Vergata"
Via di Tor Vergata, 110 - 00133 Rome, Italy
[2] Dipartimento di Statistica, Probabilitá e Statistiche Applicate
University of Rome "La Sapienza"
Piazzale Aldo Moro, 5 - 00133 Rome, Italy
caramia@disp.uniroma2.it, dellolmo@iasi.rm.cnr.it

Abstract. In this paper a fast and simple local search algorithm for graph coloring is presented. The algorithm is easy to implement and requires the storage of only one solution. Experimental results on benchmark instances of the DIMACS Challenge and on random graphs are given.

1   Introduction

The aim of the coloring problem is to assign a color to each node of a graph G such that no monochromatic edge is allowed and the number of colors used is minimum. This value is called *chromatic number* of G and is denoted as $\chi(G)$.

Studies on the coloring problem have been motivated by several applications, such as scheduling [1], timetabling [4], resource allocation, memory management and many others. Although these applications suggest that effective algorithms for solving graph coloring would be of great interest, a limited number of exact algorithms are presented in literature (see, for example: [3,21,24,28,29]). This can be due to the problem intractability confirmed by both theoretical and experimental results. In fact, the coloring problem, in the general case, is $\mathcal{NP}$-complete [16] and it remains $\mathcal{NP}$-complete also in many more particular cases such as the 3-colorability of a four regular planar graph [8], and approximating the chromatic number of a graph to within any constant factor (or even to within n^ϵ for a sufficiently small ϵ) [23].

Experimental results show that exact codes are able to solve random instances up to 70 nodes for all the densities within a time limit of one hour, while only for low densities larger instances are solved at the optimum. These values are very small if compared with the sizes of many real world applications.

In such a scenario, the importance of designing good heuristics is twofold: first obtaining good upper bounds for branch and bound schemes, and second finding colorings of large graphs arising from applications where good solutions are required quickly.

Finding a good initial upper bound in a branch and bound algorithm is one

J.S. Vitter and C.D. Zaroliagis (Eds.): WAE'99, LNCS 1668, pp. 316–329, 1999.
© Springer-Verlag Berlin Heidelberg 1999

of the most important target for all combinatorial optimization problems, and in particular for coloring, considering that the decreasing of the initial upper bound from $\chi(G) + 1$ to $\chi(G)$ results in a 40% reduction in the total number of subproblems [29].

Coloring large graphs is an important issue for heuristics as exact codes are not able to solve instances with a large number of nodes. Both classical heuristics and meta heuristics [27], i.e. single shot and multi start, have been implemented in the attempt of satisfying these requirements. These latter approaches are more recent and give good results if compared with those obtained by classical schemes: tabu search techniques [17], genetic operators [5, 11, 12] and simulated annealing [18] represent examples of these new techniques.

Due to the complexity of the problem, the computational time of these algorithms can be very high especially for large graphs (see for instance 41.35 hours to color the random graph 1000.05 [12]). But in some applications coloring algorithms are used as subroutines of more complex control and scheduling procedures where a set of limited resources has to be allocated among a set of tasks. Typically, in such problems a coloring solution corresponds to a feasible schedule where edges represent constraints among tasks (nodes) (e.g. see [2, 6, 20]). Notwithstanding the complexity of the coloring problem, in such applications a solution is required in a limited time, especially in real time control systems, where the coloring algorithm should not be the bottleneck of the whole process. Note that this was not the main objective of more sophisticated algorithms like [5, 11, 12, 17, 18, 25] where the aim of the procedures is to obtain the best solution even if it requires a significant computational effort.

The goal of this paper is to present a fast and simple algorithm for graph coloring which could be effectively employed as subroutine of complex real time systems. In particular we propose a re-engineering of TABUCOL, which represents already a good trade-off between time and solution quality, as verified also by Sewell [29].

Our approach strengthens some ideas of tabu search in order to improve its performance. In the computational analysis it appears that graph coloring is quite a destructured problem and this behavior seems to be as more observable as instances become larger. As a matter of fact, algorithms that make choices aiming at reducing the possibility of bad assignments, depend strongly from the graph structure, but in doing so they can be trapped, requiring a lot of time in the attempt to improve the solution found.

Instead, our algorithm does not consider the structure of the graph, but performs simple priority-based operations with the objective of finding and escaping quickly from local optima.

In doing so, our algorithm avoids to be trapped, and at the same time performs the exploration of disjoint neighboring solutions very quickly. We verified the experimental behavior of this algorithm both on random graphs and on DIMACS Challenge benchmark instances and as can be seen from the computational analysis, solutions have a better quality and are obtained always faster than TABUCOL.

The remainder of the paper is the following. Section 2 presents our algorithm and its main differences from previous local search algorithms. In Section 3 its performance on random graphs and benchmarks is reported and comparisons with TABUCOL are also provided. Finally, in Section 4, we list some concluding remarks.

2   Our Algorithm

Tabu search techniques are widely used in many combinatorial optimization problems. This is motivated by the good experimental performance of these algorithms that move step by step towards the optimum value of a function using a tabu list to avoid cycling and being trapped in local optima.

In general it can not be defined a prefixed tabu search scheme that fits well for all combinatorial optimization problems. In fact different way of movement through the admissible solutions are required. Moreover the achievement of an initial solution from which it starts the local search is difficult to manage in order to obtain the best performance.

Such techniques have been applied to the coloring problem in [17], giving good performance if compared with the difficulty of the problem. However, in some cases these algorithms make a large number of moves that increase significantly the overall computational time.

What we propose is a new local search algorithm, inspired from tabu search techniques, from which it maintains the possibility to move to a solution with an higher number of colors, and is improved by the absence of a tabu list substituted by an assignment of priorities to the nodes of the graph that avoids repetitions in further moves of the algorithm.

The proposed algorithm, denoted HCD, is constituted by four procedures. The *Initialization* procedure assigns to each node i a color c_i equal to its number, and a priority p_i equal to its color, i.e. $p_i = c_i$. The initial upper bound on the number of required colors (UB) is set equal to the highest color used by this initial assignment, i.e. $|V|$.

Successively the procedure *Pull − colors* starts and the node having the highest priority is selected. To this node, HCD assigns the lowest color c compatible with those assigned to its neighbors, updating its priority $p_i = c$. The possible failure of this attempt, i.e. $c_i = c$, implies the exam of the successive node having the highest priority in order to repeat the previous process. This step can be iterated at most $|V|$ times if for none of the nodes in V, the color initially assigned can be modified.

In this case the procedure *Push − colors* is called and to each node the color already assigned is changed with the highest possible color that does not exceed the value UB. If the coloring is not changed by *Push−colors*, then the procedure *Pop − colors* tries to escape from local optima assigning to the independent set formed by the nodes having associated the color one, the color UB+1. Along with this new assignment, for each node $i \in V$ is assigned a priority $p_i = 1/c_i$, and the procedure *Pull − colors* is restarted.

The priorities have been changed to avoid that, using the old priority assignment p_i=UB+1, the procedure $Pull - colors$ would have chosen one by one, as node having the highest priority, those having assigned the color UB+1, returning back to the previous solution.

The motivation of the choice of the independent set having assigned the color one is that the corresponding nodes have been updated less recently than the others, and updating them would result in a perturbation that could decrease the actual solution.

The algorithm HCD stops either when the procedure $Push - colors$ has been called more than l times, where l is a prefixed value, or when, defined a target color bound, the algorithm achieves an admissible solution with at most that number of colors.

Note that the solution found by HCD is dependent from the initial numbering of the nodes, thus to obtain a better performance, HCD could be executed after different node numberings to let the starting solution be more effective on HCD.

Algorithm HCD

Initialize

1. for each $i \in V$
    1.1. assign $c_i = i$;
    1.2. assign $p_i = c_i$;
2. set the initial upper bound UB;
3. access-time-push=0;
4. assign k=0;
5. assign $V' = V$;

Pull − colors

1. choose the node $i \in V'$ having the highest priority;
2. assign to the node i chosen, the lowest possible color c compatible with those already assigned to its neighbors;
3. assign $p_i = c$;
4. if $c = c_i$ then
    4.1. $V' = V' \setminus \{i\}$;
    4.2. increase k;
    4.3. if $k = |V|$ then
        4.3.1. if access-time-push is less than l then goto $Push - colors$ else stop;
        else goto 1;
   else
    4.4. assign k=0;
    4.5. update UB;
    4.6. goto 1;

Push − colors

1. increase access-time-push;
2. for each $i \in V$ do
    2.1. if it can be assigned a color c greater than c_i such that c is less or equal
    to UB then
        2.1.1. assign $c_i = c$;
        2.1.2. assign $p_i = c_i$;
    2.2. if the coloring has not changed then goto *Pop − colors*
    else
        2.2.1. assign $V' = V$;
        2.2.2. goto *Pull − colors*;

Pop − colors

1. assign to the nodes $i \in V$ having assigned the color one, the color $c_i = UB+1$;
2. for each $i \in V$ assign $p_i = 1/c_i$;
3. assign $V' = V$;
4. goto *Pull − colors*;

Complexity analysis

The procedure *Initialize* runs in time linear with the number of nodes.

The procedure *Pull − colors* executes for each node a neighborhood search to assign the lowest possible color compatible with those adjacent. Its complexity is then $O(|V|^2)$.

The procedure *Push − colors* has the heaviest step in 2, which implies to check for each node the colors already assigned to its neighbors; thus its whole complexity is $O(|V|^2)$.

The procedure *Pop − colors* runs in time linear with the number of vertices $|V|$.

Hence, the overall worst case complexity of the algorithm is $O(|V|^2)$.

3   Experimental Analysis

Computational results are reported in Tables 1–10. The computational analysis is mainly devoted to give theoretical insights and experimental evidence of the better performance of HCD with respect to TABUCOL. For this purpose we need to recall how TABUCOL works: it is based on partitioning the vertices of a graph into color classes attending to move vertices from one class to another. Each iteration of consists of generating a sample of neighbors; that is, different partitions in stable sets that can be obtained from the current one by moving one vertex to a different class. It then selects the neighbor partition that has the fewest conflicts (number of coloring violations), even if the neighbor has more conflicts than the current partition. The set of neighbors is restricted by a TABU list that prevents a vertex from moving back into a class that was recently a member of in a previous iteration. This helps the program struggle out of local optima.

Nevertheless, as in other combinatorial problems, TABU search has an hard parameter tuning. There are values to be fine tuned: the maximum iterations before quitting after the last improvement detected, the number (maximum and minimum) of neighbors to generate before switching to a new partition, and the tabu list size. Moreover, from the experimental analysis, TABUCOL needs a good starting solution, otherwise its performance could be poor.

In order to compare the iterations of TABUCOL with those of HCD, we define one iteration of our algorithm an attempt of changing color class to a node.

The experiments can be divided into two different classes: the former (Tables 3-7) contains the solution values and the CPU time for random and benchmarks graphs; the latter (Tables 8-10), report execution profiling of the two algorithms on hard benchmark graphs. Parameters like the number of updates, the number of routine calls and the cache miss rate are provided. As it can be observed we have divided the experimental analysis in four subsections: random graphs, benchmarks, CPU time and profiling.

Computer and implementation environments

In order to allow comparisons of algorithms which have been implemented on different machines, it is necessary to know the approximate speeds of the computers involved.

Machine: Workstation Digital Alpha Model 1000A 5/500
RAM: 256 Mb
Cache memory: 8 Mb
Computer language: 'C'
Compiler: standard Digital
Operating system: Unix 40B
TABUCOL code: available at [7]
HCD code: available from the authors under request

Table 1. Experiment environment characteristics

Graph	r100.5	r200.5	r300.5	r400.5	r500.5
CPU	0.00	0.07	0.64	4.00	15.46

User time in seconds

Table 2. CPU time obtained for the DIMACS Machine benchmarks

For this purpose we report in Table 2 the values obtained for the DIMACS Machine benchmarks by our machine (see Table 1). These benchmarks are available at the DIMACS ftp site [10], and consists of a benchmark program (DF-MAX) for solving the maximum clique problem and five benchmark graphs (r100.5, r200.5, r300.5, r400.5, r500.5).

Analysis on random graphs

Tables 3 and 4 present results on *random graphs*: we generated graphs with 50, 70, 80 and 90 nodes with an edge probability of 0.1, 0.3, 0.5, 0.7 and 0.9, and graphs with 300, 400, 500, 600 and 700 nodes with density equal to 0.1 and 0.3. For all the graphs we give average values over 10 instances. The values reported in the tables are the characteristics of the graphs (type, size (nodes, edges)) and solution values of HCD and TABUCOL.

It can be noticed that HCD, in the average, has always equal or better solutions with respect to TABUCOL. This is enforced by Table 4 where this gap between the proposed algorithm and TABUCOL increases as the number of nodes becomes higher, that is a good result for a real scenario.

Analysis on benchmarks

Tables 5, 6 and 7 report values for the benchmark graph coloring instances available at the site [9]. In particular the first refers to miscellaneous benchmark graphs: Book Graphs [19], Game Graphs [19], Miles Graphs [19], Queen Graphs [14], and Mycielski Graphs [26]. The others refer to the DIMACS Challenge benchmarks and are so divided: in Table 6 are Leighton graphs [22], in Table 7 Register allocation graphs. As for Table 3 the values reported in the tables are the characteristics of the graphs (type, size (nodes, edges)) and the solution values of HCD compared with those of TABUCOL. We have reported also the maximum clique $\omega(G)$ and the chromatic number of these graphs.

What appears is that HCD outperforms TABUCOL in all the benchmark instances tested. In particular the reported values show that in many instances both HCD and TABUCOL find the same heuristic value equal to the maximum clique. But, in some instances such as Queen5.5, Queen8.12 and Le450-25c, HCD is extremely efficient finding solutions equal to $\omega(G)$ where TABUCOL was not able. This has a great effect if HCD is used as upper bound in an exact branch and bound algorithm, because the solution would be found without branching.

Analysis on CPU time

In Tables 3 and 5 we computed the percentage of reduction of CPU time, denoted with Rct, obtained by HCD with respect to TABUCOL to achieve the best solution of this latter. The CPU time are always faster and the reduction of CPU time goes from a minimum of 10% to a maximum of 72% on random graphs, and from a minimum of 10% to a maximum of 70% on benchmarks. For all the other tables we give the CPU time of HCD.

In order give more insights on the performance of HCD, we tested our code using a simulator, and in the next subsection, Profiling analysis, we describe the experiments done.

Type	Nodes	Edges	HCD	TABUCOL	Rct %
50.01	50	110.2	3.8	3.8	10
50.03	50	367.2	6.4	6.4	28
50.05	50	608.8	9.6	9.6	44
50.07	50	841.0	13.9	14.0	54
50.09	50	1113.4	22.4	22.4	58
70.01	70	239.0	4.0	4.0	12
70.03	70	753.8	8.0	8.0	31
70.05	70	1218.4	11.8	11.9	38
70.07	70	1656.2	16.9	16.9	40
70.09	70	2177.0	28.1	28.1	59
80.01	80	310.2	4.4	4.4	35
80.03	80	961.0	8.2	8.4	50
80.05	80	1599.2	12.3	12.4	57
80.07	80	2190.4	19.0	19.2	62
80.09	80	2836.2	30.9	30.9	64
90.01	90	400.6	4.8	4.8	48
90.03	90	1285.2	8.9	9.0	52
90.05	90	2014.8	15.2	15.2	65
90.07	90	2788.0	21.0	21.4	65
90.09	90	3826.6	32.0	32.8	72

All statistics are averages of 10 test problems.

Table 3. Results on Random Graphs

Type	Nodes	Edges	HCD	TABUCOL	CPU time HCD
300.01	300	4535.2	17.2	20.8	5.2
300.03	300	12984.8	21.4	26.8	10.4
400.01	400	6034.2	18.2	21.6	7.4
400.03	400	18154.4	21.8	28.2	14.4
500.01	500	12798.2	18.4	22.8	10.5
500.03	500	35556.2	22.0	28.4	17.2
600.01	600	17956.6	19.0	24.6	15.7
600.03	600	50977.2	22.4	28.6	22.4
700.01	700	22478.6	19.8	26.0	20.8
700.03	700	67123.4	22.8	29.8	32.6

All statistics are averages of 10 test problems.

Table 4. Results on large Random Graphs

Type	Nodes	Edges	HCD	TABUCOL	$\chi(G)$	$\omega(G)$	Rct %
Myciel2	5	5	3	3	3	2	20
Myciel3	11	20	4	4	4	2	32
Myciel4	23	71	5	5	5	2	34
Myciel5	47	236	6	6	6	2	45
Myciel6	95	755	7	7	7	2	56
Myciel7	191	2360	8	8	8	2	60
Huck	74	301	11	11	11	11	34
Jean	80	254	10	10	10	10	45
Homer	561	1629	13	13	13	13	51
David	87	406	11	11	11	11	48
Anna	138	493	11	11	11	11	54
Queen5.5	25	160	5	7	5	5	70
Queen6.6	36	290	8	9	7	6	56
Queen7.7	49	476	9	11	7	7	57
Queen8.8	64	728	11	12	9	8	34
Queen9.9	81	1056	12	13	10	9	35
Queen8.12	96	1368	12	14	12	12	23
Games120	120	638	9	9	9	9	10
Miles250	128	387	8	8	8	8	22
Miles500	128	1170	20	20	20	20	52
Miles750	128	2113	32	32	32	32	55
Miles1000	128	3216	42	42	42	42	59
Miles1500	128	5198	73	73	73	73	62

Table 5. Results on Miscellaneous Graphs

Profiling analysis

In Tables 8 and 9 we report results related to the profiling of HCD when executed on Leighton and School graphs. The values have been obtained by using Digital Alpha Atom V2.17 simulator, which allows to obtain an instrumented version of the application program. This instrumented version is obtained by means of a tool, called *iprof*, which allows to perform a profiling returning the following reported values: the number of times each procedure is called as well as the number (dynamic count) of instruction executed by each procedure. We included these detailed analysis just for two graph classes, selected as the most representative for studying the algorithm performance.

Table 10 reports cache miss rates: these values are obtained by using the tool *cache* supported by Atom, which simulates execution of the algorithm in 8Kb pages and with a fully associative translation buffer. Moreover, two parameters, independent from the architecture of the computer, are shown: the number of visits and the number of updates. The visit count increases each time a node is used in a generic operation; while an update occurs each time a node changes

Type	Nodes	Edges	HCD	TABUCOL	$\chi(G)$	$\omega(G)$	CPU time HCD
Le 450-5a	450	8168	6	6	5	5	15.4
Le 450-5b	450	8169	6	8	5	5	17.2
Le 450-5c	450	16680	5	5	5	5	4.5
Le 450-5d	450	16750	5	5	5	5	4.7
Le 450-15a	450	8260	15	15	15	15	140.2
Le 450-15b	450	8263	15	15	15	15	120.0
Le 450-15c	450	17343	17	22	15	15	25.2
Le 450-15d	450	17425	17	22	15	15	25.8
Le 450-25a	450	5714	25	25	25	25	3.8
Le 450-25b	450	5734	25	25	25	25	4.2
Le 450-25c	450	9803	25	27	25	25	148.2
Le 450-25d	450	9757	27	28	25	25	152.7

Table 6. Results on Leighton Graphs

Type	Nodes	Edges	HCD	TABUCOL	$\chi(G)$	$\omega(G)$	CPU time HCD
Mulsol.i.1	197	3925	49	49	49	49	3.8
Mulsol.i.2	188	3885	31	31	31	31	3.2
Mulsol.i.3	184	3916	31	31	31	31	3.4
Mulsol.i.4	185	3946	31	31	31	31	3.5
Mulsol.i.5	186	3973	31	31	31	31	3.7
Zeroin.i.1	211	4100	49	49	49	49	3.5
Zeroin.i.2	211	3541	49	49	30	30	3.7
Zeroin.i.3	206	3540	30	30	30	30	3.6
School1	385	19095	14	14	14	14	4.8
School1.nsh	352	14612	17	21	14	14	4.4

Table 7. Results on Register and School Graphs

its current color class.

In the following we explain the notation used in Tables 8-10.

Init: Initialization procedure

Update_UB: Procedure updating the current upper bound value

Max_Prior: Procedure choosing the node having the highest priority

Lowest: Procedure assigning the lowest admissible color to a node

Highest: Procedure assigning the highest admissible color to a node smaller than UB

Pull: Procedure Pull_colors

Push: Procedure Push_colors

Pop: Procedure Pop_colors

Instr.: Total number of instructions executed

Calls: Number of times a procedure is called

%: Percentage of instructions executed by a procedure

Visits: Number of visits

Updates: Number of updates

Cache miss rate: Percentage of cache miss

Note that Lowest and Max_Prior are subroutines of Pull, while Highest is a subroutine of Push. The experiments reported are performed by fixing the number of iterations equal to 100000.

	Init		Update_UB		Max_Prior		Lowest	
Type	Calls	%	Calls	%	Calls	%	Calls	%
Le 450-5a	1	0.002	17228	11.27	9995	12.70	9995	35.71
Le 450-5b	1	0.002	17238	11.15	10005	12.62	10005	35.97
Le 450-5c	1	0.001	16320	10.13	9991	12.22	9991	38.09
Le 450-5d	1	0.002	16316	10.52	9987	12.87	9987	38.30
Le 450-15a	1	0.001	16769	7.65	9988	8.92	9988	36.95
Le 450-15b	1	0.001	16316	7.50	9987	8.92	9987	37.77
Le 450-15c	1	0.001	17220	5.28	9987	5.83	9987	42.39
Le 450-15d	1	0.001	17217	5.24	9984	5.81	9984	42.01
Le 450-25a	1	0.001	16451	6.19	10122	7.52	10122	34.52
Le 450-25b	1	0.001	16481	6.02	10152	7.31	10152	36.34
Le 450-25c	1	0.001	16772	4.48	9991	5.17	9991	41.11
Le 450-25d	1	0.001	16768	4.50	9987	5.16	9987	42.42
School1	1	0.001	16568	4.19	9988	4.91	9988	44.61
School1.nsh	1	0.001	17061	4.43	9980	4.99	9980	42.11

Table 8. Algorithm profiling on Leighton and School graphs

	Higest		Pull		Push		Pop		Instr.
Type	Calls	%	Calls	%	Calls	%	Calls	%	
Le 450-5a	7200	39.48	9995	0.100	16	0.037	16	0.024	552725190
Le 450-5b	7200	39.43	10005	0.100	16	0.036	16	0.024	559028996
Le 450-5c	6300	38.74	9991	0.097	14	0.033	14	0.021	582664140
Le 450-5d	6300	37.45	9987	0.100	14	0.035	14	0.022	561026957
Le 450-15a	6750	45.86	9988	0.070	15	0.034	15	0.022	561835698
Le 450-15b	6300	45.14	9987	0.070	14	0.025	14	0.016	787343693
Le 450-15c	7200	46.14	9987	0.047	16	0.017	16	0.011	1180136265
Le 450-15d	7200	46.50	9984	0.047	16	0.017	16	0.011	1187823734
Le 450-25a	6300	51.29	10122	0.059	14	0.021	14	0.012	962260851
Le 450-25b	6300	49.83	10152	0.058	14	0.020	14	0.012	990398113
Le 450-25c	6750	48.84	9991	0.041	15	0.015	15	0.009	1354228722
Le 450-25d	6750	47.52	9987	0.041	15	0.015	15	0.009	1347427108
School1	6545	45.98	9988	0.046	17	0.016	17	0.010	1223129778
School1.nsh	7040	48.10	9980	0.051	20	0.019	20	0.011	1092484899

Table 9. Algorithm profiling on Leighton and School graphs

	HCD	HCD	HCD	TABUCOL	TABUCOL
Type	Visits	Updates	Cache miss rate	Updates	Cache miss rate
Le 450-5a	19850397	42718	0.28	30124	0.35
Le 450-5b	18309136	46334	0.27	31298	0.35
Le 450-5c	28393804	49608	0.27	32455	0.35
Le 450-5d	31704677	49658	0.29	32566	0.36
Le 450-15a	49017250	57542	0.34	37989	0.40
Le 450-15b	49159682	56031	0.34	36524	0.40
Le 450-15c	152928220	50195	0.38	34202	0.42
Le 450-15d	121478415	50248	0.38	35004	0.41
Le 450-25a	379156898	62198	0.32	38487	0.42
Le 450-25b	367214287	65278	0.32	38228	0.42
Le 450-25c	402798244	68129	0.34	40124	0.44
Le 450-25d	427981389	68924	0.34	40289	0.44
School1	169077769	43137	0.29	39822	0.40
School1.nsh	162616670	47526	0.26	40892	0.38

Table 10. Visits, Updates and Cache miss rate on Leighton and School graphs

4   Conclusions

Tabu search techniques are widely used in all the field of combinatorial optimization problems. This can be motivated by the high experimental performance of this algorithms that move step by step towards the minimum value of a function using a tabu list to avoid cycling and being trapped in local optima. However, in some cases this algorithms make a huge number of moves that increase significantly the overall computational time and memory requirement. In this paper a new local search algorithm for graph coloring has been presented. The algorithm is easy to implement and requires the storage of only one solution. Experimental comparisons on benchmark instances and random graphs with TABUCOL show that it produces always equal or better solutions.

Further researches will be devoted to employ this routine in real time systems and in tuning the proposed method to be efficient if compared with more complex coloring techniques.

References

1. Bianco, L., Caramia, M., and Dell'Olmo, P.: *Solving a Preemptive Scheduling Problem using Coloring Technique*, in Project Scheduling. Recent Models, Algorithms and Applications, J. Weglarz (Ed.) Kluwer Academic Publishers (1998).
2. Blazewicz, J., Ecker, K. H., Pesch, E., Schmidt, G., and Weglarz, J.: *Scheduling in Computer and Manufacturing Processes*, Springer (1996).
3. Brelaz, D., "New methods to color the vertices of a graph": *Communications of the ACM* **22** (1979), 251–256.
4. Cangalovic, M., and Schreuder, J.: "Exact coloring algorithm for weighted graphs applied to timetabling problems with lectures of different lengths", *EJOR* **51** (1991) 248–258.
5. Chams, M., Hertz, A., and de Werra, D.: "Some experiments with simulated annealing for coloring graphs", *EJOR* **32** (1987), 260–266.
6. Coffman, E. G.: "An Introduction to Combinatorial Models of Dynamic Storage Allocation", *SIAM Review* **25** (1983), 311–325.
7. Culberson's Coloring WEB site:
   http://web.cs.ualberta.ca/joe/Coloring/index.html.
8. Dailey, D. P.: "Uniqueness of colorability and colorability of planar 4-regular graphs are NP-complete", *Discrete Mathematics* **30** (1980), 289–293.
9. DIMACS ftp site for benchmark graphs:
   ftp://dimacs.rutgers.edu/pub/challenge/graph/benchmarks/color/.
10. DIMACS ftp site for benchmark machines:
   ftp://dimacs.rutgers.edu/pub/challenge/graph/benchmarks/volume/mac hine.
11. Eiben, A. E., Van Der Hauw, J. K., and Van Hemert, J. I.: "Graph Coloring with Adaptive Evolutionary Algorithms" *Journal of Heuristics* **4** (1998), 25–46.
12. Fleurent, C., and Ferland, J. A.: "Genetic and hybrid algorithms for graph coloring", *Annals of Operations Research*, **63** (1995), 437-461.
13. Fleurent, C., and Ferland, J. A.: "Object-Oriented Implementation of Heuristic Search Methods for Graph Coloring, Maximum Clique and Satisfiability", Volume 26 of *Discrete Mathematics and Theoretical Computer Science* AMS (1996), 619-652.

14. Gardner, M.: *The Unexpected Hanging and Other Mathematical Diversions*, Simon and Schuster, New York, (1969).
15. Garey, M. R., and Johnson, D. S.: "The complexity of near-optimal graph coloring", *Journal of the ACM* **23** (1976), 43–49.
16. Garey, M. R., Johnson, D. S., and Stockmeyer, L.: "Some simplified NP-complete graph problems", *Theor. Comput. Sci.* **1** (1976), 237–267.
17. Hertz, A., and de Werra, D.: "Using Tabu Search Techniques for Graph Coloring", *Computing* **39** (1987), 345–351.
18. Johnson, D. S., Aragaon, C. R., McGeoch, L. A., and Schevon, C.: "Optimization by simulated annealing: An experimental evaluation; Part II, Graph coloring and number partitioning," *Operations Research* **39** (1991), 378-406.
19. Knuth, D. E.: *The Standford GraphBase*, ACM Press, Addison Welsey, New York, (1993).
20. Krawczyk, H., and Kubale, M.: "An approximation algorithm for diagnostic test scheduling in multicomputer systems", *IEEE Trans. Comput.* **C-34** (1985), 869–872.
21. Kubale, M., and Jackowski, B.: "A generalized implicit enumeration algorithm for graph coloring", *Communications of the ACM* **28** (1985), 412–418.
22. Leighton, F. T.: "A graph coloring algorithm for large scheduling problems", *Journal of Research of the National Bureau of Standards* **84** (1979), 412–418.
23. Lund, C., and Yannakakis, M.: "On the Hardness of Approximating Minimization Problems", Proc. 25th Annual ACM Symp. on Theory of Computing, (1993), 286–293 (Full version in *J. ACM* **41** **5** (1994), 960-981.)
24. Mehrotra, A., and Trick, M. A.: "A Column Generation Approach for Graph Coloring", *INFORMS J. on Computing* **8** (1996), 344–354.
25. Morgenstern, C.: "Distributed Coloration Neighborhood Search", Volume 26 of *Discrete Mathematics and Theoretical Computer Science* AMS (1996), 335-358.
26. Mycielski, J.: "Sur le Coloriage des Graphes", *Colloquim Mathematiques* **3** (1955) 161–162.
27. Osman, I. H., and Kelley, J. P.: *Metaheuristics: Theory and Applications*, Kluwer Academic Publishers, Hingam, MA (1996).
28. Sager, T. J., and Lin S.: "A Pruning Procedure for Exact Graph Coloring", *ORSA J. on Computing* **3** (1993), 226–230.
29. Sewell, E. C.: "An Improved Algorithm for Exact Graph Coloring", Second DIMACS Challenge (COLORING Papers), *DIMACS Series in Computer Mathematics and Theoretical Computer Science* (1993), 359-373.

BALL: Biochemical Algorithms Library

Nicolas Boghossian, Oliver Kohlbacher, and Hans-Peter Lenhof

Max-Planck-Institut für Informatik, Saarbrücken
{nicom|oliver|len}@mpi-sb.mpg.de
http://www.mpi-sb.mpg.de/CompMolBio

Abstract. In the next century, virtual laboratories will play a key role in biotechnology. Computer experiments will not only replace time-consuming and expensive real-world experiments, but they will also provide insights that cannot be obtained using "wet" experiments. The field that deals with the modeling of atoms, molecules, and their reactions is called Molecular Modeling. The advent of Life Sciences gave rise to numerous new developments in this area. However, the implementation of new simulation tools is extremely time-consuming. This is mainly due to the large amount of supporting code that is required in addition to the code necessary to implement the new idea. The only way to reduce the development time is to reuse reliable code, preferably using object-oriented approaches. We have designed and implemented BALL, the first object-oriented application framework for rapid prototyping in Molecular Modeling. By the use of the composite design pattern and polymorphism we were able to model the multitude of complex biochemical concepts in a well-structured and comprehensible class hierarchy, the BALL kernel classes. The isomorphism between the biochemical structures and the kernel classes leads to an intuitive interface. Since BALL was designed for rapid software prototyping, ease of use, extensibility, and robustness were our principal design goals. Besides the kernel classes, BALL provides fundamental components for import/export of data in various file formats, Molecular Mechanics simulations, three-dimensional visualization, and more complex ones like a numerical solver for the Poisson-Boltzmann equation.

1 Introduction

The fast development of new methods in Molecular Modeling (MM) becomes increasingly important as the field of Life Sciences evolves more and more rapidly. According to our experience, a researcher spends only a small fraction of his or her time on the creative process of developing a new idea, whereas the realization of the idea – the implementation and verification – is usually very time consuming. For this reason, many good ideas have not been implemented and tested yet. Although the need for shorter development times and a sound software design has been recognized and addressed in the software industry for a long time, these design goals have often been neglected while developing new software in the academic field.

In Molecular Modeling applications, the implementation from scratch is too time consuming, because the developer has to spend most of his time on the

J.S. Vitter and C.D. Zaroliagis (Eds.): WAE'99, LNCS 1668, pp. 330–344, 1999.

implementation of supporting code, only marginally related to the problem he is interested in. This supporting code usually handles file I/O, data import/export, standard data structures, or visualization and output of results. The only way to reduce the development time is to reuse reliable code. The main goals of software reuse are reduced development and maintenance times, as well as increased reliability, efficiency, and consistency of the resulting software. There are different approaches for code reuse. According to Coulange [7] and Meyer [16] the object-oriented (OO) approach is the most favorable one.

In the course of our protein docking project [14], we searched for a suitable software package that has the above-mentioned properties. First of all, we had a look at commercial MM software packages. Most of them come with either a scripting language [13], or even a dedicated software development kit (SDK) [4]. They are quite expensive and do not allow the development of free software. Additionally, these packages are solely based on the procedural programming paradigm. To our knowledge, there is no object-oriented package commercially available. Out of academia only a few class libraries have evolved for Molecular Modeling, although in chemical engineering object-oriented programming is quite common and class libraries exist in abundance. Hinsen [12] presented his *Molecular Modeling Tool Kit*, which embeds efficient (C-coded) routines for force-field-based simulations into the object-oriented scripting language Python [19]. The other existing libraries or tool kits concentrate on very specialized areas (*e.g.* PDBLib [5], a library for handling PDB files) or are only remotely related to Molecular Modeling (*e.g.* SCL [18], a class library for object oriented sequence analysis).

None of the above packages satisfies the basic requirements for rapid prototyping in Molecular Modeling, because they are either not state-of-the-art concerning software engineering or lack the required functionality. Thus there seemed a great need for a well designed, efficient package for rapid prototyping in the area of Molecular Modeling. Our answer to this is BALL. The library is implemented in C++, chosen over Java and other OO languages because of efficiency and wide acceptance. Our principal design goals were *extensibility*, *ease of use*, and *robustness*. We spent much effort on the object-oriented modeling of the fundamental concepts in biochemistry like atoms, residues, molecules, and proteins. By the use of the composite design pattern [9] and polymorphism, we were able to model the multitude of complex biochemical concepts in a well-structured and comprehensible class hierarchy, our *kernel* classes. The isomorphism between the biochemical structures and the kernel classes leads to an intuitive interface. With the *processor* concept (see Section 4), we also introduce an efficient and comfortable way to implement operations on all these kernel data structures. The kernel makes use of some basic data structures included in BALL, the so-called *foundation classes*. They provide classes for mathematical objects (*e.g.* matrices), hash associative containers, an extended string class, advanced data structures like three-dimensional hash grids, and many more. The foundation classes are partially based upon the Standard Template Library (STL) that is part of the ANSI C++ Standard [1]. A careful in-depth analysis of many typical Molecular Modeling applications identified data import/export in various formats, Molecular Mechanics simulations, and visualization of molecules as their most important functional components.

BALL provides not only an implementation of a Molecular Mechanics force field (AMBER95 [6]), but also support for generic force fields. This support includes automated reading and interpretation of the required parameter files, data structures for these parameters, and modular force field components. Using this infrastructure, it is possible to implement user-defined force fields with minimal effort. Visualization of the results is often an important part in Molecular Modeling. However, providing a portable and flexible three-dimensional visualization of molecules is quite challenging. We decided to use QT [2] and OpenGL for this purpose. This combination allows the construction of portable, GUI-based applications on all platforms. Apart from this basic functionality, we also included some more advanced features. For example, BALL also contains efficient code to solve the linear Poisson-Boltzmann equation, a differential equation describing the electrostatics of biomolecules in solution. Using this Poisson-Boltzmann solver, we are able to model the effects of water on biomolecules that have long been neglected by simpler models. The implementation of the Poisson-Boltzmann solver was the first test for BALL's rapid prototyping abilities. Making extensive use of the BALL foundation classes, its basic functionality was implemented within a week. We also included functionality to identify common structural motifs in proteins and search for these motifs in large data sets. The usefulness of BALL was shown by the implementation of an algorithm that maps two similar protein structures onto each other. Whereas the first implementation took about five months (implemented in the course of a diploma thesis [3]), we were able to re-implement the algorithm using BALL within a day. The BALL implementation even showed better performance than the original implementation.

In Section 2 we discuss the main design goals of BALL. After giving an overview in Section 3, we will describe the object-oriented modeling of the fundamental biochemical concepts in Section 4. The foundation classes are summarized in Section 5, as is the functionality of the remaining components in Section 6. Section 7 presents an example of a BALL application. The future development of BALL is then outlined in Section 8.

2   Design Goals

There were three main goals we had in mind when we designed BALL: its ease of use, its extensibility, and its robustness. Of course, *ease of use* is a very important property when it comes to rapid software prototyping. The user should be able to use the basic functionality immediately with a minimum of knowledge of the whole library, but what he learns should be globally applicable. To achieve this intuitive use, we designed the class hierarchy of the molecular data structures in a way that directly reflects the biochemical concepts and their relationship (see Section 4). This requires a consistent, minimal, complete, and well documented interface. The call for minimal interfaces is partially a contradiction to the requirements of rapid software prototyping. We break this rule and add additional functionality to the interface, if this leads to a significant decrease in source code size of standard applications (Rule: standard operations should be performed in not more than three lines). We also aim for an easily comprehensible, natural syntax. The last aspect of the ease of use, but certainly not the least, is a complete and consistent documentation. This was achieved by integrating the documentation of each class into the header files. The documentation may then be extracted in LaTeX or HTML format using the tool **doc++** [22].

Extensibility means basically two things: seamless integration of new functionality into the framework and compatibility with other existing libraries or applications. Compatibility is achieved through the use of namespaces. This guarantees no name collisions with other libraries such as the Standard Template Library (STL, now a part of the ANSI C++ standard [1]) and libraries like CGAL [8] or LEDA [15]. Compatibility with these libraries is particularly important, because computational geometry as well as graph algorithms play an important role in Molecular Modeling. The *efficiency* of our implementations has sometimes been sacrificed in favor of flexibility and especially robustness. However, time critical sections like the numerical parts (*e.g.* Poisson-Boltzmann solver, Molecular Mechanics simulations) have been carefully optimized for space and time efficiency.

Verifying the *correctness* of an implementation is very difficult in general. In the case of BALL, this is partly due to the extensive use of templates and partly due to some of the algorithms implemented. For example, it is difficult to verify whether an algorithm for the numerical solution of differential equations works correctly for all possible inputs. We decided to provide at least some basic black box testing whether every member function compiles correctly and behaves as intended. A test program is provided for each class (at least this is intended; not every test is yet completely implemented) that tests all member functions of the class. The resulting test suite is automatically compiled and run after the compilation of the library thus identifying compile-time as well as link-time problems.

Since BALL uses some of the more advanced features of the ANSI C++ Standard, such as member templates, namespaces, and run-time type identification, at the moment only a limited set of platforms is supported. However, we expect most compilers to support these features soon. Up to now, we support the following platforms: (1) i386-Linux using **egcs 1.1.x** or Kuck&Associate's KAI C++, (2) Solaris 2.6/7 (SPARC and Intel) using **egcs 1.1.x**, and (3) IRIX 6.5 using MipsPro 7.2.x. We also intend to port BALL to other platforms, like AIX, HP-UX, and Microsoft Windows.

3 Library Overview

The central part of BALL is the *kernel* (see Fig. 1), a set of data structures representing atoms, molecules, proteins, and so on. Design and functionality of the kernel classes will be described in Section 4. The kernel is implemented using the *foundation classes* that extend – and partially depend on – the classes provided by the STL (*e.g.* the vector class) and ANSI C++ (*e.g.* the string class). These three layers (STL, foundation classes, and kernel) are used by the different *basic components* of the fourth layer. Each of these basic components provides functionality for a well-defined area: *file import/export* provides support for various file formats, primarily to read and write kernel data structures. The *visualization* component BALLVIEW provides portable visualization of the kernel data structures and general geometric primitives. The *Molecular Mechanics* component contains an implementation of the AMBER95 force field and support for user-defined force fields. The *structure* component provides functionality for the comparison of three-dimensional structures, mapping these structures onto each other and searching for structural motifs. The *solvation* component primarily contains a numerical solver for the Poisson-Boltzmann equation that describes the behavior and properties of molecules in solution.

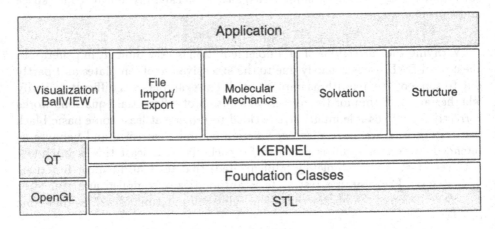

Fig. 1: Overview of the components of BALL

A typical BALL application makes use of kernel data structures, the foundation classes, and one or more basic components, for example it uses the import/export component to read a molecule from a file, performs some simulation using the Molecular Mechanics component and visualizes the result using BALLVIEW.

4 Object-Oriented Modeling of Biochemical Concepts

As its name tells, Molecular Modeling software manipulates representations of molecular systems at an atomistic level. Since these molecular systems may contain complex molecules of different kinds, one of the most difficult problems that we faced during the development of BALL was the design of a suitable class hierarchy to model all entities in these systems. Since most readers will not be very familiar with these biochemical models, we will outline them briefly. First of all, the whole "world" is built of *atoms* which are connected via *bonds*, thus forming *molecules*. Molecular Modeling usually deals with biomolecules (molecules of biological origin or relevance). Out of the many different classes of biomolecules proteins, nucleic acids (the carriers of hereditary information), and polysaccharides (sugars) are the most important ones. Common to these three classes is that their structures can be seen as a sequence (polymer) of smaller building blocks, so-called monomers. *Proteins*, for example, can be seen as chain molecules built from amino acids. We refer to the amino acids in such a *chain* as *residues*. Parts of a chain may form spatial periodic structures – like helices or strands – that are called *secondary structure* elements. *DNA* and *RNA* molecules are linear polymers built from monomers called *nucleotides*. We also introduce the term molecular *system* for a collection of molecules or biomolecules. The relationships between these entities is shown in Fig. 2.

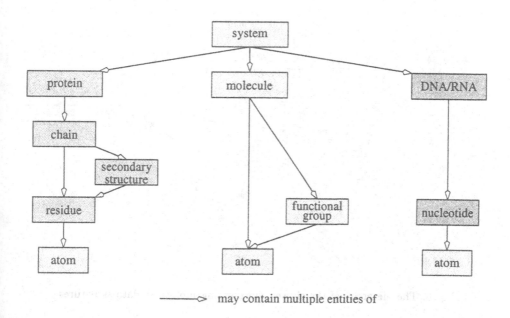

─────▷ may contain multiple entities of

Fig. 2: Hierarchical structure of molecules, proteins, and nucleic acids

Just from these brief definitions, we can deduce some basic properties of the data structures required. We recognize some **is-kind-of** relationships: a protein is a molecule, likewise a sugar. This is modeled by inheritance. Almost all of these "objects" printed in italics above represent containers: proteins contain chains, that contain secondary structures, that contain residues and so on. We also want to reach a certain uniformity between all these container classes. Most of our algorithms should operate on molecules as well as on residues or systems. We therefore model this relationship via the composite design pattern [9].

Without describing the different classes in depth, we give an overview of the molecular data structures in Fig. 3, describing our solution domain. The base class of all our molecular classes is the **Composite** class. It implements a protocol (insertion, removal, and traversal of children) and may contain an arbitrary number of children, thus representing a tree structure. **BaseFragment** specializes the concept of a composite to an object representing molecular data structures. Also derived from **Composite** are **Atom**, **Bond**, and **System** (which represents a collection of molecular objects). Molecules are represented by **Molecule**, a specialization of **BaseFragment** that may contain either **Atoms**, or **Fragments**, which usually represent groups of atoms within a molecule. **Atom**, **Fragment**, and **Molecule** are used to represent general molecules; we call them the molecular framework. Biomacromolecules are represented by specializations of these classes. There are two frameworks already implemented besides the molecular framework: the protein framework and the nucleic acid framework, both representing a specialization and extension of the molecular framework.

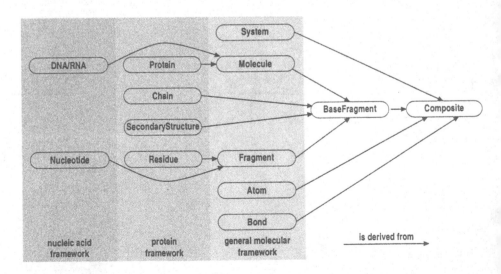

Fig. 3: The hierarchy of the classes representing molecular data structures

In the protein framework, **Protein** replaces the **Molecule**. A Protein may contain **Chains**, that may contain **Residues**. A **Residue** is a specialized variant of a **Fragment** possessing additional functionality. **Chains** may also contain **SecondaryStructures**. In the nucleic acid framework, there exist two specialized classes derived from **Molecule**: DNA and RNA. Each **DNA** or **RNA** instance may contain **Nucleotides**.

Using this class hierarchy, we have the possibility to model the specific properties of biological macromolecules. Due to polymorphism we can design algorithms that may be applied to molecules or fragments of molecules in general. One might argue that the decomposition we chose to represent the different biochemical terms and their relationships is the best. The difficulty with these biochemical terms stems from their lack of precision. For example, it is doubtful, whether a protein "is a" molecule. Every biochemist would confirm, that proteins are biomolecules. If asked for a definition of molecule, his or her answer would describe something like all atoms that are connected to each other via bonds. However, this definition does not apply to all proteins, because a protein might also be a non-bonded complex of multiple "molecules". After long discussions, we found the hierarchy we described before the most satisfying and the most intuitive one, but we do not expect it to fit all needs. Considering BALL to be a *framework* instead of a class library, this problem can easily be solved by a user-derived set of classes that satisfy these specialized needs. An example of this could be a framework representing polysaccharides.

We also introduce the concept of a **processor**. The processor performs an operation on objects and is parameterized with the class type of these objects. Each processor defines three basic methods: **start**, **operator** () and **finish**. The typical usage of a processor is to **apply** it to a kernel object. First, the processor's **start** method is called which may be used to perform initializations, but often left unimplemented. Every **Composite** can be seen as the root of a subtree. The **apply** method now iterates over all nodes in this subtree. Using run-time type identification, the actual class of each node is determined. If it matches the processor's template argument (*i.e.*, if the processor was parameterized with a base class of the node's type), the processor's **operator** () is called for that node. After the end of the iteration, **finish** is called to perform some final action (*e.g.* clearing up temporary data structures). This concept is an elegant way to define operations that can be performed on any kernel class. It is very convenient to use, because the **apply** method usually takes the place of more complicated iteration loops.

5 Foundation Classes

Besides the kernel described in Section 4, BALL also provides more fundamental data structures. We will refer to these as the *foundation classes*. Since the total number of foundation classes is roughly 100, we can only present some selected classes here.

One of the most important classes is the **String** class, that is derived from the ANSI C++ **string**. This class is especially useful if complex string handling is needed, *e.g.* in the parsing of line-based file formats. The **String** class allows case insensitive comparisons, efficient substring handling, among others. The foundation classes also provide more advanced data structures, such as hash associative containers, different types of trees, or three-dimensional hash grids (see the example in Section 7). Another part of the foundation classes provides mathematical and geometric objects like matrices, vectors, points, spheres, and so on. Object persistence was implemented for the kernel classes and for most of the foundation classes. In combination with the `SocketStream` classes, we are able to transport objects from one application to another via the network, a feature we use for the visualization of kernel objects (see Section 6).

6 The Basic Components

The Import/Export component File import and export is an important basic functionality that every application in the field of Molecular Modeling uses. The molecular structures that an application has to handle come in a multitude of different file formats and sometimes even in different variants of these formats. Since tools are available to convert between most of these formats (*e.g.* BABEL [20]), we implemented only some of the important file formats: The Brookhaven Protein Databank (PDB) format, the MOL2 format from Tripos Associates and the HIN format used by HyperChem. We also added support for Windows INI files and a special format for hierarchically organized data.

The Molecular Mechanics component Molecular Mechanics is an important method to simulate the dynamic behavior of molecules and to predict their properties (*e.g.* geometry or relative energies). It is based on classical mechanics, describing atoms as mass points connected by bonds. The behavior of these atoms can be described in terms of a *force field*, *i.e.* a set of analytical descriptions of interactions between the atoms. The force field describes the total energy for a given molecular system. The forces acting upon each atom are obtained by calculating the gradient of the energy.

A *molecular dynamics simulation* is used to examine the dynamic behavior of molecules. The forces acting upon the atoms are evaluated after discrete time steps and the displacements of the atoms are calculated from these forces. Another application is *geometry optimization*, *i.e.* the search for an atom arrangement of the molecular system with the lowest possible energy. Molecular dynamics and geometry optimization are two very important, but time consuming applications in Molecular Modeling. Efficiency is a big topic here, because molecular dynamics simulations and geometry optimizations usually require a large number of iterative evaluations of the force field. To achieve optimal performance, we had to convert our kernel data structures to more compact data structures, containing just the information we needed to evaluate the force field. This conversion is done only once before the calculation starts. After the calculation, the kernel data structures are updated with their new values. As the time spent in the force field calculations is usually much larger than the time needed to set up these internal data structures, this overhead does not matter.

Since there are so many important force fields in use, we designed this component to be modular and easily extensible. We implemented a number of support classes that free the user of some of the annoying and difficult work. We call the classes the *generic force field*. The generic force field is able to automatically read and interpret force field parameter files and even handles multiple versions of parameter sets in one file. From the parameter files it constructs predefined data structures for most of the common force field terms. It also automatically constructs the fast internal data structures and assigns atom types according to a user-defined rule set. Leaving only the implementation of the force field terms to the user, it is possible to implement a simple force field within a day. Using the generic force field, we implemented AMBER95, one of the most important force fields for modeling proteins.

The visualization Component BALLVIEW The three-dimensional visualization of molecules is a challenging subject. First of all, efficiency is a problem, because large biomolecules usually consist of several thousands of atoms. Visualization also requires a graphical user interface (GUI) that has to be portable. Considering these requirements, we decided to implement BALLVIEW using OpenGL. By compiling our kernel data structures into OpenGL display lists, we achieve optimal performance. We chose QT [2] as a platform-independent tool kit for the construction of graphical user interfaces. QT is free (at least on most platforms and for academic use), it is thoroughly tested, and provides a well designed class hierarchy.

Nevertheless, the development of a GUI is often too time consuming while testing new ideas. So we had to come up with a method to integrate a visualization component into BALL applications with just a few lines of code. We developed a stand-alone viewer that may be integrated via TCP sockets into a BALL application. If the application needs to visualize some kernel data structures (*e.g.* a protein), it sends these objects to the viewer via a stream connected to a socket, an operation requiring a single line of code. This feature uses the object persistence that we implemented for the kernel objects.

Further Components Two other components that we will only briefly discuss are the solvation component and the structure component. The solvation component is used to describe the interaction of biomolecules with solvents, usually water. These interactions may be modeled using the so-called continuum model. The electrostatic part of these interactions is described by the Poisson-Boltzmann equation, a differential equation connecting the electrostatic potential and the charge distribution in the molecule. We implemented a modified version of the algorithms described by Nicholls and Honig [17]. It solves the linear Poisson-Boltzmann equation in its finite difference form on a grid using the method of successive over-relaxation.

The structure component allows the geometric comparison of molecules, the calculation of transformations that map one molecule onto the other, and provides functions for the search for common structural motifs in proteins.

7   Example: The ProteinMapper Class

Using BALL, we have implemented several applications that stem from the area of chemical modeling and simulation. The implementation of these applications gave us the opportunity to test the usability of BALL. One of the tests that we carried out was the re-implementation of an algorithm for matching the 3D-structures of proteins, which had been implemented in the course of a diploma thesis [3]. The first implementation without BALL took more than five months; using BALL we were able to re-implement the algorithm within one day. Although implemented with a tool for rapid software prototyping the re-implementation has exhibited even better running times than the original implementation. In this section we describe the algorithm for matching the 3D-structures of proteins, present the BALL implementation of the algorithm as an example of a BALL program, and discuss some additional features of BALL.

In Section 4 we briefly introduced one of the most important classes of biological macromolecules, the proteins. Proteins are linear polymers (chain molecules) composed of twenty monomers, the so-called amino acids. The amino acid chain of every protein folds in a specific way. This folding process puts the protein into a well defined three-dimensional shape that has a strong influence on the chemical reactivity of the protein. The van-der-Waals model approximates the 3D-structure of a protein as a union of spheres, where each atom is represented by a sphere. Another simplified model represents a protein by a discrete set of points. Every amino acid is represented by the coordinates of the atom center of its "central" C_α-atom. The nature of the amino acid binding guarantees that all pairs of neighboring C_α-atoms have roughly the same Euclidean distance. Because of the high packing density of proteins and certain chemical properties the C_α-atoms form a good discrete representation of the volume of the protein.

Since the 3D-structures of proteins determine their chemical reactivity, the comparison of the 3D-structures or even parts of the 3D-structures of proteins plays an important role in the functional analysis of proteins. An abstract version of the structure comparison problem can be formulated as follows: Given two proteins P_1 and P_2 and their 3D-structures, check if P_1 is similar to P_2 or to a part of it (or vice versa). In this abstract formulation the important term "similar" is not defined. The study of the literature about structure comparison algorithms shows that a large variety of similarity definitions and algorithms for these special problem instances have been published. One of these approaches formulates the structure comparison problem as a point set congruence problem: Given a distance bound $\epsilon > 0$ and two three-dimensional point sets, S_1, the atom centers of the C_α-atoms of P_1, and S_2, the atom centers of the C_α-atoms of P_2, compute the maximal number $k \in N$, such that S_1 and S_2 are (ϵ, k)-congruent. Two point sets S_1 and S_2 are (ϵ, k)-congruent if there exists a rigid transformation M (rotation + translation) and an injective function $f : S_1 \longrightarrow S_2$ such that at least k points $p \in S_1$ are moved by the transformation M into the ϵ neighborhood of the corresponding points $f(p) \in S_2$, i.e., $d(M(p), f(p)) < \epsilon$, where $d(,)$ is the Euclidean distance. The definition of the (ϵ, k)-congruence is due to Heffernan [10], who published an algorithm for testing if point sets are

(ϵ, k)-congruent. Tests with average-sized proteins showed that this algorithm has huge running times for values of ϵ suitable for protein similarity tests. In his Master's thesis [3], Becker developed and implemented a heuristic algorithm for testing (ϵ, k)-congruence that is based on the techniques presented by Heffernan and Schirra [11]. We sketch this heuristic algorithm now.

- The algorithms determines a set T_1 of triangles between points in S_1 with edge lengths greater than l_{min} and smaller than l_{max}. The triangles of T_1 are stored in a three-dimensional hash grid G with box length ϵ. Each triangle (p_1, p_2, p_3) is stored in G according to the length $(d(p_1, p_2), d(p_2, p_3), d(p_3, p_1))$ of the "sorted" triangle edges in the grid box with indices $(\lfloor \frac{d(p_1,p_2)}{\epsilon} \rfloor, \lfloor \frac{d(p_2,p_3)}{\epsilon} \rfloor,$ $\lfloor \frac{d(p_3,p_1)}{\epsilon} \rfloor)$. This technique is called geometric hashing [21].
- The algorithm computes a set T_2 of triangles between points in S_2. For each triangle t_2 in T_2 the algorithm searches for all triangles t_1 stored in the grid G that are ϵ-similar to t_2, i.e., the absolute difference between the lengths of the ith edge of t_1 and the ith edge of t_2 is smaller than ϵ for every $i \in \{1, 2, 3\}$. For each ϵ-similar pair $(t_1 = (p_1^1, p_2^1, p_3^1), t_2 = (p_1^2, p_2^2, p_3^2))$ of triangles the algorithm computes a rigid transformation M that maps triangle t_1 onto triangle t_2. More precisely, the algorithm computes a rigid motion that maps point p_1^1 onto point p_1^2, p_2^1 onto the ray starting in p_1^2 that goes through the point p_2^2, and p_3^1 into the plane generated by p_1^2, p_2^2, p_3^2.
- For each rigid transformation M computed in the previous step the following operations will be carried out: The transformation M is applied to the point set S_1. For each point p_1^M in the transformed point set $M(S_1)$, the algorithm determines all points $p_2 \in S_2$ with $d(p_1^M, p_2) < \epsilon$. In order to determine the best injective function f for the given transformation M, the algorithm has to solve the maximum matching problem for the following bipartite graph formed from the points of S_1 and S_2. There is an edge between a vertex p_1 and a vertex p_2 if $d(M(p_1), p_2) < \epsilon$. If k is the number of edges in a maximum matching of this bipartite graph, then S_1 and S_2 are at least (ϵ, k)-congruent. The algorithm stores the transformation M_{best} that yields the maximal (ϵ, k)-congruence.

The listing below shows a short BALL program implementing the key parts of the algorithm. Several details (like include directives) have been omitted for brevity. In lines 3 through 11 we read two PDB files into two proteins P1 and P2. We then declare two STL vectors S1 and S2 parameterized with a three-dimensional vector (a BALL type: Vector3). The for loop in line 16 iterates over all atoms that are contained in P1. The overloaded operator + of AtomIterator is a shorthand for AtomIterator::isValid(). In the following line we extract all carbon atoms (carbon is defined as Element::C in the periodic system of elements PSE) with the name CA. The positions of these C_α atoms (coordinates of the atom centers) are then stored in the vector S1. Lines 20 through 22 contain the same loop for protein P2. After these preparatory steps, we use a static member function of StructureMapper to calculate the maximum k such that the two point sets S1 and S2 are (ϵ, k)-congruent. The resulting transformation M is then

written to cout. We then initialize the TransformationProcessor T using the
best transformation M. A TransformationProcessor multiplies the coordinates
of all atoms contained in a kernel object with a matrix. The transformed protein
P1 is then written to a PDB file.

```
1 int main (int argc, char** argv)
2 {
3        Protein P1, P2;
4
5        PDBFile infile (argv[1]);
6        infile >> P1;
7        infile.close();
8
9        infile.open(argv[2]);
10       infile >> P2;
11       infile.close();
12
13       vector<Vector3> S1, S2;
14
15       AtomIterator it;
16       for (it = P1.beginAtom(); +it; ++it)
17     if (it->getElement() == PSE[Element::C] && it->getName() == "CA")
18       S1.push_back(it->getPosition());
19
20       for (it = P2.beginAtom(); +it; ++it)
21     if (it->getElement() == PSE[Element::C] && it->getName() == "CA")
22       S2.push_back(it->getPosition());
23
24       int k = 0;
25       Matrix4x4 M =
26       StructureMapper::calculateEpsilonKCongruence (S1, S2, k, 0.5, 4, 8);
27
28       cout << "Could_match_" << k << "_Calpha_atoms." << endl;
29       cout << "Best_transformation_is:" << endl;
30       cout << M << endl;
31
32       TransformationProcessor T(M);
33       P1.apply(T);
34       PDBFile outfile (argv[3]);
35       outfile << P1;
36       outfile.close()
37
38       return 0;
39 }
```

The listing of StructureMapper::calculateEpsilonKCongurence below illustrates some additional features of BALL. In lines 9 and 10 the triangle sets T1
and T2 are calculated by the function StructureMapper::computeTriangles.
Using the overloaded operator<<, we insert the triangles of T1 into the hash
grid G (line 18). In lines 24–38 we use the iterator t2 to traverse all triangles in
T2. We then iterate over all triangles in the grid G (using t1) that are ϵ-similar to
t2 (for brevity, we removed the ϵ-similarity test, but the box size we chose for G
guarantees (2ϵ)-similarity). For each of the similar triangle pairs we calculate the
transformation M that maps t1 onto t2 (see description of the algorithm). In
lines 30–31 we transform all points in S1 using M and store them in transformed.
We then calculate the maximum matching of transformed and S2. The function StructureMapper::calculateMaximumMatching returns a vector of point
pairs. The number k of matched points equals the number of pairs in this vector.
If k is larger than the best matching we found up to that point (k_best), we
store it along with the corresponding transformation (line 33). Finally we return
the best transformation in line 40.

```
 1 Matrix4x4 StructureMapper::calculateEpsilonKCongruence
 2         (vector<Vector3>& S1, vector<Vector3>& S2,
 3          Size& k_best, float epsilon,
 4          float l_min, float l_max)
 5 {
 6         typedef TVector3<int>
 7                             IndexVector;
 8         typedef pair<IndexVector, Vector3>                    Triangle;
 9
10   vector<Triangle> T1 = computeTriangles(S1, l_min, l_max, epsilon);
11   vector<Triangle> T2 = computeTriangles(S2, l_min, l_max, epsilon);
12
13   HashGrid3<IndexVector>                                        G;
14                                                        M_best, M;
15                                                        k = 0;
16   vector<Vector3>                            transformed(S1.size());
17
18   k_best = 0;
19         G << T1;
20
21   vector<Triangle>::iterator                               t2;
22   HashGridBox3<IndexVector>::BoxIterator         box_it;
23   HashGridBox3<IndexVector>::DataIterator        t1;
24
25   for (t2 = T2.begin() ; t2 != T2.end(); ++t2) {
26     for (box_it = G.getBox(t2->second).beginBox() ; +box_it; ++box_it ) {
27       for (t1 = box_it->beginData() ; +t1; ++t1) {
28         M = matchPoints(S1[t1->x],        S1[t1->y],        S1[t1->z],
29                         S2[t2->first.x],S2[t2->first.y],S2[t2->first.z]);
30
31                   for (Size i = 0; i < S1.size(); i++)
32                       transformed[i] = M * S1[i];
33
34         k = calculateMaximumMatching(transformed, S2, 2 * epsilon).size();
35
36         if (k > k_best) { k_best = k; M_best = M; }
37       }
38     }
39 }
40
41   return M_best;
42 }
```

8 Conclusion

BALL is the first object-oriented application framework for rapid software prototyping in the area of Molecular Modeling. It provides rich functionality and an intuitive interface. BALL has shown its usefulness in several example applications. In each case it decreased the development time tremendously without loss of performance.

However, it is still far from complete. We are currently implementing the missing parts of the test suite to improve the robustness of BALL. After the completion of these tests, we will release a first version of the library in summer 1999.

References

1. Programming Languages – C++. International Standard, American National Standards Institute, New York, July 1998. Ref. No. ISO/IEC 14882:1998(E).
2. Troll Tech AS. QT release 1.42. http://www.troll.no/products/qt.html.
3. Jörg Becker. Allgemeine approximative Kongruenz zweier Punktmengen im R. Master's thesis, Universität des Saarlandes, 1995.
4. Cerius² modeling environment. Molecular Simulations Inc., San Diego, 1997.
5. W. Chang, I.N. Shindyalov, C. Pu, and P.E. Bourne. Design and application of PDBLib, a C++ macromolecular class library. *CABIOS*, 10(6):575–586, 1994.
6. Wendy D. Cornell, Piotr Cieplak, Christopher I. Bayly, Ian R. Gould, Kenneth M. Merz, Jr., David M. Ferguson, David C. Spellmeyer, Thomas Fox, James W. Caldwell, and Peter A Kollman. A second generation force field for the simulation of proteins, nucleic acids and organic molecules. *J. Am. Chem. Soc.*, 117:5179–5197, 1995.
7. Bernard Coulange. *Software reuse*. Springer, London, 1997.
8. Andreas Fabri, Geert-Jan Giezeman, Lutz Kettner, Stefan Schirra, and Sven Schönherr. On the design of CGAL, the computational geometry algorithms library. Technical Report MPI-I-98-1-007, Max-Planck-Institut für Informatik, Saarbrücken, February 1998.
9. Erich Gamma, Richard Helm, Ralph Johnson, and John Vlissides. *Design patterns : elements of reusable object-oriented software*. Addison-Wesley, Reading, MA, 1995.
10. Paul J. Heffernan. Generalized approximate algorithms for point set congruence. In *WADS93*, 1993.
11. Paul J. Heffernan and Stefan Schirra. Approximate decision algorithms for point set congruence. *Computational Geometry : Theory and applications*, 4(3):137–156, 1994.
12. Konrad Hinsen. The Molecular Modelling Toolkit: a case study of a large scientific application in python. In *Proceedings of the 6th International Python Conference*, pages 29–35, San Jose, Ca., October 1997.
13. HyperChem release 4.5. Hypercube Inc., 1995.
14. Hans-Peter Lenhof. New contact measures for the protein docking problem. In *Proc. of the First Annual International Conference on Computational Molecular Biology RECOMB 97*, pages 182–191, 1997.
15. Kurt Mehlhorn, Stefan Näher, Michael Seel, and Christian Uhrig. *The LEDA user manual : version 3.6*. Max-Planck-Institut für Informatik, Saarbrücken, 1998.
16. Bertrand Meyer. *Object-Oriented Software Construction*. Prentice Hall PTR, New Jersey, 2nd edition, 1997.
17. Anthony Nicholls and Barry Honig. A rapid finite difference algorithm, utilizing successive over-relaxation to solve the poisson-boltzmann equation. *J. Comput. Chem.*, 12(4):435–445, 1991.
18. Wolfgang Vahrson, Klaus Hermann, Jürgen Kleffe, and Burghardt Wittig. Object-oriented sequence analysis: SCL – a C++ class library. *CABIOS*, 12(2):119–127, 1996.
19. Guido van Rossum. Python version 1.5.1. http://www.python.org.
20. Pat Walters and Matt Stahl. BABEL version 1.6. University of Arizona.
21. H. J. Wolfson. Model based object recognition by 'geometric hashing'. In *Proc. 1st European Conf. Comput. Vision*, pages 526–536, 1990.
22. Malte Zöckler and Roland Wunderling. DOC++ version 3.2. http://www.zib.de/Visual/software/doc++/.

An Experimental Study of Priority Queues in External Memory*

Klaus Brengel[1], Andreas Crauser[1], Paolo Ferragina[2], and Ulrich Meyer[1]

[1] Max-Planck-Institut für Informatik, Im Stadtwald, 66123 Saarbrücken, Germany
{kbrengel,crauser,umeyer}@mpi-sb.mpg.de
[2] Dipartimento di Informatica, Università di Pisa, Corso Italia 40, 56125 Pisa, Italy
ferragin@di.unipi.it

1 Introduction

A *priority queue* is a data structure that stores a set of items, each one consisting of a tuple which contains some *(satellite) information* plus a *priority* value (also called *key*) drawn from a totally ordered universe. A priority queue supports the following operations on the processed set: access_minimum (returns the item in the set having minimum key), delete_min (returns and deletes the item in the set having the minimum key) and insert (inserts a new item into the set). Priority queues (hereafter PQs) have numerous important applications: combinatorial optimization (e.g. Dijkstra's shortest path algorithm [7]), time forward processing [5], job scheduling, event simulation and online sorting, just to cite a few. Many PQ implementations currently exist for small data sets fitting into the *internal memory* of the computer, e.g. k–ary heaps [23], Fibonacci heaps [10], radix heaps [1], and some of them are also publicly available to the programmers (see e.g. the LEDA library [15]). However, in large-scale event simulations or on instances of very large graph problems (as they recently occur in e.g. geographical information systems), the performance of these *internal-memory* PQs may significantly deteriorate, thus being a bottleneck for the overall application. In fact, as soon as parts of the PQs do not fit entirely into the internal memory of the computer, but reside in its *external memory* (e.g. in the hard disk), we may observe a heavy paging activity of the external-memory devices because the pattern of memory accesses is not tuned to exhibit any locality of reference. Due to the technological features of current disk systems [17], this situation may determine a slow down of 5 or 6 orders of magnitude in the final performance of each PQ-operation [1]. Consequently, it is required to design PQs which take explicitly into account the physical properties of the disk systems in order to achieve efficient I/O-performances that allow these data structures to be plugged successfully in software libraries.

A lot of work has already been done in designing I/O-efficient variants of well-known internal-memory data structures (see e.g. [22] for a survey). In all

---

* Research partially supported by EU ESPRIT LTR Project Nr. 20244 (ALCOM-IT), and by UNESCO under the contract no. UVO-ROSTE 875.631.9
[1] The internal-memory access time is currently 10^5 to 10^6 times faster than the disk access time.

J.S. Vitter and C.D. Zaroliagis (Eds.): WAE'99, LNCS 1668, pp. 345–359, 1999.
© Springer-Verlag Berlin Heidelberg 1999

these cases, the model adopted to evaluate their efficiency is the *Parallel Disk Model* proposed by Vitter and Shriver [21]. Here the memory of a computer is abstracted to consist of *two levels*: a fast and small internal memory, of size M, and a slow and arbitrarily large external memory physically partitioned into D "independent" disks. Data between the internal memory and the disks is transfered in blocks of size B (called *disk pages*). Each disk access transfers a total of DB items by retrieving one page from each disk. Algorithmic performance is measured in this model by counting (i) the number of performed disk accesses (shortly I/Os), (ii) by measuring the internal CPU time, and (iii) by counting the number of occupied disk pages. Recently, this accounting scheme was refined by separating between consecutive (bulk) I/Os and I/Os to random locations [8]. Bulk I/Os are known to be faster than random I/Os because they can profit from the prefetching and caching policies of the underlying disk system [17]. We will use the concept of bulk I/O in the experimental section to refine our analysis and better evaluate the I/O-behavior of the tested data structures.

Previous Work. It has been observed by several researchers that d-ary heaps perform better than the classical binary heaps on multi-level memory systems [16, 14]. Consequently, a variety of external PQs already known in the literature follow this design paradigm by using a *multi-way tree* as a basic structure. Buffer trees [2, 11] and M/B–ary heaps [9, 13] achieve optimal $O((1/B) \log_{M/B} N/B)$ amortized I/Os per operation, in a sequence of total N operations. Unfortunately, most of these data structures are quite complex to be implemented (the simplest proposal is in [9]), and the constants hidden in the space and I/O bounds are not negligible. For instance, all these data structures require the maintenance of some kind of child pointers and some rebalancing information which may induce a space overhead and impose to write a complex rebalancing code. Recently, starting from an idea of Thorup [19] for RAM priority queues, Brodal and Katajainen [4] designed an external-memory PQ consisting of a *hierarchy of sorted lists* that are merged upon level- or internal-memory overflows. Their main result is to achieve optimal worst-case I/O-performance. However, the complicated details of the structural design make this proposal mainly of theoretical interest. In this paper we show that this *hierarchical approach* offers some advantages over the tree-based data structures which make it appealing also in practice.

This Paper. We propose two novel external-memory heaps especially designed to offer effective I/O-performances in the practical setting. The first PQ proposal concerns with an adaptation of the two-level radix heap of [1] to the external memory. This external PQ supports monotone insertions and manages integer keys in a range-size C, thus achieving an amortized I/O-bound of $O(1/DB)$ for the **insert** and $O((1/DB)\log_{\frac{M}{(DB \log C)}} C)$ for the **delete_min**. The space requirements are optimal, and the I/O-constants are actually very small, thus driving us to conjecture good practical performances. Our second PQ proposal [2] is a simplification of [4], carefully adapted to exploit the simplicity of a collection of fixed-size lists over balanced tree structures. The resulting array-based PQ is simple to be implemented, is I/O-optimal in the amortized sense, does not impose any constraints on the priority values (cfr. radix heaps) and involves small constants in the I/O, time and space bounds. Consequently, this structure turns out to be very promising in the practical setting and thus deserves a careful experimental analysis to validate its conjectured superiority over the other PQs.

In the second part of the paper, we will perform an extensive set of experiments comparing the implementation of four external-memory PQs: one based on Buffer Trees [2], another based on B-Trees [3], and our two new proposals R-Heaps and Array-Heaps. Additionally, we will compare these PQs against four internal-memory priority queues: Fibonacci heaps, k-ary heaps, pairing heaps and internal radix heaps. Our experimental framework includes some simple tests, which are used to determine the actual I/O-behavior of **insert** and **delete_min** operations, as well as more advanced tests aimed to evaluate the CPU-speed of the internally used data structures. As a final experimental result, we will also test the performance of our proposed PQs in a real setting by considering special sequences of **insert/delete_min** operations which are generated via a simulation of the Dijkstra's shortest path algorithm. These artificial sequences will allow us to check the behavior of our PQs on a "non-random", but application driven, pattern of disk accesses. We think that our variegate spectrum of experimental results gives a good picture of the I/O-properties and practical behavior of these external-PQs, thus being helpful to anyone interested to use such data structures on very large data sets.

2   External Radix Heaps

Our first external-heap proposal consists of a simple and fast data structure based on internal two-level radix heaps [1] (shortly *R-heaps*). Let C be a positive integer constant and assume that the element priorities are positive integers. R-heaps work under the following condition: *Upon insertion, any priority value must be an integer in the range* [min, min + C], *where* min *is the priority value of the last element removed from the heap via a* **delete_min** *operation (*min = 0 *if the queue was empty).*

Hence the queue is *monotone* in the sense that the priority values of the deleted elements form a *nondecreasing* sequence. This requirement is fulfilled in many applications, e.g. in Dijkstra's shortest path algorithm. The need for integer priorities is not severe, since by interpreting the binary representation of a floating point number as an integer does not change the ordering relation [12]. R-heaps with $C = 2^{64}$ can therefore also be used for 64 bit floating point numbers.

The structure of an external R-heap is defined as follows. Let r be an arbitrary positive integer (also called *radix*) and choose the parameter h to be the

---

[2] Independently, Sanders [18] used similar ideas for implementing PQs in caching hierarchies.

minimum integer such that $r^h > C$ (i.e. $h = \lceil \log_r (C+1) \rceil$). Let k be the priority of an arbitrary element in the queue and let $k_h k_{h-1} \ldots k_0$ be its representation in base r (denoted by $(k)_r$). Similarly, let $m_h m_{h-1} \ldots m_0$ be the representation of min in base r (denoted by $(\min)_r$). If an element with priority k belongs to the queue then $k - \min < r^h$. Consequently $k_h = m_h$ or $k_h = m_h + 1$.

In detail, our external R–heap consists of three parts (see also Figure 1):

- h arrays of size r each. An array entry is a linear list of blocks called a *bucket*. Let $\mathcal{B}(i,j)$ denote the bucket associated with the j-th entry of the i-th array, for $0 \le i < h, 0 \le j < r$. For each bucket we maintain the first block (disk page) in main memory. This constrains r to satisfy the relation $h \cdot r \cdot B \le M$.
- a bucket $\mathcal{N}$ containing all the elements having priority k with $k_h = (m_h + 1) \bmod r$.
- an internal memory priority queue $\mathcal{Q}$ containing all indices of the non-empty buckets. These indices are ordered lexicographically, i.e., $(i,j) < (i',j')$ if either $i < i'$ or, $i = i'$ and $j < j'$. $\mathcal{Q}$ never stores more than $h \cdot r$ indices.

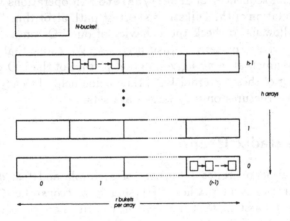

Fig. 1. The structure of the external radix heap.

Insert: In order to insert a new element with priority k in the external R–heap, we first compute the least significant $h + 1$ digits of $(k)_r$, thus taking $O(h)$ time and no I/Os. Then we insert k into the bucket $\mathcal{N}$, whenever $k_h \neq m_h$; otherwise, we insert k into the bucket $\mathcal{B}(i, k_i)$, where $i = \max(\{l \mid m_l \neq k_l, 0 \le l \le h\} \cup \{0\})$ If bucket $\mathcal{B}(i,j)$ was empty before the insert operation, we also insert the index (i,j) into $\mathcal{Q}$ taking $O(\log (rh))$ time.

Lemma 1. *An* insert *operation takes amortized* $O(1/B)$ *I/Os and* $O(h + \log r)$ *CPU time.*

Proof. This follows from the discussion above. Notice that $O(h + \log(hr)) = O(h + \log r)$. The constant in the I/O term is one. □

Delete_min: If the bucket $\mathcal{B}(0, m_0)$ is not empty, we delete an arbitrary element from this bucket. This takes $O(1/B)$ amortized I/Os and $O(\log(hr))$ time. Otherwise we use the internal priority queue $\mathcal{Q}$ to find the first non-empty bucket. This is either a bucket $\mathcal{B}(i, j)$, for some i and j, or the bucket $\mathcal{N}$. In both the two cases, we scan the non-empty bucket and determine the new minimum element, thus setting the new value for min. Since the minimum has changed, we reorganize the elements in the selected bucket according to the rule exploited for the **insert** operation. We update $\mathcal{Q}$ accordingly and then return any element stored in the bucket $\mathcal{B}(0, m_0)$.

Lemma 2. *A* **delete_min** *operation takes* $O(h/B)$ *amortized I/Os and* $(h \log(rh))$ *amortized CPU time.*

Proof. Each element can be redistributed at most h times, namely once for each of the h arrays. This amounts to 2 scans per redistribution, implying the I/O-bound. For the CPU time we observe that $O(\log(hr))$ time is required to find the first non-empty bucket (by using $\mathcal{Q}$) and this is also the internal time for moving a single element. Since each element is moved at most h times, the final time bound also follows. □

It remains to determine the appropriate values for r and h that allow the R–heap data structure to work efficiently. The only constraint we previously imposed on these parameters was that $r \cdot h \cdot B + O(r \cdot h) \leq M$, i.e. the first block of every bucket and the internal queue fit in internal memory. Notice that the second additive term is nearly neglectable because $1 \ll B$.

Now, since $h \approx \log_r C$, it suffices to choose the maximum value of r such that the constraint above holds, that is (where $m = M/B$):

$$r \log_r C \approx m \quad \Rightarrow \quad r/\log r \approx m/\log C \quad \Rightarrow \quad r \approx (m/\log C)\log(m/\log C)$$

Theorem 1. *Let* $m = M/B$ *and let* $r = (m/\log C)\log(m/\log C)$. *An* **insert** *into an external R–heap takes amortized* $O(1/B)$ *I/Os and* $O(\log_r C + \log r)$ *CPU time. A* **delete_min** *takes* $O((1/B)\log_r C)$ *amortized I/Os and* $O(\log_r C \cdot \log(r \log_r C))$ *amortized CPU time.*

External radix heaps are efficient in applications where C is small. Examples are time-scheduling, where typically C is the time at which an event takes place, or in Dijkstra's shortest path computations where C is an upper bound on the edge weight. For typical values of M, B and C, $\log_r C$ is two or three.

As far as the external space consumption is concerned, we observe that only *one* disk page can be non-full into each bucket (by looking at a bucket as a stack). But this page does not reside on the disk, so that there are no partially filled disk pages. We can therefore conclude that:

Lemma 3. *An external R–heap storing N elements occupies no more than N/B disk pages.*

The complexities on the D-disk model can be easily derived by observing that a scan operation on x contiguous elements striped over the disks requires $O(x/DB)$ I/Os.

3   External Array-Heaps

Our second heap structure is a simplification of [4] and consists of two parts: an internal heap $\mathcal{H}$ and an external data structure $\mathcal{L}$ consisting of a collection of sorted arrays of different lengths. $\mathcal{L}$ is in turn subdivided into L levels $\mathcal{L}_i$, $1 \leq i \leq L$, each consisting of $\mu = (cM/B) - 1$ arrays (called *slots*) having length $l_i = (cM)^i/B^{i-1}$. Each slot of $\mathcal{L}_i$ is either empty or it contains a sorted sequence of at most l_i elements. The following property is easy to see: *The total size of* $(\mu + 1)$ *slots containing* l_i *elements each is equal to* l_{i+1}.

Elements are inserted into $\mathcal{H}$; if $\mathcal{H}$ gets full, then $l_1 = cM$ of these elements are moved to the disk and stored in sorted order into a free slot of $\mathcal{L}_1$. If there is no such free slot (overflow), we merge all the slots of $\mathcal{L}_1$ with the elements coming from $\mathcal{H}$, thus forming a sorted list which is moved to the next level $\mathcal{L}_2$. If no free slot of $\mathcal{L}_2$ does exist, the overflow process is repeated on $\mathcal{L}_2$, searching for a free slot in $\mathcal{L}_3$. We continue until a free slot is eventually found.

The **delete_min** operation maintains the invariant that the smallest element always resides in $\mathcal{H}$. Therefore $\mathcal{H}$ needs to be refilled by deleting some appropriate blocks from the sorted slots of $\mathcal{L}$ whenever the minimum element is removed from the heap.

We describe now in more detail two versions of this heap; one is a simplified non-optimal structure intended to be fast in practice, while the second proposal is more complicated but reaches I/O-optimality.

3.1   A Simplified Version

The internal heap $\mathcal{H}$ is conceptually divided into two parts, H_1 and H_2. H_1 stores the newly inserted elements (at most $2cM$ in total), whereas H_2 stores the smallest B elements of each non-empty slot in (any level of) $\mathcal{L}$. Thus, we keep less than $cM(2 + L)$ elements in internal memory. The analysis shows that $L = O(\log_{M/B}(N/B))$, thus the internal memory condition $\mu \cdot B \cdot L \leq M$ can only be fulfilled if L is a constant. For practical values of M, B and N we have $L \leq 4$, so that $c = 1/6$ satisfies the condition above.

Insertion. A new element is initially inserted into $\mathcal{H}$. If $\mathcal{H}$ gets *full*, the greatest cM elements of $\mathcal{H}$ form the set S which is moved to the level $\mathcal{L}_1$ (the smallest B elements among them stay in internal memory). This insertion may influence the L-levels in $\mathcal{L}$ in a cascading way. In fact at a generic step, the sorted sequence S has been originated from level $\mathcal{L}_{i-1}$ (initially $\mathcal{L}_0 = \mathcal{H}$) and must be moved to the next level $\mathcal{L}_i$. If $\mathcal{L}_i$ contains an empty slot, then S is stored into it. Otherwise a further overflow occurs at $\mathcal{L}_i$. Two situations may arise: (i) at least 2 slots can be merged into a single one (they lost their elements via **delete_mins**) thus freeing a slot that can be used to store S; otherwise (ii) all slots of L_i plus the elements from S are merged into a new sorted sequence S'. This is moved to the next level, thus repeating the loop above (now S' plays the role of S). It is simple to prove that no more than L iterations of the loop suffice to find a free slot, thus breaking the loop. Similarly, it is simple to prove that $L = O(\log_{M/B} N/B)$ over a sequence of N operations (consider the case of N inserts).

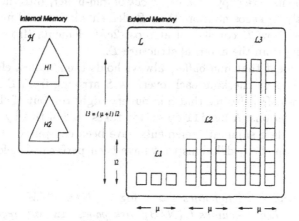

Fig. 2. The array heap

Deletion. The smallest element is removed from the internal heap $\mathcal{H}$. If it was the last one in a slot-buffer j of some level $\mathcal{L}_i$, we load the next B elements from this slot into H_2. This way, the corresponding buffer is refilled with the next smallest B elements of that slot. In order not to use unreasonably many partially filled slots, we check the sizes of the slots after each load operation: *if two slots j and j' of $\mathcal{L}_i$ contain less than $l_i/2$ elements each, they are merged into a single slot.* Consequently, every level $\mathcal{L}_i$ contains at most one slot which is non-empty and contains less than $l_i/2$ elements.

Theorem 2. *Under the assumption that L is a constant, the* insert *and* delete_min *operations take $O(1/B)$ amortized I/Os. The total required space in external memory is $(2X/B) + L$ pages[3], where X is the current number of elements in the heap. The total required internal space is $cM(2 + L)$.*

For practical values of M, N and B it is $L \leq 4$, thus c can be chosen to be $1/6$.

---

[3] Recall that we assume to implement each slot via a (static) array. To get this space bound we need to employ a sort of "rebuilding" step that reconstructs the heap from scratch at fixed time intervals. In practice, however, it is reasonable to assume that we have at least an half-filled slot per level, so that the bound $4X/B$ holds and the "rebuilding" step is unnecessary.

3.2   An Improved Version

We introduce a slight modification in the external data structure $\mathcal{L}$ that reduces its internal space usage thus making it useful also when less stringent conditions are imposed on L (i.e. it is not a constant), while still keeping its simplicity. We associate with each level $\mathcal{L}_i$ a special slot of length l_i, called min-buffer$_i$. It always contains the smallest elements in the slots of $\mathcal{L}_i$ as a sorted sequence of at most l_i elements. We copy *only one block* of min-buffer$_i$ into the internal heap $\mathcal{H}$. Consequently $\mathcal{L}_i$ keeps no more than B elements in internal memory, instead of $B\mu$ elements. Now $\mathcal{H}$ consists of at most $2cM$ elements (from H_1) plus LB elements coming from the external structure $\mathcal{L}$.

The requirement that min-buffer$_i$ always holds the smallest elements within $\mathcal{L}_i$ can be fulfilled by checking each overflow S arriving from $\mathcal{L}_{i-1}$ against the elements in min-buffer$_i$. Notice that min-buffer$_i$ might run out of elements due to a sequence of deletions. A **Refill** operation takes care of this by merging the μ slots in $\mathcal{L}_i$ until their smallest l_i elements have been computed. The merge cost can be charged on the l_i deletions that have been performed before min-buffer$_i$ got empty.

Theorem 3. *Under the hypothesis that* $\log_{M/B} N/B \le M/B$, *the new variant of the array heap occupies* $O(N/B)$ *disk pages,* **insert** *requires* $O((1/B) \log_{M/B} N/B)$ *amortized I/Os and* **delete_min** *requires* $O(1/B)$ *amortized I/Os.*

4   Experiments

External-Memory Library. We implemented external radix heaps, array heaps, buffer trees and B-trees using an external-memory library of algorithms and data structures called **LEDA-SM** [6] (an acronym for "LEDA for Secondary Memory") [4]. This library is an extension of the internal-memory library *LEDA* [15] and is *portable, easy to use* and *efficient*. The library **LEDA-SM** consists of a collection of efficient data structures and algorithms explicitly designed to work in an external memory setting. The system underlying **LEDA-SM** reflects a realistic view of the external memory model [21]: The internal memory is directly provided by the internal memory of the computer; whereas the D abstract disks are modeled by the file system, which also provides proper tools for implementing the low-level I/O via block transfers. Each disk is modeled with a *single file* and it is divided into logical blocks of a fixed size B (disk pages). The size of this file is fixed, thus modeling the fact that real disk space is bounded. Since **LEDA-SM** uses the file system, it explicitly takes advantage of the underlying *I/O-buffering* and *read-ahead* strategy at no additional implementation effort [5]. The programmer

---

[4] There exist other interesting external memory libraries, like TPIE [20] and VIC* (see http://www.cs.dartmouth.edu/~thc/ViC).

[5] As the file system is allowed to buffer disk pages, some of the disk requests can immediately be satisfied by the I/O-buffer. This can notably speed up the real performance of external-memory algorithms.

can still keep track of the number of disk accesses (bulk, random and total) performed by his/her algorithms, because the system allows the explicit counting of the number of both writes and reads to the D disks. LEDA-SM is implemented in C++ as a set of template classes and functions.

Experimental Setup. We compare eight different PQ implementations, namely array heaps, external radix heaps, buffer trees, B-trees, Fibonacci heaps, k-ary heaps, pairing heaps and internal radix heaps. The first four priority queues are explicitly designed to work in external memory, whereas the last four ones are LEDA-implementations of well-known internal-memory priority queues. B-trees are a natural extension of (a, b)-trees to secondary memory where we increase the fanout of each internal node to $O(B)$ so that the entire node fits into a single disk page. Insert and delete_min require $O(\log_B N/B)$ I/Os in the worst case. Buffer-Trees are also an extension of (a, b)-trees but they are intended to reach better amortized I/O-bounds for the insert and delete_min operations, namely $O((1/B)\log_{M/B}(N/B))$ amortized I/Os. The fanout of the internal nodes is $O(M/B)$ thus reducing the height to $O(\log_{M/B}(N/B))$. Each internal node has a buffer of size $O(M)$. Instead of storing a newly inserted element directly in the correct leaf position, the element is put into the root-buffer. When the root buffer is full, we empty it and move its elements to the appropriate buffers of the internal nodes in the next level. The smallest element now resides somewhere in the buffers of the nodes on the leftmost path. Delete_min is handled by storing the smallest $O(M)$ elements in an internal data structure. Newly inserted elements are checked against the ones stored in the internal structure; they are either put into the internal structure or into the buffer tree. If the internal structure runs out of elements due to a sequence of delete_mins, then we empty all buffers on the leftmost path and take the next smallest $O(M)$ elements.

We use the following three tests to analyze the performance of these PQs:

Insert-All-Delete-All. We perform N insert followed by N delete_min operations. This test allows us to compare the raw I/O-performance of the various PQs. The keys are integers, randomly drawn from $[0, 10^7]$.

Intermixed insertions and deletions. This test is used to check the I/O- and the CPU-speed of the internal-memory part of the external PQs. We first insert 20 million keys in the queue and then randomly perform insert and delete_min operations. An insert occurs with probability 1/3 and a delete_min occurs with probability 2/3.

Dijkstra's shortest-path algorithm. We simulate the Dijkstra's algorithm in internal memory on a large graph (using a large compute server). We take the sequence of insert and delete_min operations executed by the algorithm on its internal PQ. This sequence is then used to test our external PQs, thus allowing us to study the behavior of the external heaps in the case of "non-random", but application driven, update operations.

The tests were performed on a SPARC ULTRA 1/143 with 256 Mbytes of main memory and a local 9 Gbytes fast-wide SCSI disk. The internal memory priority queues use swap-space on the local disk to provide the necessary working space and a page size of 8 kbytes. The external memory priority queues used a disk block size of 64 kbytes [6]. Each external memory data structure uses about 16 Mbytes of main memory for the in-core data structures. Using only 16 out of 256 Mbytes of internal memory leaves enough space for the buffer cache of the file system. In order to have a better picture of the I/O-behavior of the experimented data structures, we decided to estimate also the actual "distribution" of the I/Os so to understand their degree of "randomness". This is a very important parameter to be considered since it is very well known [17] that accessing one page from the disk in most cases decreases the cost of accessing the page succeeding it, so that "bulk" I/Os are less expensive per page than "random" I/Os. This difference becomes much more prominent if we also consider the reading-ahead/buffering/caching optimizations which are common in current disks and operating systems. In order to take into account this technical feature, without introducing new parameters that would make the analysis much more complex, we follow [8] by choosing a reasonable value for the *bulk load size* and by counting the number of random and total I/Os. Since the transfer rates are more or less stable (and currently large) while seek times are highly variable and costly (because of their mechanical nature), we choose a value for the bulk size which allows to hide the *extra cost* induced by the seek step when data are fetched. For the disk used in our experiments, the average `t_seek` is 11 msecs, the `disk_bandwidth` is 7 Mbytes/sec. We have chosen `bulk_size` = 8 disk pages for a total of 512 kbytes. It follows that `t_seek` is 15% of the total transfer time needed for a bulk I/O. Additionally, the bulk size of 512 kbytes allows us to achieve 81% of the maximum data transfer rate of our disk while keeping the service time of the requests still low. Using a page size of 64 kbytes, we still keep the service time for random blocks reasonable small and the throughput rate reasonable high.

4.1   Experimental Results

We comment below the experimental results obtained for the tested priority queues. B-trees are implemented as a B^+-tree where the keys are stored in the leaves, and copies of them are put into the internal nodes as routers. We use path caching to speed up `insert` and `delete_min` operations on them.

External radix heaps are implemented in the natural way, a Fibonacci heap is used to store the indices of non-empty buckets. Array heaps are implemented in the simple (non-improved) variant because we can meet the memory requirements imposed by that structure, i.e. we can store the smallest block from *each* slot in internal memory (see Section 3). The slots are implemented as linked lists of blocks. This simplifies the `reload`-operation during a `delete_min` and leads to a better disk space usage. Allocated disk blocks for a slot are consecutive in order to perform bulk I/Os during insert. The in-core data structures H_1 and

---

[6] This value is optimal for request time versus data throughput.

H_2 are implemented as follows: H_1 is a specialized binary heap while H_2 is implemented via arrays. Buffer trees use the LEDA data type *sorted sequence* to implement the in-core ordered set (see also [2]). As also observed by [11], we do not restrict the internal buffer size in order to speed up operation insert.

Table 1 summarizes the running time in seconds and the number of I/Os on the insert-all-delete-all test.

Results for Buffer Trees. The standard buffer-tree is designed to be a lazy search tree. Our implementation follows an idea proposed in [11] that uses big unbounded buffers instead of buffers of size M, as proposed by [2]. This allows to save much time in the costly rebalancing process which does not occur as often as in the standard implementation. Another speed-up in our implementation is induced by the use of a simple in-core data structure (**sorted sequence** of LEDA) which reduces the CPU time. As a result, the insert operation is very fast as shown in Table 1. However, delete_min is slower than that of radix heaps and array heaps, and this is due to the costly rebalancing step. Earlier experiments [6, 11] have already shown that the internal buffer size heavily influences the performance of buffer trees. Unfortunately, the optimal value of the buffer is machine dependent and must be found experimentally.

If we sum up the time required for the insert and delete_min operations, we find that buffer trees are three times slower than array heaps and six times slower than radix heaps (see Table 1). They perform approximately four (resp. three) times the number of I/Os executed by radix heaps (array heaps), see Table 1. The 75% of these I/Os are random I/Os, which are mainly executed during the delete_min operation and are triggered by the costly rebalancing code. Overall, buffer trees are the second best "general-purpose" PQ, when no restriction is imposed on the priority value.

Results for B-Trees. B-Trees are not developed to work well in a priority queue setting. The worse insert performance of $O(log_B N)$ I/Os leads to an experimentally large number of I/Os, large running time, and hence to an overall poor performance. The B-tree executes only random I/Os during the various operations. delete_min can be speeded up by caching the path leading to the leftmost child. However, the final performance does not yet reach any of the other external priority queues.

Results for R-heaps. External radix heaps are the fastest integer-based PQ. Their simple algorithmic design allows to support very fast insert and delete_min operations. There is no need to maintain an internal data structure for the minimum elements as it is required for array heaps and buffer trees. This obviously reduces the CPU time. If we sum up the time required for the insert and delete_min operations, radix heaps are about 2.5 times faster than array heaps and six times faster than buffer trees (see Table 1). They execute the smallest number of I/Os (see Table 1), and additionally only 15 % of these I/Os are random. These random I/Os mainly occur during the operation delete_min.

Unfortunately, there are two disadvantages incurred by radix heaps: they cannot be used for arbitrary key data types, and the queue must be monotone. Consequently their overall use, although very effective in practical applications, is slightly restricted.

Results for Array Heaps. The array heap obviously needs a tricky in-core data structure to differentiate between newly inserted elements (structure H_1 is implemented by a modified binary heap) and between minimum elements coming from the external slots (structure H_2 is implemented by a set of arrays). This leads to a slowdown in the CPU-time for the operation insert, which is actually substantial. Indeed insert is up to ten times slower than in radix heaps or in buffer trees even if, in the latter case, array-heaps execute a smaller number of both random and total I/Os. Hence, on our machine the insert behavior of array-heaps is not I/O-bounded, but the CPU-cost is predominant (see Section 5). On the other side, array heaps achieve a slightly better performance on the delete_min operation than radix heaps, and they result up to nine times faster than buffer trees. In this case, there is no costly rebalancing operation as it instead occurs in buffer trees. In conclusion, if we sum up the time for both update operations, we find that array heaps are more than three times faster than buffer trees and 2.5 times slower than radix heaps. Consequently, array heaps are the fastest "general-purpose" priority queue among the ones we have tested. If we look at the I/O-behavior, we see that array heaps perform only slightly more I/Os than radix heaps, and only $\approx 2\%$ are random. These random I/Os are quite evenly distributed between insert and delete_min operations.

Internal-Memory Priority Queues. Our tests considered standard PQs whose implementations are available in LEDA. All these data structures are very fast when running in internal memory, and among them the best choice are again the radix heaps. However, there is a big jump in their performances (see Table 1) when the item set is so large that it cannot be fit into the internal memory of the computer. In this case, the PQs make heavy use of the swap space on disk. This jump occurs between 5 and 7.5 million items (Fibonacci heaps) or 10 and 20 million items (radix heaps, pairing heap and k-ary heap). The early breakdown for Fibonacci heaps is due to the fact that a single item occupies a large amount of space: about 40 bytes for a 4 byte priority value and a 4 byte information value. As we expected, none of the tested internal-memory PQs is a good choice for large data sets: the heavy use of pointers causes these data structures to access the external-memory devices in an unstructured and random way, so that hardly any locality can be exploited by the underlying caching/prefetching policies of the operating system.

Mixed Operations. The test on an intermixed sequence of insert and delete_min operations is used to check the speed of the internal data structures used in the best external PQs. The results of this test are summarized in Table 2.

We see again that radix heaps are superior to array heaps because they do not need to manage a complicated internal data structure, which reduces CPU-time, and also execute much less random (and less total) I/Os. This leads to a speed up of a factor 1.5. Buffer trees are more complicated and rebalancing is costly. The buffer tree is about six times slower than array heaps and seven times slower than radix heaps. They execute about nine times more I/Os and nearly all of them are random.

Test for Dijkstra's Algorithm We considered d-regular graphs with N nodes and $3N$ edges. Edge weights are integers drawn randomly and uniformly from the interval $[1, 1000]$. Again radix heaps execute significantly less random (and total) I/Os (see Table 2). We immediately see from this example that external radix heaps (see also[1]) perfectly fit in Dijkstra's shortest path algorithm because they can profit from bounded edge weights.

5   Conclusions and Further Research

We have compared eight different priority queue implementations: four of them were explicitly designed for external memory whereas the others were standard-internal queues available in LEDA. As we expected, all in-core priority queues failed to offer acceptable performance when managing large data sets. The fastest PQ turns out to be the radix heap: its simple structure allows to achieve effective I/O and time performances. Unfortunately, radix heaps are restricted to work on integer keys and monotone heaps. In the general case, array heaps become the natural choice. In the light of their good performance, we therefore plan in the future to re-engineer this data structure concentrating our attention on its internal part in order to hopefully speed up the operation insert (recall that it was CPU-bounded). Finally, we also plan to experimentally find the best choice for the parameters involved in the buffer-tree design, thus aiming at a more careful comparison of these data structures.

Acknowledgments

We are grateful to the referees for their deep and constructive comments.

References

1. R. Ahuja and K. Mehlhorn and J.B. Orlin and R.E. Tarjan, 'Faster Algorithms for the Shortest Path Problem', *Journal of the ACM*, 37:213-223, 1990.
2. L. Arge, 'The Buffer Tree: A new technique for optimal I/O-algorithms', *Workshop on Algorithms and Data Structures*, LNCS 955, 334-345, 1995.
3. R. Bayer and E. McCreight, 'Organization and Maintenance of Large Ordered Indices', *Acta Informatica*, 1:173-189, 1972.
4. G. S. Brodal and J. Katajainen, 'Worst-case efficient external memory priority queues', *Scandinavian Workshop on Algorithm Theory*, LNCS 1432, 107-117, 1998.
5. Y. Chiang, M. T. Goodrich, E. F. Grove, R. Tamassia, D. E. Vengroff, and J. S. Vitter, 'External-memory graph algorithms', *ACM-SIAM Symposium on Discrete Algorithms*, 139-149, 1995.

6. A. Crauser and K. Mehlhorn LEDA-SM, extending LEDA to Secondary Memory. *Workshop on Algorithmic Engineering*, 1999 (this volume).
7. E.W. Dijkstra. 'A note on two problems in connection with graphs.' *Num. Math.*, 1:269–271, 1959.
8. M. Farach and P. Ferragina and S. Muthukrishnan. 'Overcoming the memory bottleneck in Suffix Tree construction', *IEEE Symposium on Foundations of Computer Science*, 194–183, 1998.
9. R. Fadel and K.V. Jakobsen and J. Katajainen and J. Teuhola, 'External heaps combined with effective buffering', *Proc. Computing: The Australasian Theory Symposium*, 1997.
10. M.L. Fredman and R.E. Tarjan, 'Fibonacci heaps and their use in improved network optimization algorithms.', *Journal of the ACM*, 34:596–615, 1987.
11. D. Hutchinson, A. Maheshwari, J. Sack and R. Velicescu, 'Early experiences in implementing buffer trees', *Workshop on Algorithmic Engineering*, 92–103, 1997.
12. 'IEEE standard 754-1985 for binary floating-point arithmetic', reprinted in *SIG-PLAN 22*, 1987.
13. V. Kumar and E.J. Schwabe, 'Improved Algorithms and Data Structures for Solving Graph Problems in External Memory', *IEEE Symposium on Parallel and Distributed Processing*, 169-177, 1996.
14. A. LaMarca and R.E. Ladner, 'The influence of caches on the performances of heaps', *Tech. Report 96-02-03, UW University*, 1996. To appear in *Journal of Experimental Algorithmics.*
15. K. Mehlhorn and S. Näher. LEDA: A platform for combinatorial and geometric computing. *Communication of the ACM*, 38:96–102, 1995.
16. D. Naor, C. U. Martel and N. S. Matloff, 'Performance of Priority Queue Structures in a Virtual Memory Environment', *The Computer Journal*, 34(5):428–437, 1991.
17. C. Ruemmler and J. Wilkes, 'An introduction to disk drive modeling', *IEEE Computer*, 27(3):17–29, 1994.
18. P. Sanders. 'Fast priority queues for cached memory', *ALENEX '99. Workshop on Algorithmic Engineering and Experimentation*, LNCS, 1999 (to appear).
19. M. Thorup. 'On RAM priority queues', *ACM-SIAM Symposium on Discrete Algorithms*, 59–67, 1996.
20. D.E. Vengroff and J.S. Vitter. 'Supporting I/O-efficient scientific computation using TPIE', *IEEE Symposium on Parallel and Distributed Processing*, 74-77, 1995.
21. J.S. Vitter and E.A.M. Shriver, 'Optimal Algorithms for Parallel Memory I:Two-Level Memories', *Algorithmica*,12(2-3):110-147, 1994.
22. J. Vitter. External memory algorithms. Invited Tutorial, *ACM Symposium on Principles of Database Systems*, 1998. Also Invited Paper in *European Symposium on Algorithms*, 1998.
23. J.W.J. William. Algorithm 232 (heapsort), In *Communications of the ACM*, 347-348,1964.

Insert/Delete_min time performance of the external queues (in secs)				
N [$*10^6$]	radix heap	array heap	buffer tree	B-tree
1	6/24	18/11	56/34	11287/259
5	17/97	74/63	148/309	66210/1389
10	35/178	353/89	201/882	-
25	85/372	724/295	311/2833	-
50	164/853	1437/645	445/6085	-
75	246/1416	2157/1005	569/9880	-
100	325/1957	2888/1408	*	-
150	478/3084	4277/2297	*	-
200	628/4036	5653/3234	*	-

Insert/Delete_min time performance of the internal queues (in secs)				
N [$*10^6$]	Fibonacci heap	k-ary heap	pairing heap	radix heap
1	3/32	4/33	3/19	3/11
2	6/73	8/75	6/45	5/27
5	17/208	21/210	14/126	11/71
7.5	172800*/-	32/344	22/207	18/124
10	-/-	43/482	30/291	23/162
20	-/-	172800*/-	172800*/-	172800*/-

Random/Total I/Os for external queues			
N [$*10^6$]	radix heap	array heap	buffer tree
1	44/420	24/720	228/668
5	422/3550	120/4560	16722/21970
10	1124/8620	168/9440	35993/47297
25	2780/21820	570/29520	93789/123285
50	7798/56830	1288/66160	190147/249955
75	12466/89370	2016/102480	286513/376625
100	17736/124740	2776/139760	*
150	27604/192500	4216/210080	*
200	38284/211570	5712/284320	*

Table 1. *Experimental results on the insert-all-delete-all test. In the time performance tables, the notation a/b indicates that a (resp. b) seconds were taken to perform N insert (resp. N delete_min) operations. A '-' indicates that the test was not started, x^* indicates that the test was stopped after x seconds and a * indicates that the test was not started due to insufficient disk space.*

Time performance on mixed operations			
N[$*10^6$]	radix heap	array heap	buffer-tree
50	544	770	4996
75	609	945	5862
100	619	1027	6029
Random/Total I/Os on mixed operations			
50	2935/19615	22325/26997	153321/177201
75	5128/26752	24256/28384	171615/196647
100	5782/30094	24220/28380	171658/196578

Time performance on the Dijkstra's test			
N[$*10^6$]	radix heap	array heap	buffer tree
15	441	812	1524
20	1266	2030	5602
25	2750	4795	-
Total/Random I/Os			
15	422/5054	160/7800	33359/40887
20	2630/20790	528/23920	156155/176275
25	5810/48650	1600/80640	-

Table 2. *Execution Time and I/Os (random/total) for the mixed operation test (left table) and for Dijkstra's test applied on a random graph consisting of N nodes and 3N edges (right table).*

Author Index

Lecture Notes in Computer Science

For information about Vols. 1–1584
please contact your bookseller or Springer-Verlag